Glassy Metals: Magnetic, Chemical, and Structural Properties

Editor

Ryusuke Hasegawa, Ph.D.
Materials Laboratory
Corporate Research and Development Center
Allied Corporation
Morristown, New Jersey

CRC Press, Inc.
Boca Raton, Florida

Library of Congress Cataloging in Publication Data

Main entry under title:

Glassy metals.

Bibliography: p.
Includes index.
1. Metallic glasses. I. Hasegawa, Ryusuke,
1940—
TN693.M4G58 1983 666'.15 82-14707
ISBN 0-8493-5746-2

This book represents information obtained from authentic and highly regarded sources. Reprinted material is quoted with permission, and sources are indicated. A wide variety of references are listed. Every reasonable effort has been made to give reliable data and information, but the author and the publisher cannot assume responsibility for the validity of all materials or for the consequences of their use.

Direct all inquiries to CRC Press, Inc., 2000 Corporate Blvd., N.W., Boca Raton, Florida, 33431.

International Standard Book Number 0-8493-5746-2

Library of Congress Card Number 82-14707
Printed in the United States

PREFACE

It was in 1960 that $Au_{75}Si_{25}$ was made noncrystalline directly from liquid state by a rapid-quenched technique. This work by Pol Duwez and his collaborators was the first to demonstrate that liquid-like states can be realized in solid metals and alloys. The subsequent discovery of the stable ferromagnetic $Fe_{80}P_{13}C_7$ glassy alloy by Duwez and S. C. H. Lin in 1967 and the development of chill casting techniques to synthesize glassy materials in ribbon form have opened up a new avenue of materials research and application.

Because the atomic structure of a glassy or amorphous alloy is nonperiodic, it is not uniquely defined. However, an average over the entire structure is reflected in diffraction, magnetic, electrical, chemical, and mechanical properties. Based on these data, structural modelling by computer simulation is possible as presented in Chapter 1. This is useful in understanding the local structure-property relationship in metallic glasses, which is the main theme of this book.

One of the simplest theoretical methods of treating magnetic properties is the mean field theory. Because of structural disorder, this approach described in Chapter 2 has been reasonably successful in glassy alloys. To obtain a more microscopic picture, however, one needs to know accurate electronic structure of the system of interest. The atomic and electronic structures are intimately related, so that a detailed knowledge of the electronic structure should provide additional insight into the atomic arrangement of glassy solids. On the other hand, the relationship between the electronic structure and the properties of the materials is crucial to an understanding of phenomena such as electron transport, superconductivity, magnetism, chemistry, and mechanical behavior. For this purpose, as in crystalline materials, efforts are on determining density of state functions. These subjects are discussed in Chapter 3.

The most well-studied noncrystalline metallic system is the transition metal-metalloid base alloys, mainly because of their technological significance. Chapter 4 summarizes the infuence of structural disorder on the magnetic properties of dilute and concentrated alloys. Local moment formation, long-range magnetic ordering, and local short-range order and magnetism are extensively discussed. One system from this class of materials, i.e., Fe–B–Si, is found suited for low frequency (50/60 Hz) magnetic devices such power transformers. This and related subjects are summarized in Chapter 5. A number of glassy alloys based on Fe–Ni–B and Co–Fe–B exhibit magnetic permeabilities comparable with or superior to those of crystalline supermalloys. These materials are, therefore, suited for use in high frequency magnetic devices. This item is included in Chapter 6 in which various devices other than those used in low frequencies are considered using metallic glasses.

Rare-earth-element-containing noncrystalline alloys can be synthesized by liquid quenching techniques. Most of these alloys, however, have relatively low magnetic transition temperatures ($\leq$300 K) and have limited technological importance. Transition metal-rich rare-earth-element-containing alloys, however, can be made glassy by sputtering techniques. These alloys are in thin film form and uniaxial magnetic anisotropy can be induced along the direction perpendicular to the film plane. These films, therefore, can be used in magnetic bubble and magnetooptic devices as discussed in Chapter 7.

Amorphous solids are, because of their structures, free of defects most commonly found in crystalline solids such as grain boundaries and stacking faults. Of special interest are many glassy alloys prepared by liquid quenching which are in single phase and free of defects such as second phases or precipitates. These materials are then considered as ideally, chemically homogeneous alloys which should be resistant against corrosion attack. This is the subject of the last chapter of this book.

The editor wishes to express his sincere thanks to all the contributors for this book, and hopes that the publication of this book will promote interest for those who are studying this expanding area of interesting research and will help those who wish to learn the current state of knowledge.

Many thanks are due to Mr. B. J. Starkoff, President of CRC Press, for his interest in the present book and continued support to its completion, to many of the editor's colleagues for many helpful suggestions and to Miss Cathleen Storzum for her always cheerful assistance in preparing the entire manuscript.

Ryusuke Hasegawa
Morristown, New Jersey
May, 1982

THE EDITOR

Ryusuke Hasegawa, Ph.D., is Research Associate of the Materials Laboratory at the Corporate Research and Development Center of Allied Corporation, heading a group in glassy metals research.

Dr. Hasegawa obtained his B. Eng. and M. Eng. in electrical science from Nagoya University in Japan, and his M.S. in electrical engineering and the Ph.D. in materials science from California Institute of Technology in Pasadena in 1968 and 1969, respectively. He was Research Fellow in Materials Science at California Institute of Technology and Post-Doctoral Fellow at IBM Thomas J. Watson Research Center before he assumed his position as Staff Physicist at Allied Corporation in 1975.

Dr. Hasegawa is a member of the American Physical Society, American Association for the Advancement of Science, Materials Research Society and Sigma XI, and is a senior member of the Institute of Electrical and Electronics Engineers. He is currently Chairman of the Magnetics Society of the IEEE Princeton Section. He was Co-chairmen of the Second International Conference on Amorphous Magnetism held at Rensselaer Polytechnic Institute in 1976. He currently serves as a member of the Steering and Publications Committee of the annual Magnetism and Magnetic Materials Conference sponsored by the American Institute of Physics. He was a Fulbright Scholar during 1964-69.

Dr. Hasegawa holds seven U.S. patents, is the author or co-author of more than 90 papers and was a co-editor of the book *Amorphous Magnetism II*. His current research interests center around the relationship between atomic- and micro-scale structure and properties in noncrystalline solids.

CONTRIBUTORS

Richard Boll, Ph.D.
Head of the Section on Magnetism
Research and Development Division
Vacuumschmelze GmbH
Hanau, West Germany

D. S. Boudreaux, Ph.D.
Corporate Research and Development Center
Allied Corporation
Morristown, New Jersey

Jacques Durand, Sc.D.
Laboratory of Solid State Physics
University of Nancy I
Nancy, France

Koji Hashimoto, D. Eng.
The Research Institute for Iron, Steel, and Other Metals
Tohoku University
Sendai, Japan

Hans Rainer Hilzinger, Dr. rer.nat.
Head of the Laboratory for Special Metallurgy
Research and Development Division
Vacuumschmelze GmbH
Hanau, West Germany

William L. Johnson, Ph.D.
Associate Professor of Material Science
California Institute of Technology
Pasadena, California

Takehito Kaneyoshi, Dr. Sc.
Research Associate
Department of Physics
Nagoya University
Nagoya, Japan

Tsuyoshi Masumoto, D. Eng.
Professor
The Research Institute for Iron, Steel, and Other Metals
Tohoku University
Sendai, Japan

Ronald F. Soohoo, Ph.D.
Professor of Electrical Engineering and Computing Science
Department of Electrical and Computer Engineering
University of California at Davis
Davis, California

Michael A. Tenhover, Ph.D.
Senior Research Physicist
Sohio Research Center
Warrensville Heights, Ohio

Hans Warlimont, Prof. Dr. rer.nat.
Director of Research and Development
Vacuumschmelze GmbH
Hanau, West Germany

TABLE OF CONTENTS

Chapter 1

STRUCTURE OF METALLIC GLASS ALLOYS

D. S. Boudreaux

TABLE OF CONTENTS

I. INTRODUCTION

Metallic glass alloys constitute an important new class of materials. Since their discovery in the 1960s by Duwez[1] and co-workers, they have advanced from laboratory samples to an article of commerce produced in ribbon form by the kilometer.[2] There are a number of techniques by which the alloys can be produced[3] in their glassy (noncrystalline and thermodynamically unstable) state and these give rise to a variety of interesting and useful properties. The potential uses cover a wide spectrum. Many alloys exhibit high strength and good ductility which make them suitable as reinforcing agents. Others exhibit unique magnetic properties, while some are exceptionally resistant to environmental degradation. Some alloys are expected to possess a surface chemistry not previously known.

All physical and chemical properties ultimately depend on atomic interactions. Mechanical behavior is usually understood in terms of crystalline defects. Magnetic response is often described as due to the local symmetry of magnetically active species in condensed matter. Surface chemistry is conventionally related to the arrangement of d-orbitals at the surface of a material. In a glass, all of the long-range order and some of the short-range order which make such explanations possible is missing. We are only beginning to understand the arrangements of atoms in metallic glasses but, at present, there are no known ways to relate these to conventional representations of material properties.

Metallic glass alloys exhibit a range of properties which extend beyond those of crystalline materials.[1] One of our major goals is to understand those found. Even better, however, would be the ability to systematize the search for new alloys which controlled sets of properties. Ultimately, one would like to predict new and useful properties of this exciting new class of materials. There are two parts to this problem. The first is that the details of the atomic structure and useful methods of describing them must be known. The second has to do with modifying or adapting the considerable machinery of solid state physics to deal with nonperiodic structures. It appears at present that the effort is somewhere between steps one and two. Some consensus has developed about the elementary aspects of atomic structure, but there is a good bit more work to be done. Only preliminary efforts have been completed to relate glass structure to electronic properties. Attempts are made to utilize the successful theory of liquid metals to describe transport phenomena but it stirs up considerable controversy. This review will be restricted in attention to questions related to the atomic structure.

Information about structure at the atomic level is difficult to obtain for a glass. Diffraction is the primary tool for structure elucidation, but it loses most of its power as repeat patterns disappear; it is the constructive interference from these patterns that produce such precise fingerprints of crystalline arrays of atoms or molecules. However, diffraction remains the best tool available. Consider a collimated beam of X-rays incident on a sample of material and a sheet of film in a plane behind the sample perpendicular to the beam. For crystalline samples, the film would record a distribution of spots related to the symmetry of the arrangement of atoms scattering the X-ray photons. For a glass, one only records a set of diffuse rings on the film with intensity diminishing from the most central outward. More typically an experimentalist would do a step scan of diffracted intensity along one of the radii of the diffuse ring pattern; the distance from the center of the pattern can be calibrated to measure the length of the vector by which the photon momentum is changed (the scattering vector). In Section II, it will be shown how these measurements can be related to the distribution of distances between pairs of atoms and how this reflects on the short-range order of the atomic distribution.

The density of metallic glasses is a quantity with significant meaning. Universally, it is only a few percent below the value for a crystalline material with similar composition. The clear implication is that the spatial packing of atoms is not very different in the glass from the crystal and that the interactions between near-neighbor pairs will be very similar. Thus certain kinds of short-range order known to exist in the crystalline state should be expected to be preserved. Recently some new tools have been brought to bear on the question of short-range structure in the glassy state. EXAFS (Extended X-ray Absorption Fine Structure)[4] is a direct probe of the short-range distribution of neighbors. Electrons created by the primary absorption at one atomic site are scattered back to that site by neighboring atoms. The returning electron modulates the absorption characteristics of the original atom in a detectable way. The information then relates to the number of neighbors, at which distance(s) they occur and their type. The technique has thus far been applied to only a few materials of interest.[5,6] Hyperfine field distributions at magnetic sites have been measured using Mössbauer spectroscopy and related to the distribution of magnetic neighbors of a typical site;[7] unfortunately, this kind of measurement cannot be applied to a wide variety of alloys because there are a limited number of appropriate isotopes of glass-forming metal elements. Finally, attempts are being made to view the structure directly using ultrahigh resolution electron microscopy.[8]

The ultimate goal is to establish an idealized concept or abstraction of atomic structure in a metallic glass which would be simple enough to apply to real alloy systems. At present, there are no known techniques which deal simply with dense collections of atoms in unordered, but not completely random, arrangements. It is difficult to find concepts which may be borrowed and modified to fit the situation at hand. Most work has been theoretical in style and centered around "building" models that may be tested against those properties which can be determined experimentally. Some of this work will be reviewed here and sheds light on the details of the atomic structure.

A number of proposals have been made from widely differing starting points. An early thought was that the materials were microcrystalline, i.e., tiny crystallites, randomly oriented and separated from one another by regions whose structure is completely amorphous. This idea is still raised from time to time, but is not generally accepted.[9] Cargill[10] has shown the diffraction from microcrystalline arrangements of atoms is inconsistent with observation. An aesthetically appealing idea is that metallic glasses are composed of a random packing of atomic clusters; Gilman[11] has advanced some important arguments in favor of this idea. The clusters may be the kind of arrangements found in crystalline materials or they may be arrangements which are forbidden by symmetry in periodic systems. Some important ramifications of these ideas will be reviewed, however, because of the inherent complexity, most work has consisted of studying models which start from the packing of spheres. Historically, physical assemblages of ball bearings were analyzed and later simulated on a computer where both hard and soft spheres could be studied. Interactions of hard spheres is a problem with a considerable history in statistical mechanics; attempts have been made to apply this work to metallic glass structure, but with limited success.

II. DIFFRACTION FROM NONCRYSTALLINE MEDIA

Much of the discussion of the atomic structure of glassy material, in general, and of metallic glass alloys, in particular, is made in terms of data from diffraction experiments, using X-ray, electron, neutron, or a combination of radiations. The measured *interference function*, described conceptually above, but more precisely by Wagner,[12] and the *radial distribution function* (RDF) deduced from it are frequently discussed. In this section these ideas are developed. The usual starting point is the experimental data, but the discussion is more transparent if one starts from a conceptualized model of the atomic structure and derives the measured quantities.

A *model* is a specific set of points in space: $\{\mathbf{r}_L\}$, at which atoms are located. The number density is the number of atoms per unit volume in the model and is given by

$$1/V \sum_{L=1}^{N} \delta(\mathbf{r} - \mathbf{r}_L) \tag{1}$$

δ is the Dirac delta function and V is the volume containing all N atoms. In addition, the average number density is usually denoted by

$$\rho_0 \equiv N/V \tag{2}$$

In correlating the properties of a given model with diffraction data, the primary point of interest is not the absolute positions of the atoms or in local values of the density but rather in correlations between positions of atoms. A most useful quantity in this regard is $\rho_L(\mathbf{r})$, which is defined as the number density of atoms at a point **r** relative to the position of atom L. Since the glass is not an ordered structure $\rho_L(\mathbf{r})$ will vary from one site L to another. The angular dependence of ρ will, in general, be completely random as L is varied. Thus a more useful quantity is introduced:

$$\rho_L(r) \equiv \langle \rho_L(|\mathbf{r}|) \rangle_{\theta,\phi} \tag{3}$$

the average is taken over a thin spherical shell a distance r from site L. In practice, this is done by counting neighbors in such a shell and dividing by the volume of the shell. In a model with a finite number of atoms, care must be taken in treating sites near the surface of the model as they have neighbors on only one side.

There will still be variations in $\rho_L(r)$ from site to site, especially for small values of r. For example, an eight or ninefold coordinated site will have quite different values of atomic density when r is of the order of the near-neighbor separation. In addition there are no experimental means of determining the particular neighbor distributions of a given site L, but rather measurements sample the average site surroundings. Averaging once more over all sites L produces

$$\rho(r) = \langle \rho_L(r) \rangle_L \tag{4}$$

This quantity measures the average number density of neighbors at a distance r from any site in the model.

The generalization of the quantities discussed above to alloys containing more than one type of atom is straightforward. The atomic density of type i atoms is

$$1/V \sum_{L=1}^{N} \delta(\mathbf{r} - \mathbf{r}_L^i) \tag{5}$$

where $\mathbf{r}_L^i$ are the position vectors of the N_i atoms of type i. The total number density is then

$$1/V \sum_{i=1}^{n} \sum_{L=1}^{N} \delta(\mathbf{r} - \mathbf{r}_L^i) \tag{6}$$

for an alloy with n components. In addition, it is useful to define

$$\rho_{0i} \equiv N_i/V \tag{7}$$

the average density of type i sites. $\rho(\mathbf{r})$ is generalized to $\rho_{ij}(\mathbf{r})$, i.e., the average number density of type j atoms a distance **r** away from a type i atom. The averages are performed on each type pair separately.

The *radial distribution function* can now be defined for models with only one kind of atom as

$$RDF(r) \equiv 4\pi r^2 \rho(r) \quad (8)$$

it is a measure of the average number of neighbors found at a distance r from any atom chosen as origin. Oscillations in this function occur about a parabolic base line, so a more convenient function, the *reduced radial distribution function*, is conventionally defined:

$$g(r) \equiv 4\pi r\,[\rho(r) - \rho_0] \quad (9)$$

$$= 1/r\,[RDF(r) - 4\pi r^2 \rho_0] \quad (10)$$

The reason for the additional factor of 1/r will become clear shortly. For a multicomponent alloy the definition of the RDF becomes

$$RDF(r) = 4\pi r^2 \rho(r) \equiv 4\pi r^2 \sum_{i=1}^{n} \sum_{j=1}^{n} w_{ij}\, \rho_{ij}(r) \quad (11)$$

The correlations between each type pair are added up to get the total radial distribution of atoms, thus the double summation. w_{ij} are weighting factors and relate to the fact that experimentally the correlation between pairs of one type may be "more detectable" than between pairs of another type; the precise definition of w_{ij} involving scattering amplitudes will be made later.

All of the above information can, in principle, be extracted from diffraction experiments and, in the following, the relationship to the raw data is sketched. Consider atoms of type i as primary scattering centers. The total distribution of sites is given by

$$\rho_{ij}(r) + \delta(\mathbf{r} - \mathbf{r}^i) \qquad \{ i=1,2,3,\ldots N_1 \} \quad (12)$$

i.e., the distribution of neighbors plus distribution of central sites of type i. Note that averaging has not yet been performed. This quantity is usually discussed as a generalized Patterson function describing correlations among atom positions.[12,13] Consider an incident wave scattered first off the central atom (of type i) then off another atom, say, of type j. If the scatterers are separated by a distance which is some integral multiple of the incident wavelength, then the scattering can occur cooperatively (the secondary scattering event is in phase with the primary one). In general, the fraction of incident radiation of wavelength λ scattered through wave vector **K** is given by the Fourier transform of the distribution of scattering sites. The transform calculates the number of pairs whose interatomic separation has a component along **K** which is nλ in length

$$\int \{\rho_{ij}(r) + \delta(\mathbf{r} - \mathbf{r}^i)\}\, e^{-i\mathbf{K}\cdot\mathbf{r}}\, d\mathbf{r} \quad (13)$$

The relative scattering strengths of the members of the pair are included via their scattering amplitudes f_i, which do in general depend on **K**.

$$f_i(\mathbf{K}) f_j^*(\mathbf{K}) \int \{\rho_{ij}(r) + \delta(\mathbf{r} - \mathbf{r}^i)\}\, e^{-i\mathbf{K}\cdot\mathbf{r}}\, d\mathbf{r} \quad (14)$$

Finally a sum is performed over all types j; the result is multiplied by the total number of primary scattering sites, N_i, and summed over all types of primary sites.

$$\sum_i \sum_j N_i f_i(\mathbf{K}) f_j(\mathbf{K}) \int \{\rho_{ij}(\mathbf{r}) + \delta_{ij}\,\delta(\mathbf{r} - \mathbf{r}^i)\}\, e^{-i\mathbf{K}\cdot\mathbf{r}}\, d\mathbf{r} \tag{15}$$

The Kronecker symbol, δ_{ij}, is added to avoid including the primary sites in the sum on j of types of secondary sites.

This equation includes what is known as volume scattering, i.e., all atoms scattering in phase simultaneously; it is an effect which is generally not observed. It is conventional to subtract out the volume contribution. It can be expressed as the scattering from all type i atoms off the average background of type j atoms:

$$-\sum_j N_i f_i f_j \int \rho_{oj}\, e^{-i\mathbf{K}\cdot\mathbf{r}} d\mathbf{r} \tag{16}$$

Combining this with the total scattering gives a quantity referred to as the observable scattering

$$I_o(K) = \sum_i \sum_j N_i f_i f_j \int \{\rho_{ij}(\mathbf{r}) - \rho_{oj} + \delta_{ij}\delta(\mathbf{r} - \mathbf{r}^i)\}\, e^{i\mathbf{K}\cdot\mathbf{r}}\, d\mathbf{r} \tag{17}$$

Next the assumption of scattering isotropy is made, i.e., it is assumed that the averaging process has been performed and that $\rho_{ij}(\mathbf{r})$ becomes $\rho_{ij}(r)$. It is now possible to evaluate some of the integrals. The term involving the delta function is trivial and is brought out front. The expression is then rewritten with the remaining volume integral explicitly expanded as

$$I_o(K) = \sum_i N_i f_i^2 + \sum_i \sum_j N_i f_i f_j \int_0^\infty r\,dr\ \{\rho_{ij}(r) - \rho_{oj}\} \int_0^\pi \sin\theta\, d\theta\, e^{-iKr\cos\theta} \int_0^{2\pi} d\phi \tag{18}$$

the ϕ integral is trivial and the θ integral can be transformed to

$$I_o(K) = \sum_i N_i f_i^2 + \sum_i \sum_j N_i f_i f_j \int_0^\infty r\,dr\ \{\rho_{ij}(r) - \rho_{oj}\}\, 2\pi \int_{-1}^{+1} e^{-iKr\cos\theta}\, d(\cos\theta) \tag{19}$$

Changing variables leaves a simple integral evaluation

$$= \sum_i N_i f_i^2 + \sum_i \sum_j N_i f_i f_j \int_0^\infty r\,dr\ \{\rho_{ij}(r) - \rho_{oj}\} \frac{2\pi}{ikr} \int_{-iKr}^{iKr} e^u\, du \tag{20}$$

with the final result

$$I_o(K) = \sum_{i=1}^n N_i f_i^2 + \sum_{i=1}^n \sum_{j=1}^n N_i f_i f_j [1/K] \int_0^\infty 4\pi r\ \{\rho_{ij}(r) - \rho_{oj}\} \sin Kr\, dr \tag{21}$$

It is desirable to further transform this expression so as to eliminate explicit reference to the number of atoms, which is, in general, unknown. This is accomplished by expressing the scattered intensity on a per ideal atom basis, where the ideal atom has unit scattering amplitude. In order to simplify the resulting expressions the atomic concentrations are introduced,

$$c_i \equiv N_i/N \tag{22}$$

It follows that

$$\rho_{0i} = c_i\rho_0 \tag{23}$$

and weighting functions are defined as follows:

$$w_{ij}(K) = \frac{c_i c_j f_i(K)\, f_j(K)}{\langle f(K) \rangle^2} \tag{24}$$

Here

$$\langle f(K) \rangle = \sum_{i=1}^{n} c_i f_i = 1/N \sum_{i=1}^{n} N_i f_i(K) \tag{25}$$

is the scattering amplitude of the average atom in the alloy. The intensity per atom I(K) is then derived by dividing $I_0(K)$ by $N\langle f\rangle^2$. In addition it is conventional to normalize I(K) so that in the absence of pair correlations, the result is unity; this is accomplished by subtracting $[\langle f\rangle^2 - \langle f^2\rangle]/\langle f\rangle^2$ from $I_0(K)$:

$$I(K) = 1 + \sum_{i=1}^{n}\sum_{j=1}^{n} w_{ij}/K \int_0^\infty 4\pi r \,\{\rho_{ij}(r)/c_j - \rho_0\} \sin Kr\, dr \tag{26}$$

This is the final expression which can be fitted to experimental data. It allows for the fact that f_i and hence the w_{ij} are K dependent. In the most general case these quantities are complex,[13] but for clarity this case has not been treated in the above derivation. The K dependence of the w_{ij} is usually ignored in the analysis of X-ray and electron scattering from metallic glass alloys. The K dependence is not a problem in neutron scattering because $K >> r_{nucleus}$; this implies that the atoms act as point scatterers and that the f_{ij} are constants.

The interference function I(K) can be related to the RDF and g(r) functions in the following way. It is convenient to define the *reduced interference function*

$$F(K) \equiv K\,[I(K) - 1] \tag{27}$$

$$F(K) = \sum_i \sum_j w_{ij} \int 4\pi r \,\{\rho_{ij}(r)/c_j - \rho_0\} \sin Kr\, dr \tag{28}$$

Since $\sum_i \sum_j w_{ij} = 1$ by definition, this can be written

$$F(K) = \int g(r) \sin Kr\, dr \tag{29}$$

where

$$g(r) \equiv 4\pi r \left[\sum_i \sum_j w_{ij}\rho_{ij}(r)/c_j - \rho_0\right] \tag{30}$$

is the reduced radial distribution function for multicomponent systems. The basic Fourier transform relationship established between F(K) and g(r) can be used to convert from reciprocal to coordinate space, i.e., to deduce real space relationships from experimental data.

In testing the theoretical models to be described in the remainder of this paper, one proceeds the other way: g(r) is determined from the coordinates of the atoms in the model, then F(K) or I(K) is calculated in order to compare to data measured from an alloy the model is designed to simulate. The calculation of g(r) is easy once the $\rho_{ij}(r)$ are known, they are determined by examining the surroundings of each type i atom for type j neighbors at a distance r. In particular, the space around atom i is partitioned into spherical shells and the number of type j atoms in each shell is counted. Corrections are made for the finite size of the model[14] to maximize the statistical significance. The data from all type i sites are averaged[15] to produce $\rho_{ij}(r)$. It is not convenient to display the ρ_{ij}'s directly but rather the quantities

$$p_{ij} \equiv \rho_{ij}(r)/\rho_{oj} \tag{31}$$

called *partial pair correlation* functions, are plotted. The normalization provides that the asymptotic dependence of each p_{ij} is unity for large r.

Observe that a single diffraction experiment is incapable of yielding the individual p_{ij} because only their weighted sum contributes to I(K). It is possible to take advantage of the fact that the scattering amplitudes, f_i, depend on the wavelength of the probing radiation; in effect the weighting factors can be manipulated. Therefore, it is necessary to use data from n(n + 1)/2 experiments each using a different radiation to determine the complete set of pair correlations for an n component alloy. Waseda[16] has studied the Fe–P alloy system in this way. Others[17,18] have combined X-ray and neutron sources in analyzing the partial pair correlations in the Co–P system. Note that in neutron diffraction studies isotopic substitution could be used. Egami[19] has pioneered the use of an energy-dispersive X-ray diffraction experiment which permits, in principle, the extraction of a complete set of characterizations from a single experiment, this is possible because a range of photon energies is present in the scattered beam and several experiments are effectively combined. To date it has only been used to collect conventional interference functions, but to higher values of K which effects improved spatial resolution.

An interesting, although not yet very popular, kind of analysis of statistical data derived from experimental data is based on the position-dependent fluctuations of concentration and number density of atom types.[20] Various correlations between pairs of these quantities can be derived directly from interference functions measured with various wavelength radiations. They can also be related to $p_{ij}(r)$ and g(r) in a straightforward way; Wagner[12] has carefully carried out these derivations. Part of the reason for the limited use of this kind of analysis is that the full set of p_{ij} have not generally been available. An increase in the use of number and concentration correlations is expected as experimental methods improve and as attempts are made to correlate atomic structure models to physical phenomena more naturally related to such quantities, e.g., domain structures in a magnetic alloy. An interesting and recent example[21] is the analysis on the $Ti_{70}Ni_{30}$ alloy suggesting a preference of Ti atoms in the first neighbor shell of Ni sites. The observation is made possible by special characteristics of the Ti nucleus to neutron scattering and corresponds very naturally to the concentration fluctuation-type analysis.

III. DEVELOPMENT OF CONTEMPORARY IDEAS

The earliest research in the field dealt primarily with inventing techniques of glass formation and discovering compositions of matter which would permit attainment of the metallic glass state.[1,3] As samples were routinely examined by X-ray diffraction to determine whether a glassy structure had been formed, a pattern of extraordinary similarity was recognized. The first comparisons were made among interference functions; Figure 1 presents a collage

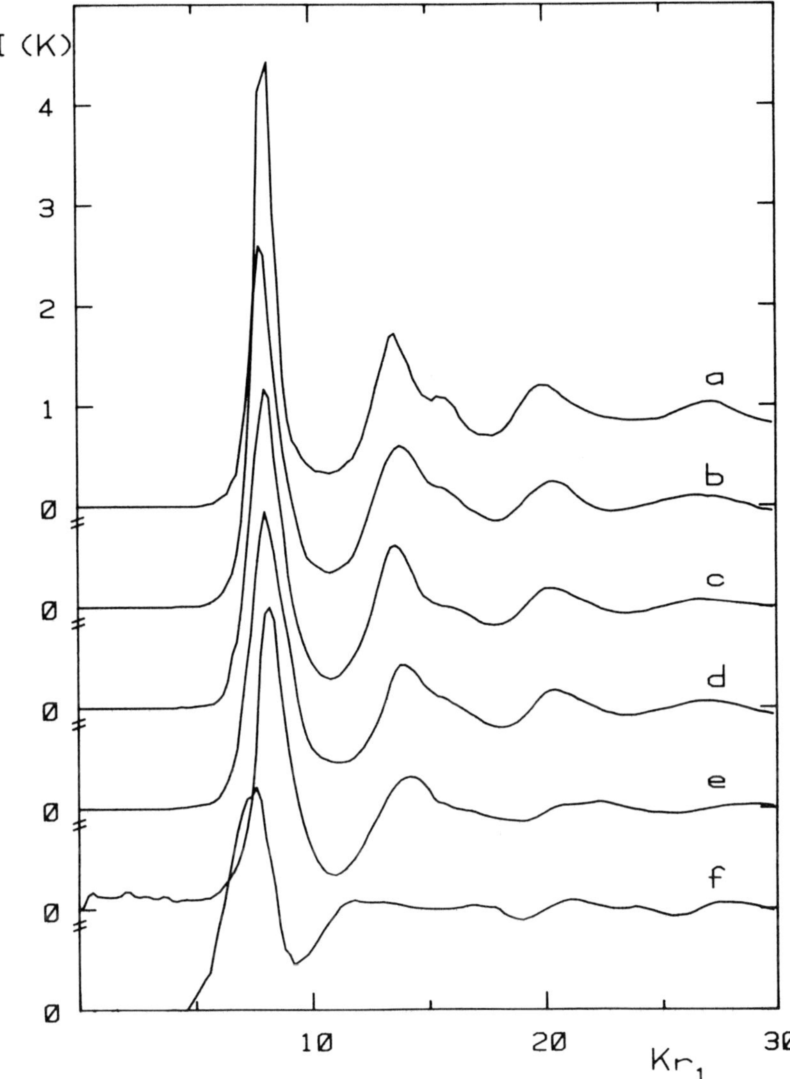

FIGURE 1. X-ray interference functions (the length of the scattering vector K is converted to a universal dimensionless variable Kr_1, where r_1 is the position of the first peak in the radial distribution function) for six metallic glass alloys. (a) $Fe_{80}P_{13}C_7$,[13] (b) $Ni_{80}P_{20}$,[31] (c) $Fe_{80}B_{20}$,[63] (d) $Pd_{80}Si_{20}$,[25] (e) $Cu_{57}Zr_{43}$,[23] and (f) $Gd_{18}Co_{72}$.[71] The first five exhibit striking similarities and the details of their structure are relatively well characterized; the last is not yet fully understood.

of such data and the similarities are obvious except for the Gd–Co case which exhibits some additional complexity at large K. This undoubtedly motivated the calculation and comparison of radial distributions. The latter is not done routinely because data must be collected with extreme care for use in the Fourier transform, a step-scan with high signal/noise ratio must be taken to as large a value of K as possible. Spurious features in F(K) will cause structure of unphysical nature in the g(r) and if there are insufficient points at large K, the effective truncation of the Fourier integral will have the same effect.

Figure 2 shows the reduced radial distribution functions corresponding to the data in Figure 1. The first sharp peak in each curve corresponds to the shell of near neighbors; its position indicates the distance between near neighbors and its area is proportional to the number of

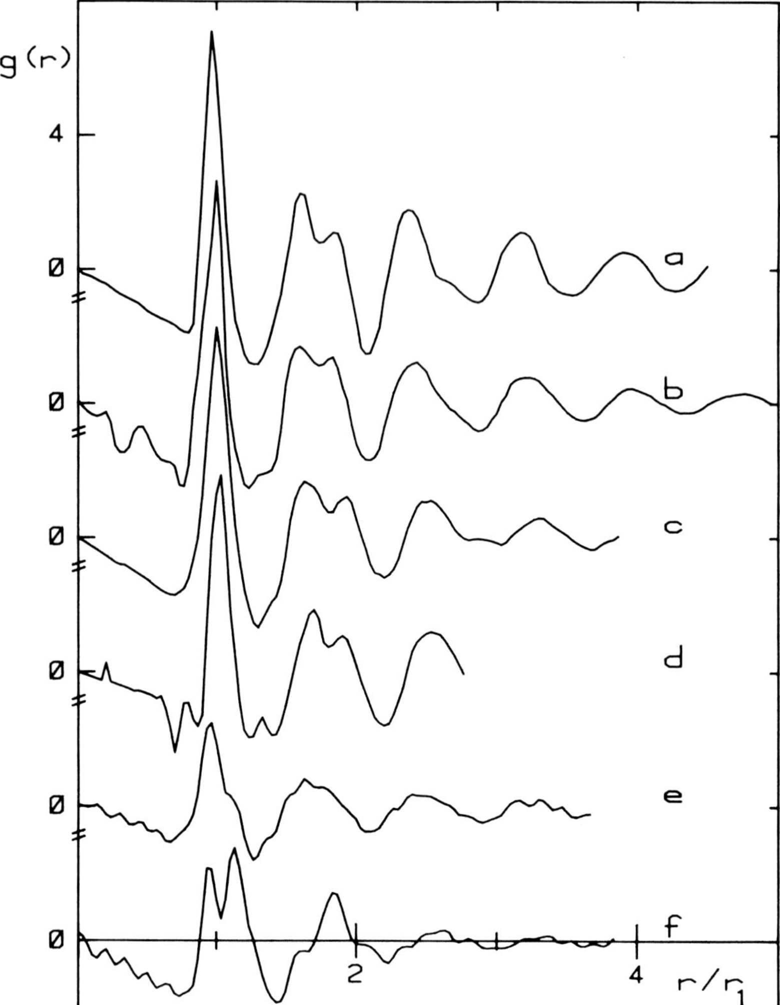

FIGURE 2. The reduced radial distribution functions calculated from the data in Figure 1. The units of the abscissa are again dimensionless: r/r_1 where r_1 is the actual position of the first peak measuring the near-neighbor separations. Each curve is a sum of contributions, but only in (f) are there two observable near-neighbor pairs (Co—Co and Gd—Co) due to the values of the scattering strengths of the alloy components. The split second peak is seen in almost all metallic glasses. The wiggles before the first peak are due to truncation of the Fourier transform integral and measure the error induced by it.

such neighbors, i.e., the *coordination number*. For all of the alloys represented in the figure this shell contains more than one kind of atom, in general, occurring at different near neighbor distances. This is quite clear for the Gd–Co alloy where the first peak is split because the scattering factors for Gd and Co are similar in magnitude and because the Gd–Co and Co–Co spacings are sufficiently different to be resolved;[13] the structure in I(K) for large K is responsible for this additional structure in g(r) for small r. There is a shoulder on the first peak of the Zr–Cu data for the same reasons, but Zr and Cu are closer to the same size and the w_{ij} for the Cu–Cu pair is much larger than for the Cu–Zr pair. X-ray diffraction does not show any structure in the first neighbor shell for the remaining transition metal-metalloid type glass alloys because large scattering factors and composition cause the metal components to dominate.

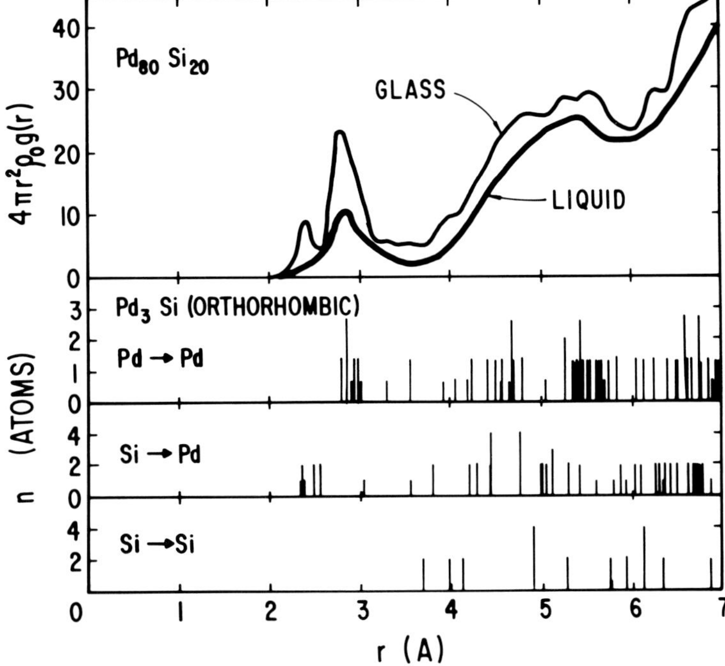

FIGURE 3. Comparison of the RDF of $Pd_{80}Si_{20}$ in the glass and liquid states. In this case high resolution neutron diffraction[25] resolved the Pd–Si and Pd–Pd near-neighbor peaks in the glass. The liquid has much less structure. For comparison a representation of diffraction lines observed from crystalline Pd_3Si is shown. Some of the short-range features in the glass can be seen to be similar to those in the crystal, but are broadened significantly.

The second peak in g(r) describes the shell of next-nearest neighbors; it has received considerable attention because it is split into two substructures. There are no simple explanations for this almost universal effect. It is not due to the multicomponent nature of the glass because it is also seen in elemental glasses[22] and is seen as well in the partial pair correlations of, e.g., Cu–Cu in ZrCu.[23] Some relatively simple pictures of its origin have been advanced[13,24] (see also Figure 9), but it is generally recognized as being due to a combination of a number of factors associated with the three-dimensional packings of spheres. Basically one can think of the second neighbors as fitting into pockets formed by triangular arrangements of atoms in the first neighbor shell; there are two classes of pockets, one "deeper" than the other. Another reason for the discussion of the second peak is that it has served as a rather good test of proposed models of the atomic structure, as will be outlined shortly. The remaining point of significance in Figure 2 is the rapid diminution of correlations at large r, the broad features simply fade away to indicate a rapid decay of order; i.e., at sufficiently large distances there is equal likelihood of finding atom centers or not finding them.

It is natural to compare the structural correlations measured on glassy metals to that of liquid metals of the same composition to see whether the glass can be viewed simply as a quenched liquid. Figure 3 makes this comparison and in addition shows the degree of structure observed in the crystalline state. In this case the scattered radiation was from a neutron source[25] and has sufficient resolution to separate the Pd–Pd and Pd–Si near-neighbor shells. Clearly the glass contains more structure than the liquid. In particular, the short-range order of the crystalline state (as exhibited by groups of lines for small r) is still seen in the glass but disappears in the liquid. Thus it is not just a question of density decrease on going from the glass to the liquid state.

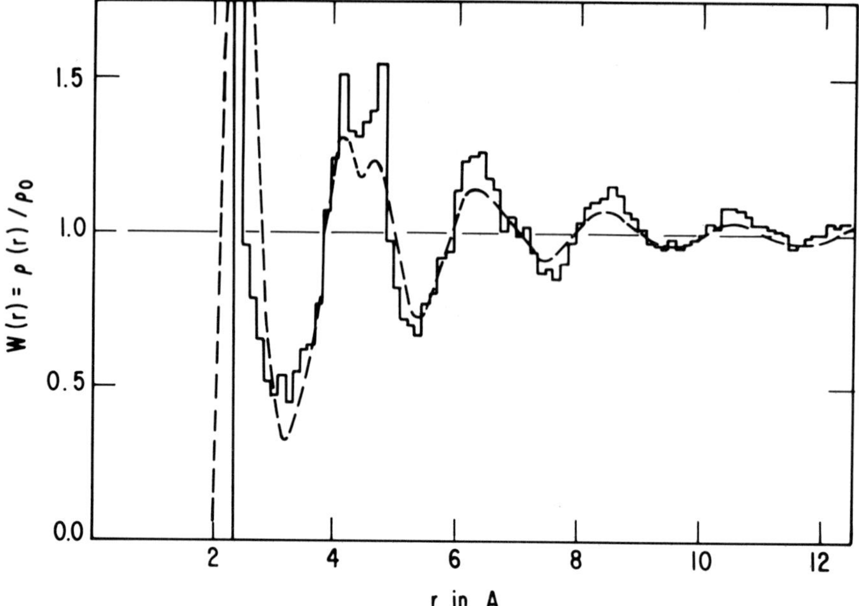

FIGURE 4. The measured pair correlation function for $Ni_{80}P_{20}$[31] (dashed line) is compared to that determined by Finney from a model of packed ball bearings[29] (histogram). The ball bearings are impenetrable and so there are no pairs inside the abrupt edge of the first peak. The second peak splitting is reproduced although with the wrong height as is the remaining periodicity.

Cohen and Turnbull[26] suggested that some earlier work by Bernal[27,28] might be appropriate as a model for the glass atomic structure. Bernal's work was concerned with dense random packings (DRP) of hard spheres which were subsequently analyzed in considerable detail by Finney.[29] The models of Bernal were constructed by placing ball bearings into a rubber bladder and kneading until maximum density was achieved; finally molten wax was poured in so that the bladder could be removed. Analysis consisted of painstakingly separating the model and measuring the coordinates of each ball removed. From the list of coordinates the appropriate statistical correlations are calculated directly. Cargill[31] compared the high resolution RDF determined by Finney with an RDF measured on a NiP alloy, see Figure 4. The agreement is striking, but the model does not contain spheres of appropriate size to represent the smaller P atoms. Polk suggested that the phosphorus occupied sites formed by "cages" of near-neighbor Ni atoms (the cage is a polyhedron with Ni atom vertices and is called a Bernal or canonical hole). Finney had determined the distribution of hole shapes and Polk, using these results was able to determine that about 20% of the cages in the Bernal model were of adequate size to contain metalloid atoms — this 20% is important because it was generally known that transition metal-metalloid glass alloys were formed most easily at such a composition.[2] In addition the idea satisfied the expectation that metalloid-metalloid neighbors should not occur since they are not seen in crystalline intermetallics. There have been some more detailed studies of Bernal's and other's models by Whittaker,[32] Frost,[33] and Yamamoto and Doyama[34] all finding only minor variations on this theme. In addition there have been criticisms[35] of the idea on grounds that near-neighbor coordination numbers are not correctly predicted. Later studies have corrected these deficiencies and have found further weaknesses. However the basic conception has proved very popular and has had a deep impact on the understanding of metallic glass structure.

A number of other efforts to understand the atomic structures of metallic glasses have originated from a defective crystal lattice point of view, in which the glassy structure is derived from the crystalline one by the introduction of high densities of defects such as grain boundaries,[10] dislocations,[36] or disclinations.[37] Conceptually, these approaches are quite similar since grain boundaries can be regarded as dislocation or disclination arrays and dislocations are equivalent to disclination dipoles. They differ, however, in the manner in which disorder, in the form of strain, is distributed in the structure.

Microcrystalline models, in which the disorder is more or less localized in the grain boundaries, have generally failed to predict experimental RDF data.[10] In the disclination-dislocation model, proposed by Morris,[37] the disorder is partially distributed as lattice curvature and partially localized in defect cores. This model permits calculation of reasonable values for both the heat of crystallization and the low temperature yield stress; it is further supported by disclination-like structures of atoms in the models of Boudreaux and Gregor[15] observed directly in stereo projection using computer graphics and as identified by Yamamoto and Doyama.[34] In addition, Rivier[38] has shown that disclinations as line defects are topologically stable in disordered condensed matter. While no attempt has been made to compute RDFs for disclination containing structures, Koizumi and Ninomiya[36] have shown reasonable agreement between experimental data and pair distributions calculated for a dislocation model and have also shown that such models contain many Bernal polyhedra arising from the dislocation core structures. Defective lattice models of the glassy structure can be considered to be complementary to DRP models and are particularly useful in considering phenomena involving correlated atomic motions.

Considerable success has been had in the description of simple fluids through the study of the statistical mechanics of interacting hard spheres. The pair correlation function of the interacting system is decomposed into a direct and an indirect part. The direct part is given the name *direct correlation function,* c(r), while the indirect part is defined via the integral equation[39]

$$p(r) = 1 + c(r) + \rho_0 \int [p(|\mathbf{r} - \mathbf{r}'|) - 1]\, c(r')\, d\mathbf{r}' \qquad (32)$$

The indirect part represented by the integral is the influence at **r** propagated through a third atom located at **r**′ summed over all **r**′; i.e., the influence of the remaining atoms in the model. So-called Percus-Yevick theory results when this equation is solved subject to the conditions:

$$\begin{aligned} p(r) &= 0 \text{ for } r < d \;:\; \text{hard spheres} \\ c(r) &= 0 \text{ for } r > d \;:\; \text{short-range interactions} \end{aligned} \qquad (33)$$

Here d is the diameter of the interacting spheres. It is not the purpose of this paper to review the application of this theory, but it has been quite successful. Weeks[40] has extended the theory to multicomponent systems and has made some calculations. The agreement with measured g(r), e.g., is not very good. One source of trouble is the unknown behavior of some of the series which must be summed. For densities appropriate to metallic glass alloy systems, the convergence rules known to apply at fluid densities, are not known to be valid. For this reason as well as the fact that more realistic interactions are very hard to include, this area of investigation has not received much attention.

The most successful approaches to the question of structure are associated in one way or another with the sphere packing ideas discussed above. The most straightforward conceptualization of glass formation from assemblages of spheres is the molecular dynamics scheme. Basically, a system of interacting particles is described by a set of pair potentials. Each particle is given a random position and momentum consistent with the temperature of the system and their subsequent motion is considered to be Newtonian. A computer is used to

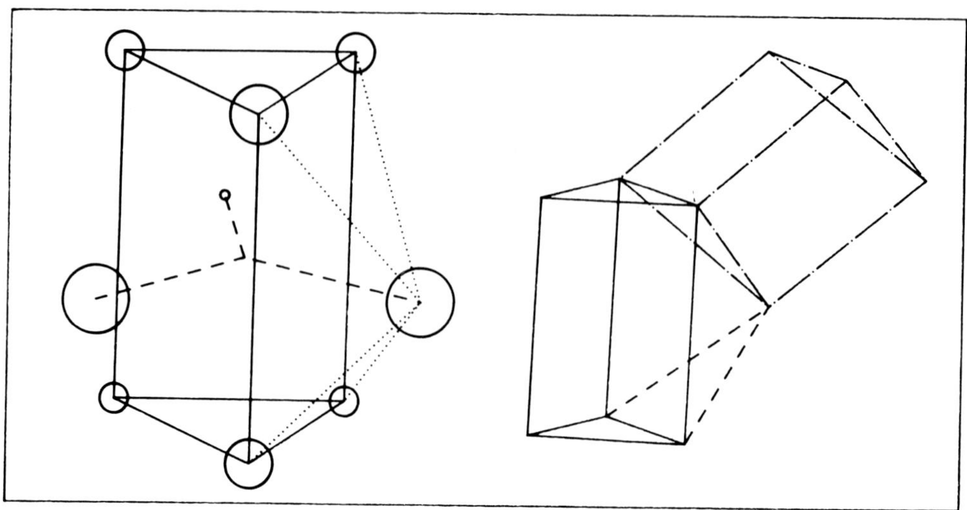

FIGURE 5. Left: A trigonal prismatic unit with a metalloid at the center (not shown) surrounded by transition metal atoms (spheres whose size implies proximity to viewer). Gaskell[43] has pointed out that such units are common in crystalline intermetallic alloys and are expected to be present in glasses as well. Each rectangular face is shown capped by another transition metal atom which forms triangular faces to which other such units can join in building the glass structure as shown on the right.

solve the set of coupled second order differential equations resulting from Newton's second law at a sequence of times. Pictorially, one is creating a series of snapshots of the atom positions. Some very interesting work has been done in this area[41,42] but is restricted by computational requirements to relatively small or periodic systems. It has not yet been possible to study systems of sufficient size to make adequate comparisons of the structure representation with diffraction data.

Extensions of the sphere packing idea have also been advanced. Each scheme attempts to study the packing of a unit which is designed to preserve some short-range order. Gilman[11] has studied correlations between viscosity data, relative strengths of atomic interactions, and eutectic compositions to conclude that the liquid state of a glass alloy already consists of molecular units which typify the short-range order in the glass. He proposed that such units be the basis for construction of an atomic model. A random packing of relatively complex units is a formidable computational problem because of the number of degrees of freedom in the problem. Gaskell[43] took the concept quite literally and supplemented the task of computing atomic coordinates by simultaneously building a physical model to approximately correspond to the set of coordinates being calculated in the computer. Human perception made the process possible. He chose the regular trigonal prism as the basic unit such as is found in the local ordering in crystalline Pd_3Si;[43] Si sites are at the geometric center of a very regular trigonal prism. Figure 5 illustrates the basic unit as well as a typical connection to neighboring units. Note that face caps are required to occur randomly in order to raise the composition to 80% Pd (the glass composition). The limitation in the procedure is that it is not possible to continue the parallel process of computer and physical models to more than a few hundred atoms. In general, Gaskell concluded that models built with certain demands on the short-range geometry of groups of atoms do yield very reasonable results; although there are somewhat sharper features in calculated statistical functions than observed, these may be due to finite size effects.

Briant and Burton[24] argued in a similar way that the basic local ordering in a glass should be an icosahedral arrangement of metal atoms. An icosahedron can be viewed as a unit with 5 planes containing 1/5/1/5/1 vertices, respectively; the vertices in the two "5" planes form regular pentagons which are in a staggered (rotated 36°) configuration with respect to each other and whose planes are perpendicular to a line formed by the vertices of the other layers. The structure has twelve C_5 axes and is thus symmetry-forbidden as a unit cell structure for a three-dimensional crystal. However, extensive theoretical studies[44] of crystal nucleation clusters have shown that the basic icosahedron and its larger analogues are among the most stable configurations of atoms known. Briant and Burton further show that diffraction characteristics of icosahedra are not inconsistent with those of metallic glasses. However an adequate test of the idea requires that a relatively large model be studied and this presents the same problem as that presented by Gilman's hypothesis.

Most of the features discussed individually above are now built into the best models of glass structure. It would appear that the reasoning behind most of the work which has developed over the last decade is sound but the models which contain the expected features and properties are constructed from an entirely different point of view.

IV. COMPUTER-SIMULATED STRUCTURES

Motivated primarily by the interesting comparison of radial distributions between the Bernal[27,28] model and the Ni–P experiments of Cargill[31] (see Figure 4), several groups began to explore the possibility of simulating dense random packings of spheres in a computer. The geometry of packing of hard spheres is considerably simpler than that of polyhedra and makes the task feasible. Discussing the atomic structure in terms of sphere packings necessitates separating metallic glass alloys into two classes: transition metal-metalloid type and metal-metal type. The former exist over a rather narrow composition range near 20% metalloid whereas the latter form glasses over broad composition ranges usually centered around the 50-50 composition.[3,13] There is very little overlap between the sets of metal elements which form one type of glass as opposed to the other. It is believed that the primary difference between the two classes is a chemical ordering[45] which the metalloid atom forces. The metal-metalloid bond is at least partially covalent in character[46] which is responsible for the strong short-range order expected by Gilman[11] and Gaskell.[43] From a sphere packing point of view, the main difference is not that there is more than one size sphere in the model but that near-neighbor metalloid pairs are energetically forbidden.

The first computer simulations were by Adams and Matheson[47] and Bennett.[48] They worked with only one size hard sphere and considered the effect of various packing algorithms on the structure of assemblages of four to five thousand spheres. The best results were obtained when each additional atom was forced to touch three existing atoms in the structure in such a way that the addition was closest to the model center. The second peak of the RDF was indeed found to be split but the ordering of the intensity of the two subpeaks was wrong. Figure 6 compares the pair correlations measured by Bennett on a 3999 atom model with the results of Finney[29] shown earlier. Clearly the outer half of the second peak is too intense and at a somewhat shorter distance than measured by Finney. Worse however, was that the density of the resulting structures monotonically decreased with model size and even near the center, where density is maximum, it did not approach that of a real glass.

Sadoc, Dixmier, and Guinier[49] were the first to explore the effects of multiple sphere sizes in computer modelling. The interaction was still expressed only as touching hard spheres, but the packing algorithm has been shown to be nonspace filling.[50] In addition, they apparently used equal weight scattering amplitudes for the various atomic species to calculate the RDF and I(K) curves, i.e., $f_{TM}(K) = f_M(K) = \text{const}$. The work is widely quoted because the comparison between the calculated I(K) and that measured from NiP alloys is quite good. It should be recognized, however, that the model suffers from a number of internal deficiencies which make it unacceptable for use in discussing structurally related phenomena.

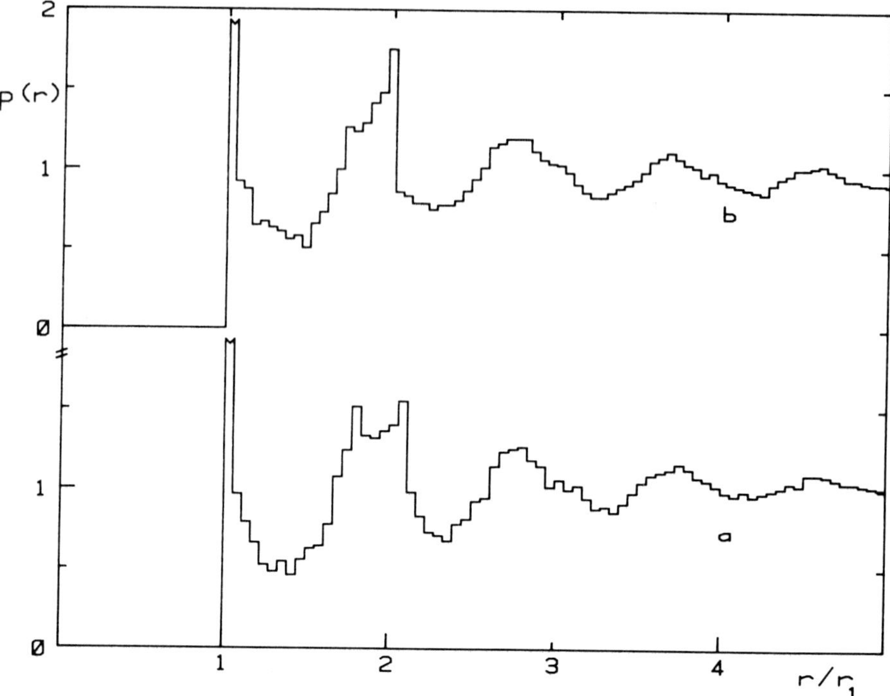

FIGURE 6. (a) The data of Finney[29] from Figure 4 is compared to (b) that deduced by Bennett[48] from a computer-generated cluster of 3999 hard spheres of equal size. The shape of the second peak is poorly reproduced and it occurs at the wrong distance as do the remaining peaks.

Ichikawa[51] was concerned with the incorrectly predicted shape of the second peak of the RDF in hard sphere models. He introduced the notion of *tetrahedral perfection* and a parameter, $0 \leq k \leq 2$, by which it is controlled. As each sphere is added it forms a tetrahedron with the three sites its touches as base; the degree of regularity (equal length sides) of these tetrahedra is, in general, random. By specifying a value of k in his construction algorithm he could reject as possible sites, all sphere positions which would have a degree of regularity less than desired; $k = 2$ requires all sites to be perfectly regular tetrahedra. He was able to resolve the problem with the second peak shape with $k \sim 1.2$. However, this was a compromise because it introduced a bit too much structure in peaks at large r. More problematic was that the monotonic density decrease was not corrected. Apparently the result of increased tetrahedral perfection is to correct some of the deficiencies of the short-range order which existed in the first models. This concept is still used[52] occasionally to build models which are approximately correct at small r; it results in an enormous computational simplicity by avoiding an energetic relaxation step.

Energetic relaxation was first published by von Heimendahl[53] and by Koskenmaki.[54] They considered portions of hard sphere models to be interacting via pairwise potentials of the Lennard-Jones and simple harmonic types, respectively. Still only including one size sphere, an improvement of second peak shape was observed. The positions of third and successive peaks, however, were at distances smaller than observed. Neither of these papers considered the effect of relaxation on density.

Boudreaux and Gregor[55] were the first to recognize the cause of the monotonic density decrease in the hard sphere models and produced algorithms which corrected the problem and also served to generate models of multicomponent alloys. After reproducing the model of Bennett, an analysis of pair correlations was carried out and analyzed before the angular

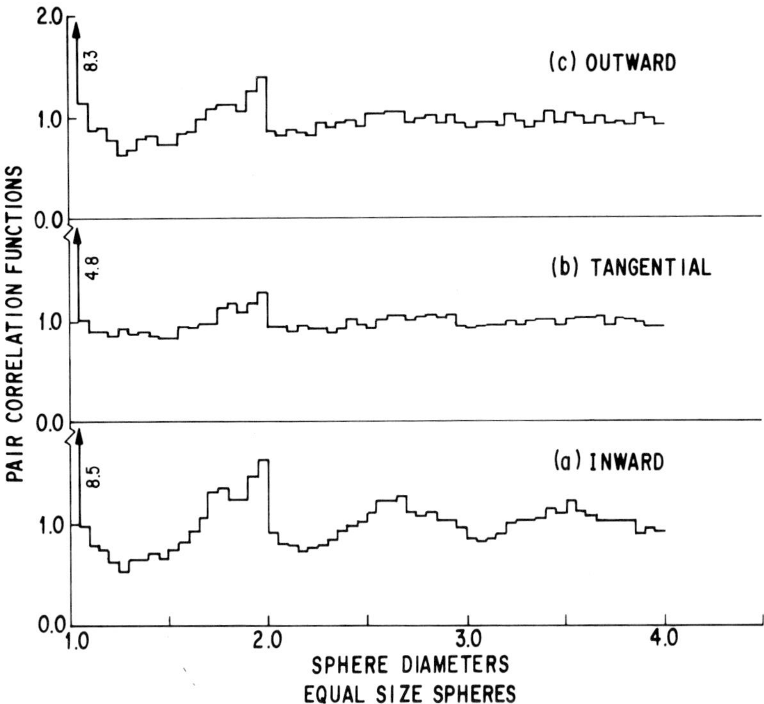

FIGURE 7. Pair correlations calculated from neighbors located in selected conical regions in a model of equal size hard spheres.[55] The regions are identified in reference to the geometric center of the model, e.g., *inward* labels the histogram depicting the probability of finding neighbors in a direction toward the center. *Tangential* implies directions within 15° of the plane normal to a radius from the center. The short-range order in this model is unphysically anisotropic.

averages over θ and ϕ were performed. A marked anisotropy was found in the structure (see Figure 7). Averaging pair correlations over selected solid angles showed that the model structure has almost no correlation beyond the first neighbor shell except when the solid angle is directed toward the center of the model from any atom considered as origin. The packing algorithm used by Bennett[48] and by Adams and Matheson[47] is very directional. In fact, it was found to become more directionally selective as the models became larger; this is the effect associated with the decrease of density with model size. Similar studies were carried out on multicomponent hard sphere models[15] but the directional packing problem was only magnified. In addition it was found that the control of alloy composition as a function of model size was not trivial. Spheres representing metalloid species are significantly smaller than those representing metals and as a result they can occupy pockets closer to the center of the model which are forbidden to the larger spheres without overlap. One cannot simply decide which species to add by use of a random number scheme because each addition is in effect adjusting the composition of the model locally. In the present spherical models, an addition modifies the composition in a spherical shell through r_a, "the radius of addition". Since r_a is dependent on the size sphere being added, composition is being controlled by different rules for different species. A more complex algorithm[15] corrects this problem but requires that the composition be recomputed for each trial site or tentative addition.

Boudreaux and Gregor[55] also explored the tetrahedral perfection scheme for multicomponent systems. The general conclusion is that it always produces too much long-range order. Energetic relaxation of homogeneous hard sphere models was carried out by the same

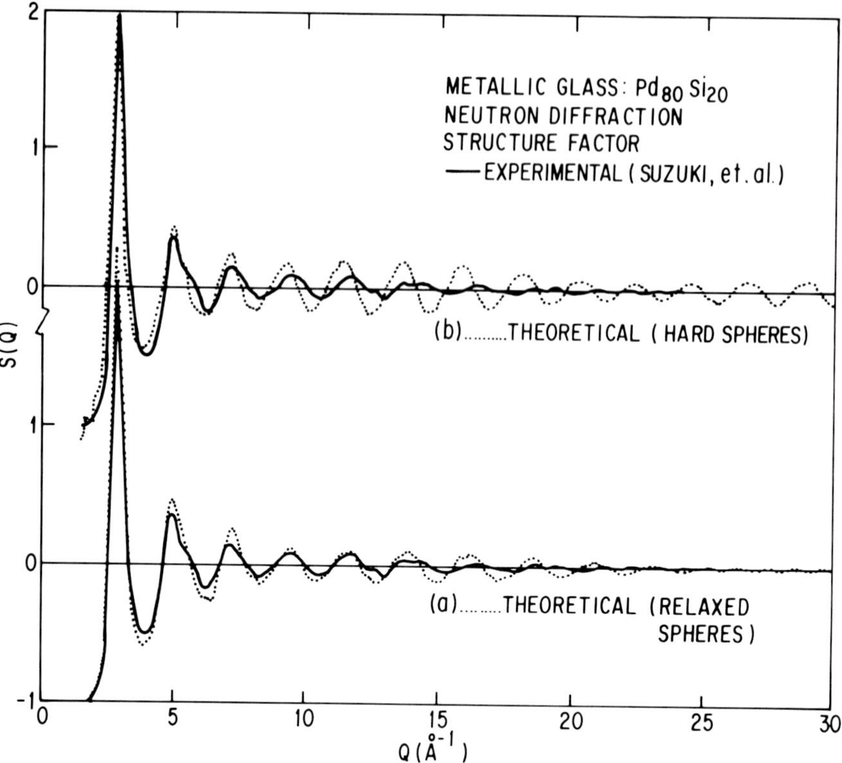

FIGURE 8. The effects of energetic relaxation on a two size sphere model[55] of $Pd_{80}Si_{20}$ compared to high resolution neutron diffraction measurements.[25] Here the reduced interference function has been calculated by Fourier transform from the RDF of the model structure and the comparison to untransformed data eliminates the effect of truncation errors.

authors.[15] In order to obtain adequate statistics for comparison of the large r peaks of the RDF to experimental data, at least 2000 spheres must be included in the relaxation. In addition, the n-component nature of the alloy requires that the relaxation be controlled by the total energy which is the sum of $n(n+1)/2$ different pair interaction potential energy functions. A conjugate gradient scheme[56] was worked out which insured a monotonic approach to the local energy minimum (the global minimum is presumably the crystalline state). The results of this process are models which fit all of the measured characteristics of metallic glass alloys. The most difficult comparison is directly to the measured I(K) curve. Figure 8 exhibits the theoretical curves superimposed on Suzuki's[25] high resolution neutron diffraction experiments on $Pd_{80}Si_{20}$. Note that the fit is good over the entire range of scattering lengths (reciprocal distances). The comparison to the interference functions requires a Fourier transform of g(r) calculated from the simulated model. Theoretically the integration must be carried all the way to infinity but practically may be terminated at some point beyond which g(r) is zero (no further correlations). A model may be constructed large enough to avoid any errors introduced by this effective truncation, but the same is not true in the inverse transformation of experimental data to g(r). The data exist only over a domain of K values fixed by the apparatus being used and frequently the domain does not span the entire nonzero range of the interference function. The effect of such truncation errors is quite important and is the reason why it is preferable to compare theory and experiment in reciprocal space.

A Lennard-Jones form, $\epsilon\,[(\sigma/r)^{12} - (\sigma/r)^{6}]$, of the pair interaction potential was used to

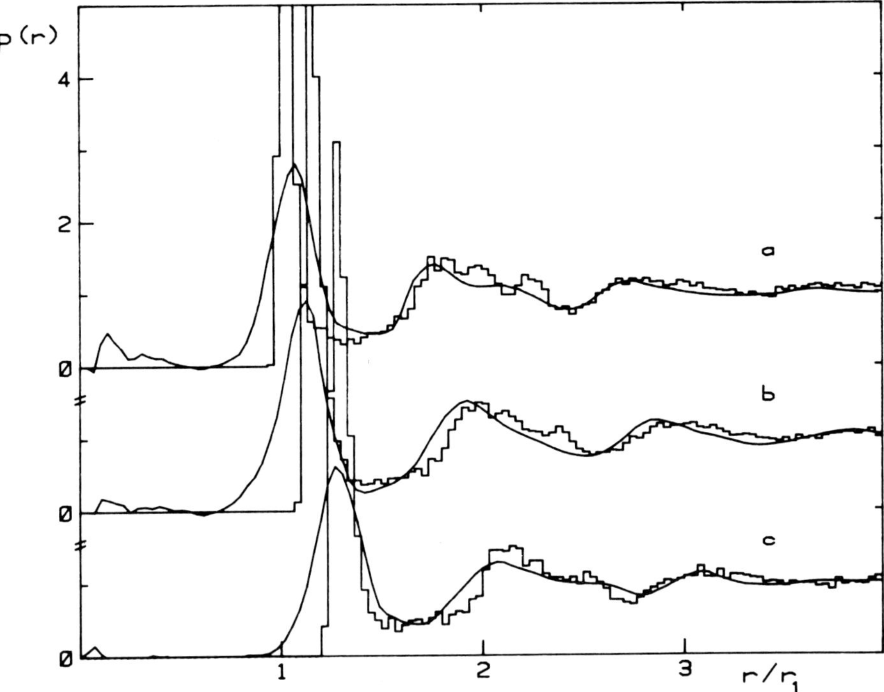

FIGURE 9. Partial pair correlations for $Cu_{57}Zr_{43}$ glass. The curves are experimentally[23] determined and the histograms are calculated from a model. (a) Cu–Cu, (b) Cu–Zr, and (c) Zr–Zr. The model was constructed without restrictions on neighbor types and relaxed under equal strength pair potentials of the Lennard-Jones form.

produce the above results. The range, σ, of each potential function was set so that its minimum occurred at a distance corresponding to the radius of the original hard sphere (this choice can be modified); the strength parameters, ϵ, were arbitrary but scaled so that the metal-metalloid interaction strength was twice as strong as the metal-metal interaction (reflecting an intuitive expectation of relative bond strengths). A very slight modification of the form of the potential[15] was made to insure proper numerical behavior of the conjugate gradient scheme of energy minimization; this involved forcing the value and derivative of each potential to zero at two large sphere diameters, the distance beyond which the effect of interactions were truncated. Other forms of the potential were experimented with including changes of exponent values in the standard Lennard-Jones form,[57] variation of relative strengths of pair potentials, and various strength Morse-type potentials. The effect was simply a broadening of the features of the RDF and pair correlation functions due to the broader range of interaction allowed by the form of other potentials. At the present time, experimental information is not sufficiently precise to select one form of the potential over another.

Another approach to computer simulation of the hard sphere structure is exemplified by the work of Kobayashi, Maeda, and Takeuchi.[58] The approach is simpler in that random positions of spheres in a unit cube are considered as long as overlap does not occur. A relaxation step follows to generate the final structure. Such models are restricted to the metal-metal type glass system because there is no capacity to simulate the correlations which prohibit metalloid-metalloid near neighbors. Figure 9 illustrates calculated pair correlation functions for a Zr–Cu alloy (by author) and compares them to measured[23] correlations. The agreement is satisfactory except for the second peak of the Zr–Zr curve; however, it is important to point out that the longer range correlations are more important than appears from this comparison for reasons described above in the discussion of truncation errors.

V. MODELING MATERIAL PROPERTIES

An acceptable model structure must simulate measured material characteristics, such as the interference and radial distribution functions. Some examples were given in the last section. There are of course many other properties which must also be reflected by a model, e.g., it must have the correct density and possess the proper number and type of neighbors in the first coordination shell. Even more important is the fact that such properties vary with alloy composition and these variations must also be correctly predicted. Simulation of effects due to variations in temperature is also highly desirable, but is much more difficult to achieve. This section reviews the present level of achievement in meeting these goals.

X-ray diffraction studies of the effect of composition variation in binary metallic glass alloys were carried out very early by Cargill[59] on the Ni–P system, and by Logan[60] on the Fe–P system. In both cases the interference and radial distribution functions exhibit very little variation; it was noted, however, that the position of the first peak of the RDF shifted inward as metalloid content was reduced. The shifts are small but quantitatively significant; they are reflections of changes in the periodicity of the peak structure of the interference functions which is accurately measured. The dominant contribution to the first peak is due to metal-metal near-neighbor pairs and the shift reflects a decrease in the size of the metal atom. The simplest view is that charge transfer between metal and metalloid is composition dependent and that as more metalloid is added, the average metal atom has more charge transferred to it. For a 5% increase in metalloid content, the measurements indicate a peak shift of 1% for Ni–P and 3% for Fe–P. All of this is consistent with the magnetic properties of these alloys;[61,62] the charge transfer is primarily from metalloid to the d-band[46] of the transition metal species. Fe–P is ferromagnetic and has empty d-states to accept the charge whereas Ni–P is paramagnetic and its d-band is nearly full. The shift in the position of the first peak cannot be expected to be simulated *a priori* by models of the type described in the last section because the position of the peak is determined by the position chosen for the minimum in the pair potential function. In fact the best models are produced by empirically adjusting the potentials so that the minima fall at positions corresponding to measured peak positions; when measurements are not available, the model builder must use data from the structure of crystalline material of nearly the same composition. It should be noted that more recent experiments have shown that there are other features of the various distribution functions which vary with composition.

High resolution neutron diffraction studies[63] have been done on a series of Pd–Si glass alloys (Figure 10) and find two important changes in the reduced radial distribution functions as the composition is changed. There are relative changes in the heights of the two subpeaks of the first neighbor shell, one corresponding to Pd–Pd correlations and the other to Pd–Si correlations; this is due primarily to the fact that the amount, c_i, of each type element is varying and thus so is its scattering strength, w_{ij}. Thus as Si content is increased the first peak grows in height. More subtle changes occur in the shape and position of the second peak, see Figure 10. Note that the inner subpeak of the second neighbor shell moves to shorter distances as metalloid content is increased. The data from models built by the author[64] reflect this trend nicely, but more important is the fact that the models can be studied for details not accessible experimentally. The exact origin of this peak shifting is isolated to the Pd–Pd correlations. The partial pair correlations of Pd–Pd interactions calculated from models are the same as those displayed in Figure 11 for Fe–P glass alloys; the two alloy systems are simulated by identical computer-generated structures.[55] Models have been built for other points on the composition scale to add further clarity to the picture. Note the same shifting of the inner half of the second subpeak. No simple geometric picture has been devised to explain this observation except for the obvious fact that an increased number of small spheres in the first neighbor shell must create deeper pockets in that shell into which

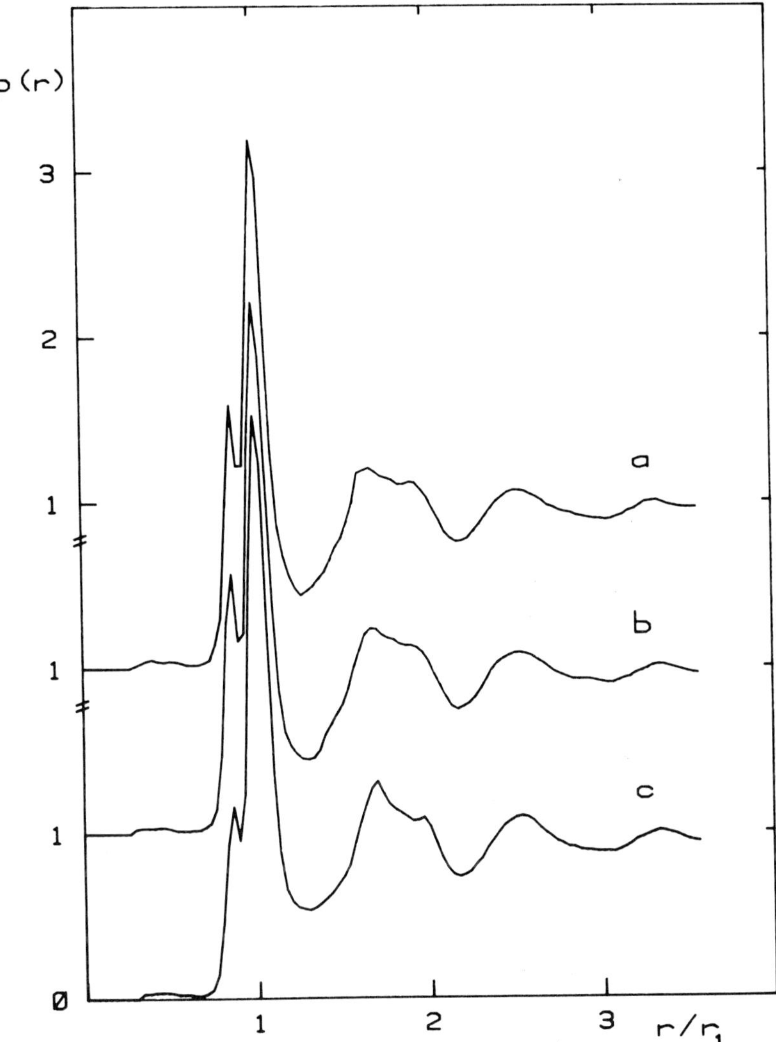

FIGURE 10. Total pair correlation functions for Pd–Si glass as a function of composition.[63] (a) $Pd_{78}Si_{22}$, (b) $Pd_{80}Si_{20}$, and $Pd_{85}Si_{15}$. Neutron diffraction resolves the Pd–Pd and Pd–Si pair contributions to the first peak. Note that the Pd–Si near-neighbor peak grows and that the inner half of the second peak shifts inward as metalloid content increases.

next-neighbor Pd sites can pack closer to the central site. This behavior of the pair correlations has been related[64] to the well-known effect of hardness increase with metalloid addition. The hardness increase is related to increased bond strength noted during the energetic relaxation; basically Pd atom pairs sample lower portions of the potential energy surface at higher metalloid content.

A number of attempts at a direct measurement of partial pair correlations have been made[16-19,65] as described in Section II. Such experiments require radiation of three separate wavelengths or a combination of radiation sources for a binary alloy and the difficulty is compounded by the fact that the data reduction requires a knowledge of the w_{ij} factors for each wavelength radiation used (and in particular their K dependence which is usually not measured or unknown). Procedures are improving to permit more accurate results and the data provide a very sensitive test of the correctness of any model structure. For example,

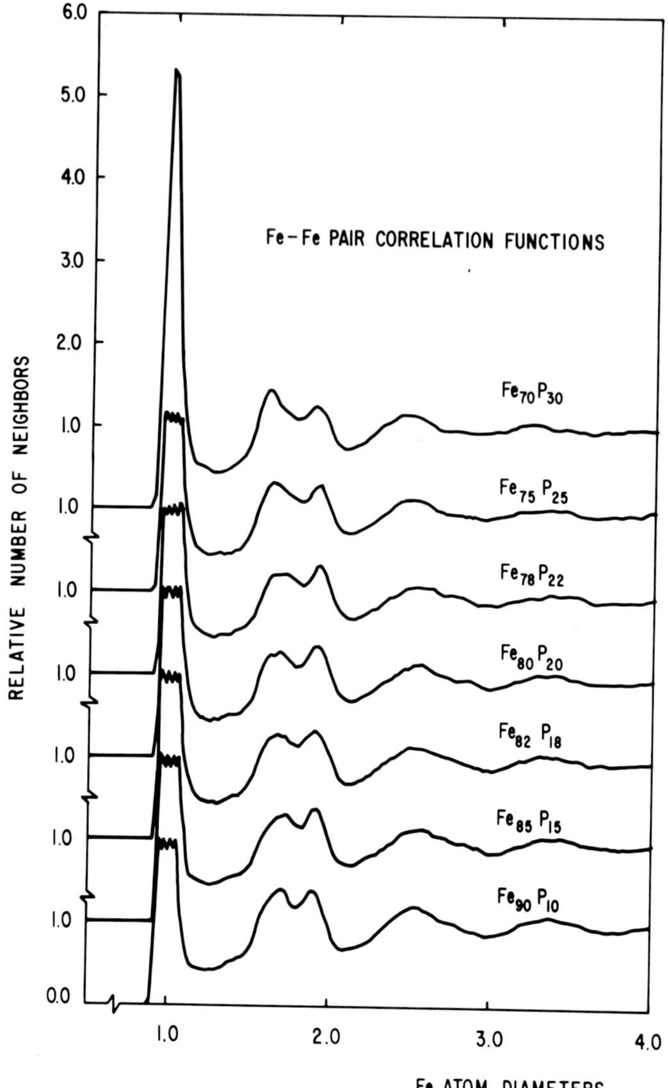

FIGURE 11. Fe–Fe partial pair correlations in models[64] with changing composition. The models serve equally well for Pd–Si glass alloys. Note the shifting inward of the inner half of the second peak as metalloid content increases. The same effect was observed experimentally and illustrated in Figure 10. The observation leads to an understanding of the hardness variation in glassy alloys (see text).

Figure 12 shows a comparison of theory[64] and experiment[65] for an $Fe_{75}P_{25}$ glass. Because $w_{P\text{-}P}$ is very small compared to $w_{Fe\text{-}Fe}$ and $w_{Fe\text{-}P}$, the P–P pairs could not be adequately resolved by the experiment. The only disagreement in the Figure is with that portion of the data (Fe–P correlations) which has the largest uncertainty. A very similar comparison holds with the models of Fujiwara and Ishii[66] who used a slightly modified procedure to relax the structure. To date there are no published results on partial pair correlations as a function of composition.

The *coordination number* of an atom site is defined as the number of near neighbors of a given type which surround it. In a disordered structure one expects variation in coordination

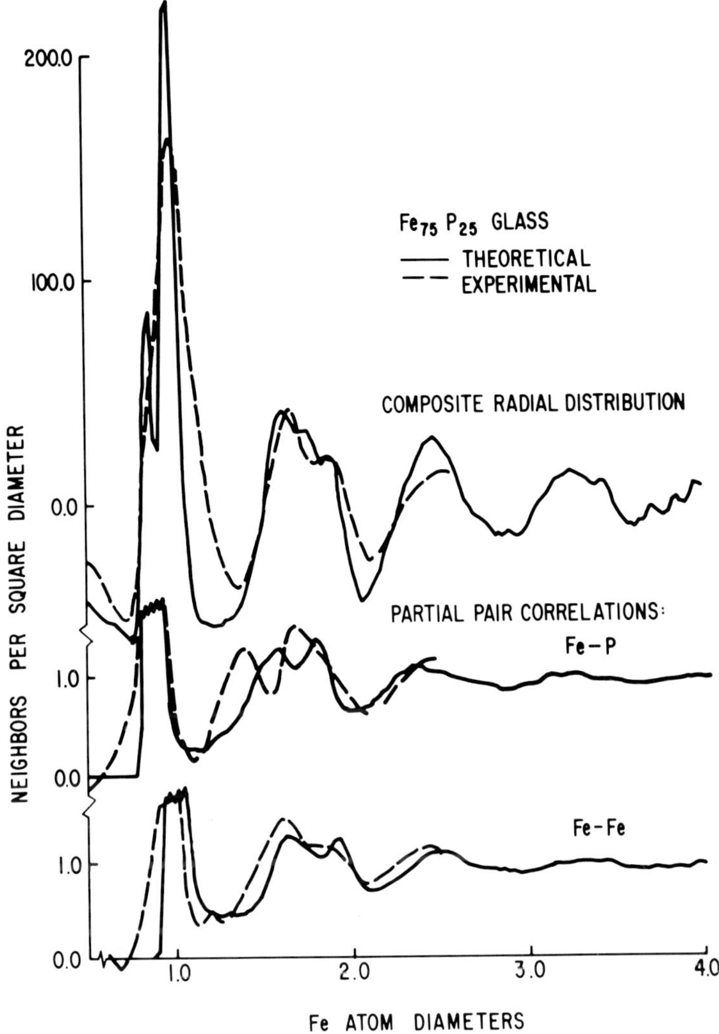

FIGURE 12. Comparison of reduced radial distribution function and partial pair correlations between theory[64] and experiment.[65] The Fe–Fe and Fe–P contributions to g(r) are resolved in the calculation from model structures. The only disagreement is in the position of the second peak of the Fe–P curve.

numbers from site to site. A statistically averaged value can be deduced from RDF data; an integral, measuring the area under the curve, yields the number of atom sites between the limits of integration, (r′,r″)

$$N = \int_{r'}^{r''} RDF(r)\, dr \tag{34}$$

In practice one usually calculates the number of atoms in the first coordination shell by integrating to the first minimum beyond the first peak. For a multicomponent alloy, the complete set of partial pair correlations are required to extract the coordination numbers because the RDF contains weighted contributions (although approximations can be made if the weight factors are known). For this reason, the coordination is most easily studied from

Table 1
COORDINATION NUMBERS

Composition	70/30	76/24	78/22	80/20	82/18	85/15	90/10
Pd-Si							
(Fe-P)							
TM—TM	10.34	10.43	10.72	10.22	10.49	10.59	10.65
TM—M	3.29	2.52	2.19	2.05	1.78	1.44	0.85
Total	13.63	12.95	12.91	12.27	12.27	12.03	11.50
M—TM	8.40	8.21	8.26	8.44	8.36	8.46	8.21
Fe—B							
TM—TM		10.97		10.79		11.25	
TM—M		1.94		1.51		1.06	
Total		12.91		12.30		12.31	
M—TM		6.89		6.41		6.56	
Pd—Ge							
TM—TM				10.23			
TM—M				2.18			
Total				12.41			
M—TM				9.29			

Experimental

$N_{Pd\text{-}Pd}$ in Pd Ge	= 10.23	$N_{Ge\text{-}Pd}$	= 9.29
$\langle N_{Fe\text{-}Fe} \rangle$ in Fe P	= 10.49	$\langle N_{P\text{-}Fe} \rangle$	= 8.33
$\langle N_{Fe\text{-}Fe} \rangle$ in Fe B	= 10.99	$\langle N_{B\text{-}Fe} \rangle$	= 6.62

$N_{Pd\text{-}Pd}$ in Pd_{80} Si_{20} (SUZUKI)	= 10.4
$N_{Pd\text{-}Si}$ in Pd_{80} Si_{20} (SUZUKI)	= 1.8
	12.2
$N_{Fe\text{-}Fe}$ in Fe_{75} P_{25} (WASEDA)	= 10.4
$N_{Fe\text{-}P}$ in Fe_{75} P_{25} (WASEDA)	= 2.6
	13.0
$N_{P\text{-}Fe}$ in Fe_{75} P_{25} (WASEDA)	= 8.1
$N_{Ga\text{-}Pd}$ in Pd_{78} Ge_{22} (HAYES)	= 8.6 ± 0.5

models. Table 1 exhibits data comparing coordination numbers calculated from various simulated structures and comparison to some experiments. When these data were first published[64] the most striking observation was that the coordination of the metalloid site by metals was essentially constant as composition varied. It was suggested that some sort of molecular unit of reasonable regularity surrounded the metalloids and that these units always formed during the relaxation with excess metal squeezing between the units. The assertion agreed with the physical expectations of Gaskell[43] and of Gilman[11] and does turn out to be the case as will be described in the next section.

Studies of theoretical models provide a much richer source of information because the distributions of the coordination numbers of particular type sites can be calculated. Each sphere in the model is examined and a list of its near neighbors prepared. The explicit identification of near neighbors always involves a decision about the distribution of ‘‘bond lengths’’; the only criterion which can be used for identification is the distance between an atom and a neighbor. If the distance is short enough the neighbor is identified as ‘‘near’’. The hard sphere touching distance (sum of sphere radii) is not adequate because during relaxation some spheres pull slightly off of others yet remain as near neighbors topologically. Thus a tolerance is added to the sum of radii; the distance to the first minimum beyond the first peak in a pair correlation function is much too far. Empirically, 10% was found[69] to be an appropriate tolerance. From the list of near neighbors of each atom an occurrence frequency distribution is then easily deduced. As a check on such predictions from models,

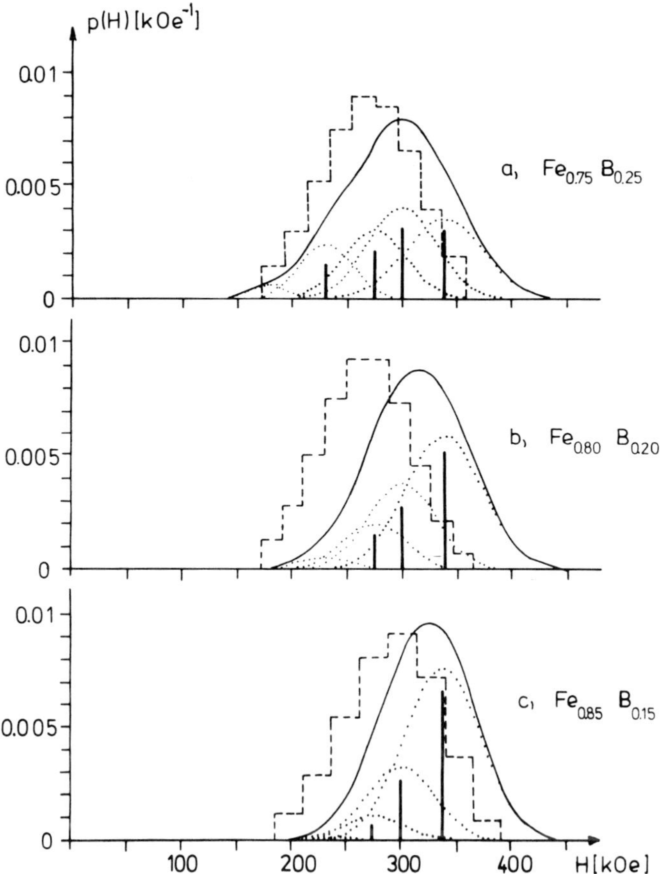

FIGURE 13. Hyperfine field distributions in Fe–B glass alloys. Dashed histograms are experimental data.[7] The solid curves are the sum of dotted gaussians representing the contributions of various type Fe sites in the model structure; the vertical bars are at the estimated hyperfine field of an Fe site with a given coordination and the length of the bar relates to the probability of finding such a site. The general trends are right but the models require improvement.

hyperfine field distributions for iron alloys can be calculated (making some assumptions about the relationship between iron site hyperfine fields and the number of metalloid near neighbors) and checked by Mössbauer spectroscopy. Comparative studies as a function of composition were made on alloys from the Fe–B glass system.[7] Figure 13 summarizes the results; while the general shift of the calculated peak position with composition is correct, the position and shape are not fit very well. Some improvements to the model are clearly required, but have not been found to date; it is speculated[67] that the model is emphasizing a particular form of short-range order while the actual Fe–B alloy system may be a mixture of two forms. The forms may be related to the orthorhombic and tetragonal phases of crystalline Fe_3B.

The relatively recent EXAFS experiment[5,6] shows great promise for resolving some of the puzzles of metallic glass structure. The results of Hayes et al.[5] on $Pd_{80}Ge_{20}$ corroborate the interatomic distances deduced from diffraction studies and provide conclusive evidence that chemical ordering does exist. No metalloid-metalloid pairs were found bringing further emphasis to the fact that structural models based on metal atoms alone are not adequate. A rather different but very interesting experiment[6] using the same technique probed temperature effects in a multicomponent glass containing about 40% of both Fe and Ni. It was clearly observed that the effects of temperature on the near environment of Ni were very different from Fe. The implication is that the bonding is significantly different and responds differently to thermal perturbation and agrees with the deduction made from the early composition studies.[59-62]

Density is one of the most difficult quantities to determine accurately from model structures. Basically one has to count the number of atoms in a specific volume, multiply by their atomic weight, and with the help of Avogadro's number calculate the density. This is easy in crystalline materials where one can determine the exact number of atoms in a unit cell, but in a glass there is no such cell. Using various volumes containing many atoms the density can be determined, but one always has a problem with atoms on the bounding surface — do they belong inside or outside! The best models yield densities within 5% of actual values by such computations but are always on the low side. The resolution of the problem lies in dividing space in such a way that an exact count of atoms can be made in some volume. The Voronoi polyhedral construction[29] does exactly this. The smallest polyhedron determined by planes which perpendicularly bisect all vectors from a given atom to all others defines the so-called Voronoi polyhedron. It contains exactly one atom and the collection of such polyhedra fills space exactly as does the unit cell or Wigner-Seitz cell. The volume of each Voronoi polyhedron allows one to define the exact density at a point; an average then gives the density of the alloy being simulated by the structure. Finney[68] has done the complete polyhedral analysis on several Pd–Si and Fe–P models and the density as a function of composition has been determined.[69] Figure 14 shows the comparison to actual density. It is important to note that the dependence of metal sphere size on composition as determined by Logan[60] is essential to producing the correct values; i.e., the scaling of the unit spheres of the model to the real world is composition dependent. The Fe–B alloy system exhibits a breaking curve which was originally thought[70] to be due to microstructural variations; it may be due to the presence of two forms of short-range order[67] as described above, the break occurring upon exhaustion of one form.

Emphasis thus far has been on the transition metal-metalloid type glass alloys. Studies of the metal-metal glasses have been thought to be a little easier since chemical ordering is not expected and since many of the alloys have constituent atoms which would be described by spheres of nearly the same "size". Thus interpretations of experimental data in terms of the single size sphere models are frequently thought to be adequate. For example it is easy to understand the shape of the first peak in the RDF for a $Gd_{18}Co_{82}$ alloy (see Figure 15a); the three subpeaks are due to Co–Co, Co–Gd, and Gd–Gd pairs, respectively. As the Gd content is reduced (Figure 15b), the Gd–Gd pairs decrease in number so that the third feature does not appear. However it is clearly necessary to simulate more subtle effects in order to reproduce all of the features measured;[71] Cargill has built models using Ichikawa's tetrahedral perfection scheme that reproduce most of the experimental results.[71] Even more recently, neutron studies on Ti-containing glasses[72-74] have suggested that there may indeed be some kind of "chemical" short-range ordering (CSRO) in the structure. Ti is a very useful material in a neutron scattering experiment because radiation scattered from Ti is 180° out of phase with that from most other alloy constituents and can emphasize certain features which would not otherwise be observable.[72,73] While the interpretation is not completely clear, there is definitely a peak in F(k) prior to the usual first peak. Fourier transformation of F(k) with and without the pre-peak[72] shows a decrease in Cu–Cu correlations in the first neighbor shell and an increase in the second. Combined X-ray and neutron studies[73] and neutron and EXAFS studies[74] provide a more detailed suggestion; namely that chemical short-range order provides for a greater than average coordination by opposite atomic species. This is another area where model builders must attempt to simulate the facts, and thereby suggest the details

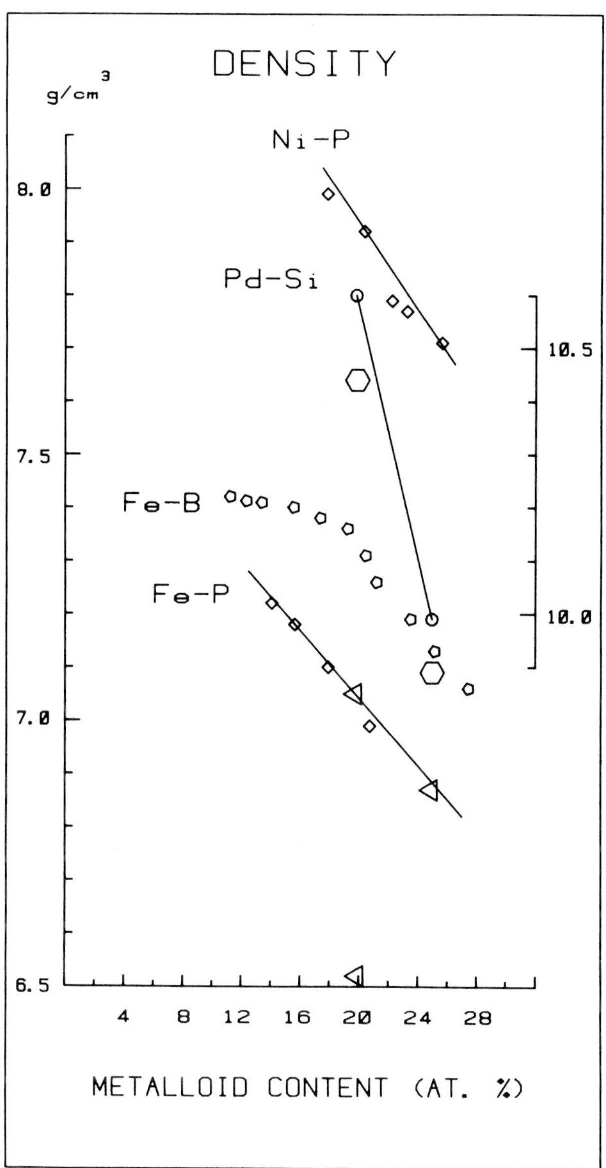

FIGURE 14. Density of metallic glass alloys as metalloid content is varied. The left scale is for the Fe and Ni-based glasses and the right scale for those of Pd. Theoretical points for Fe–P glasses are shown by triangles; the point at 6.52 g/cm^3 for $Fe_{80}P_{20}$ is calculated using the same Fe–Fe separation as $Fe_{75}P_{25}$. The density is brought into agreement by shortening the Fe–Fe length as observed by Logan.[60] The calculated values for Pd–Si are shown by hexagons.

of the atomic structure. A clear problem is that one would expect the CSRO to be nearly identical in Ti–Cu and Zr–Cu alloy systems. If deductions from neutron measurements are correct, why do the models of Zr–Cu, containing no special CSRO, fit X-ray data so well (see Figure 9)?

The simulation of thermal effects in the glass structure is very much in its infancy. One example is the work of Bennett and Finney[75] who performed a molecular dynamics study on 1000 atoms from the central portion of a single size sphere model. The computations were too complex to obtain results of the desired statistical accuracy but certain very sugges-

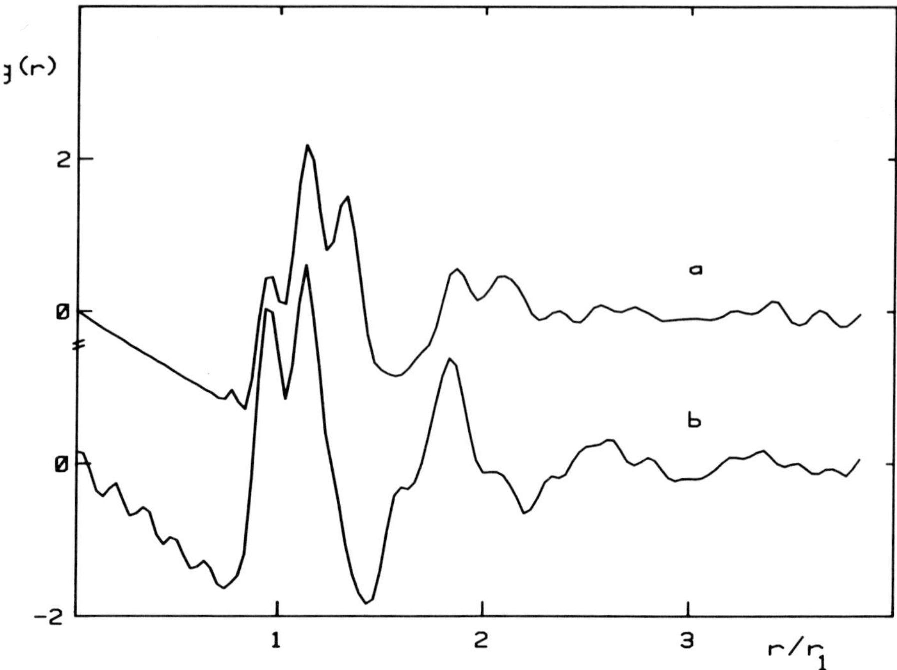

FIGURE 15. Reduced radial distribution functions for (a) $Gd_{32}Co_{68}$ and $Gd_{18}Co_{82}$ as measured by Cargill.[71] The dependence of the variation in structure of the composite first peak is easily understood (see text), but the remaining structure is difficult to interpret and has not yet been satisfactorily modeled.

tive observations were made. The splitting of the second peak disappeared above the glass transition temperature and its shape changed on heating, the outer subpeak moved to smaller r values and decreased in prominence. This is qualitatively similar to measurements made by Waseda et al.[76] on an Fe–P–C alloy after annealing at various temperatures. The theoretical and experimental data are compared in Figure 16. These results are only illustrative of much more work which will be needed to clarify the detailed behavior of atom motion as a function of temperature.

The relationship of the local structure variations to magnetic phenomena — in particular magnetic anisotropy — is a subject of considerable importance because of the application of magnetic alloys as low loss core materials. This subject is too big to add to the present review, but the reader is referred to other chapters of the present volume and to the review of Cochrane, Harris, and Zuckermann.[77] The models studied in the latter are not the most sophisticated, but significant relationships are established which are useful in understanding the magnetic properties of amorphous alloys, especially those containing rare-earth elements.

VI. STRUCTURAL CONCEPTS DEDUCED BY MODELING

The most significant contribution to the understanding of metallic glass properties to come from modelling is the importance of short-range order in the atomic structure. Studies on models have clarified and confirmed intuitive expectations[11,43] derived from properties of liquid and crystalline materials. Although almost all of the models built and studied have started from a sphere-packing approach, several analyses have uncovered significant order in the immediate vicinity of any given site. The constancy of metalloid coordination number described earlier is a hint of this phenomenon, as are the Mössbauer[7] measurements. Machizaud, Kuhnast, and Flechon[78] have suggested that careful diffraction studies on Ni–P glasses

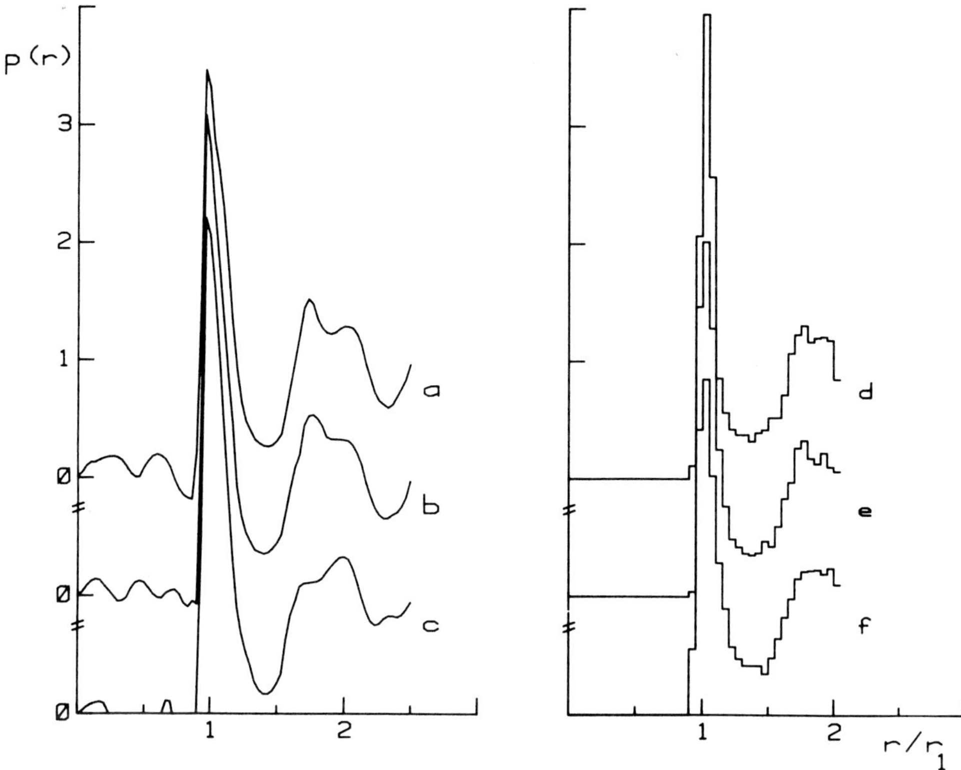

FIGURE 16. Pair correlation functions for an Fe–P–C glass annealed[76] at 330°C for (a) 0 min, (b) 1300 min, and (c) 1700 min. The shape of the second peak should be compared to the molecular dynamics snapshot models[75] at T* = (d) 0.13, (e) 0.33, and (f) 0.36.

show evidence of regular icosahedral clusters as implied earlier by Briant and Burton.[24] Any regularity in local atomic environments provides an extremely useful starting point for describing the physics and chemistry of metallic glasses in more detail and so it is quite important to discover and to catalog such order. This section reviews the first few attempts to do this on appropriately constructed models.

The coordination numbers in Table 1 were calculated by the integration technique (see Equation 34) and suffer from lack of detail and inaccuracies. In particular the upper limit of integration for any given site could either include too few or too many atoms as near neighbors, and further the number is an average with no clue as to the shape of the distribution it is associated with. Motivated by the observation that metalloid coordination is composition independent, further analyses were carried out to more explicitly show the distribution of site coordinations.[69] The vicinity of each atom was searched to find neighbors within the hard sphere touching distance ±10% as explained in the last section. A summary of the near-neighbor data is presented in Figure 17 for various models of the Fe–B alloy series. The first column details the metalloid environment while the second and third separate the metal environment into two parts, metalloid neighbors and metal neighbors. The average metalloid coordination is six as reported in Table 1 but it is now clear that more than half of the metalloid sites actually exist at the average coordination. Even more striking is the fact that the distribution sharpens about the average near the center of the known composition range suggesting that a more stable structure is favored by more uniform short-range order. The distribution of B sites about Fe centers is understood as having a decreasing average as less B is included in the alloy; the data of Figure 17(b) were used to determine the hyperfine

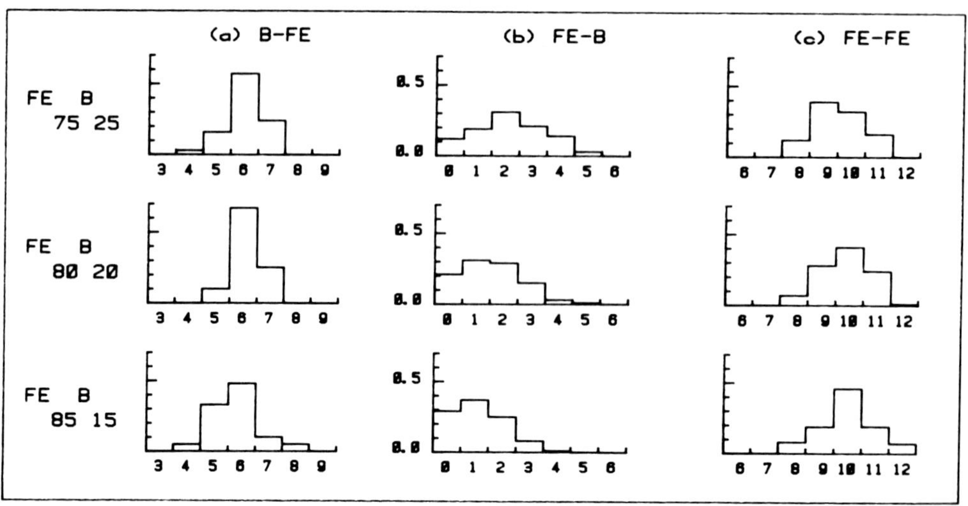

FIGURE 17. Probability distribution of the number of near-neighbors in models of Fe–B glass alloys. (a) Fe atoms around B atoms, (b) B atoms around Fe atoms, and (c) Fe atoms around Fe atoms. Near-neighbors are defined as atoms within the hard sphere touching distance ± 10%.

field distributions shown in Figure 13. The distributions in the last column sharpen about the average and the average moves to higher values as the composition approaches pure iron which in the fcc structure is 12.

Several approaches have been used to elucidate the geometric significance of the short-range order preferences deduced by near-neighbor counting. Yamamoto and Doyama[34] (YD) and Maeda and Takeuchi[79] (MT) have each studied the local order in single size sphere models (of amorphous iron). Both have performed Voronoi polyhedral analysis on models produced in a slightly different way and find evidence for clustering of atoms into an arrangement best described by the regular icosahedron. In fact, MT find that nearly every atom in their model belongs to one of the Kasper's polyhedra[80] especially the 13 atom icosahedron. They suggest, as did Briant and Burton,[24] that the icosahedral arrangements are responsible for the splitting and shape of the second peak in the pair correlation function; their reasoning is based on the various intersphere separations in the regular icosahedron. YD in addition perform a "cavity analysis" which is roughly similar to the Bernal hole study performed by Finney.[29] While YD did not concern themselves with regularity of local geometry, they did find regular occurrences of tetragonal dodecahedra and capped trigonal prisms (see Figure 5a) which they found chained together. These chains can be regarded as dislocation cores and provide evidence that relaxed DRPS models possess at least some of the characteristics of the defective lattice models.

Boudreaux and Frost[69] sought to clarify the regularities in the number of near neighbors in a different way. Their studies were carried out on models of *binary* alloys in distinction to the work described in the last paragraph and contain more complications but should have a better correspondence to actual glass structure. In order to avoid any possible bias in the identification of near neighbors on the dependences of the distributions shown in Figure 17, the following study was performed. The metalloid atoms were removed from the model, leaving behind vacancies. The size of these vacancies should be related in some way to the local geometry of the environment of the metalloid site. As a measure of the distribution of sizes, a sphere was inscribed in each hole of the metalloid free model and the distribution of sphere diameters plotted. Figure 18 exhibits data for two Pd-based alloys. There are two peaks in each curve, the one corresponding to the larger inscribed spheres relates to the metalloid vacancies. The other and larger peak describes the distribution of smaller voids,

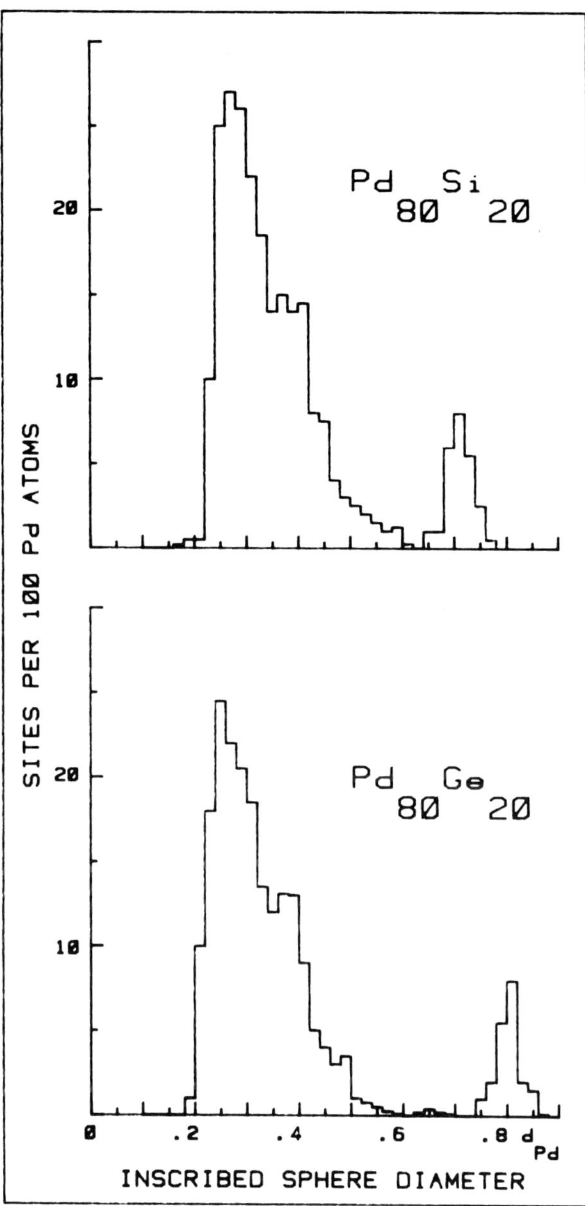

FIGURE 18. Calculated hole size distributions for Pd-based glass models. The metalloids were removed before analysis and the holes in the remaining structure are characterized by the size of an inscribed sphere (measured in units of Pd atom diameters). There are clearly two classes of small holes as indicated by the substructure on the right of the first peak. The second peak shows the distribution of hole sizes previously occupied by metalloid atoms.

which one would expect in any sphere packing, but which are not large enough to be occupied by metalloid atoms. Not surprisingly the distribution of B vacancy sizes in Fe–B alloy models shows significant sharpening near the center of the composition range (see Reference 69 for details). The Pd data shown in Figure 18 did show an unexpected effect, namely the clear shoulder on the right of the peak describing unoccupied voids. While further exploration of this effect is required there are clearly structural details not yet understood. It is speculated that the present data are connected with the so-called "two level system" which has been invoked to explain low temperature transport phenomena. Any atom occupies the common vertex of several polyhedra surrounding the small voids in a structure; a slight motion of one vertex (inhibited by the potential barrier in the two level system) could cause an inversion of a pair of small voids each having a different size described by the two subpeaks of the size distribution.

To assess the local order in more geometric detail, the Fe–B models were studied using computer graphic techniques to inspect directly the immediate surroundings of any given atomic site.[69] The polyhedra observed are not exactly regular but occur with varying degrees of distortion. For example, a continuous range of distortions is available between the octahedron and the trigonal prism; thus it is necessary to establish explicit procedures which define whether a specific neighbor polyhedron is one shape or another. Two approaches were identified and studied. The first cataloged the geometries of polyhedra formed by the network of near-neighbor connections (using some specified tolerance) among the atoms in the first neighbor shell of the sites at the center of the polyhedra. The second method used the directions of the set of vectors pointing to near neighbors to determine the shape of the polyhedra. Some of the results of the second scheme will be discussed in the following paragraphs.

Primary concern is with the geometry of the metal atoms surrounding the smaller metalloid atoms. In the Fe–B models most boron atoms have six iron near neighbors. The most regular shapes with six vertices are the octahedron and the trigonal prism, although several distorted shapes may be defined in between these two. The metal atoms can be thought of as being loose to move out of position to produce a range of distorted shapes. An octahedron can be created by twisting a trigonal prism along an axis through the triangular faces until one triangle is oriented 60° from the other.[69] Using this notion it is easy to devise a method which determines whether the surroundings of a particular boron site are more nearly octahedral or trigonal prismatic. The coordinates of the site to be investigated and the coordinates of its near neighbors are selected from a model and the positions of the atoms are displayed on a CRT screen at points corresponding to their (x,y) coordinates, with symbols whose size is scaled by their z coordinates (larger implies closer to viewer). The figure could be rotated about any axis permitting the consideration of each triangular face in turn. A second triangular face was sought nearly parallel to the first. For six vertices, if the two triangles had nearly congruent xy projections the polyhedron was called a trigonal prism. If one triangle was "upside down" with respect to the other, the polyhedron was called an octahedron. For more than six vertices the procedure becomes more complex, but it is usually possible to visually identify a distorted regular polyhedron possibly with atoms in face capping or edge-bridging positions.

An interesting trend was found in the polyhedra surrounding the boron atoms. Most of the shapes in $Fe_{75}B_{25}$ were identified as trigonal prisms (as in the crystalline form of the compound); as boron content was reduced a more frequent occurrence of octahedra was observed. Typical examples are shown in Figure 19. The trigonal prism in Figure 19(a) is distorted on one of its rectangular faces to allow for the almost near-neighbor iron atom which caps the face. The interatomic forces under which these structures were produced were twice as strong between iron and boron than between iron and iron, and clearly favors the capping of the prism and in some cases distortion of a face to allow closer proximity of the capping iron to the attractive boron site. Figure 19(b) shows an octahedrally coordinated boron site from $Fe_{85}B_{15}$. The reason for the more common occurrence of octahedra when less boron is present is due to packing optimization; but studies of the more difficult matter of packing of the polyhedra themselves as suggested by Gilman[11] and Gaskell[43] have not yet been completed. An intuitive explanation is that the trigonal prism presents only five

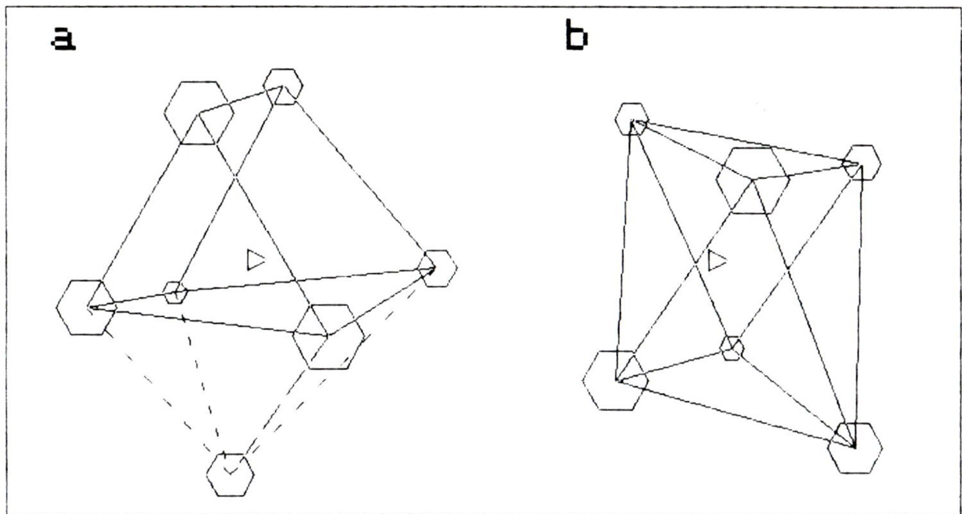

FIGURE 19. Geometry of short-range order around B sites in Fe–B glass alloys. The boron is at the center of the structure shown by the triangle. The surrounding iron atoms are drawn smaller if farther away from viewer. (a) Trigonal prismatic coördination in $Fe_{75}B_{25}$. (b) Octahedral coordination in $Fe_{85}B_{15}$.

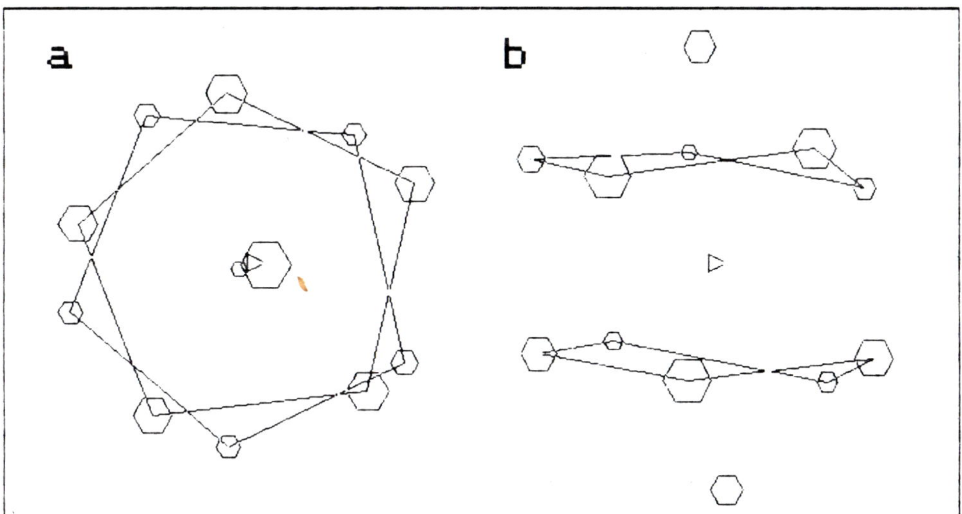

FIGURE 20. Icosahedral units found around Fe (triangle) sites in Fe–B alloy models. (a) "top view" shows staggered pentagonal rings and central line of atoms. (b) Same unit rotated 90° about the x-axis to show the planarity of the rings.

faces for next-nearest-neighbor iron sites, while the octahedron presents eight. In the alloys with small boron content, each boron site must accommodate more iron sites in its vicinity and this is favored by octahedral short-range symmetry.

Iron site surroundings were also considered in the computer graphic studies and there one could almost always identify an icosahedral arrangement of neighbors when the coordination number was 12 (the average value, see Figure 17b and c). Figure 20 shows an icosahedrally coordinated iron site from $Fe_{80}B_{20}$. It is shown in two views so that the regularity of the shape can be appreciated. Figure 20(a) shows the staggered pentagonal rings and the three atoms on the axis nearly aligned while Figure 20(b) shows a 90° degree rotation about the x-axis to illustrate the planarity of the rings.

The degree of regularity in the glass structure was not really expected a few years ago, but will clearly play an increasing role in simplifying the understanding of the physics of these complex materials. Abstraction from the actual atomic arrangement to simpler, more regular pictures will make progress possible in certain areas such as the calculation of electronic structures. Understanding from model building will then permit the application of ideas from abstract results to real glass alloys.

ACKNOWLEDGMENTS

I wish to thank G. Fish and R. Hasegawa for their careful reading of this manuscript and for helpful suggestions. R. C. Morris generously assisted in the preparation of the discussion of defective lattice models. I. Vincze has provided some valuable insights prior to their publication and I have profited greatly from discussions with him.

REFERENCES

1. **Duwez, P.,** The structure and properties of alloys rapidly quenched from the liquid state, *Trans. Am. Soc. Met.*, 60, 607, 1967.
2. **Gilman, J. J.,** Metallic glass alloys, *Phys. Today*, 28, 46, 1975.
3. **Chaudhari, P., Giessen, B. C., and Turnbull, D.,** Metallic glasses, *Sci. Am.*, 242, 98, 1980.
4. **Sayers, D. E., Lytle, F. W., and Stern, E. A.,** Extended X-ray absorption fine structure, *Adv. X-ray Anal.*, 3, 248, 1970.
5. **Hayes, T. M., Allen, J. W., Tauc, J., Giessen, B. C., and Hauser, J. J.,** Short-range order in metallic glasses, *Phys. Rev. Lett.*, 40, 1282, 1978.
6. **Wong, J., Lytle, F. W., Greegor, R. B., Lieberman, H. H., Walter, J. L., and Luborsky, F. E.,** EXAFS studies of the dynamic disorder of Fe and Ni in metallic glasses, in *Proc. 3rd Int. Conf. Rapidly Quenched Metals*, Cantor, B., Ed., The Metals Society, London, 1978, 345.
7. **Vincze, I., Boudreaux, D. S., and Tegze, M.,** Short-range order in Fe-B metallic glass alloys, *Phys. Rev. B*, 19, 4896, 1979.
8. **Gaskell, P. H., Smith, D. J., Catto, C. J. D., and Cleaver, J. R. A.,** Direct observation of the structure of a metallic alloy glass, *Nature (London)*, 281, 465, 1979.
9. **Cochran, W.,** Scattering properties of amorphous structures, *AIP Conf. Proc.*, 177, 1974.
10. **Cargill, G. S., III,** Structural investigation of non-crystalline Nickel-Phosphorus alloys, *J. Appl. Phys.*, 42, 12, 1970.
11. **Gilman, J. J.,** Structure of ferrous eutectic liquids, *Philos. Mag. B*, 37, 577, 1978.
12. **Wagner, C. N. J.,** Direct method for determination of atomic-scale structure of amorphous solids (X-ray, electron and neutron scattering), *J. Non-Cryst. Solids*, 31, 1, 1978.
13. **Cargill, G. S., III,** Structure of metallic alloy glasses, *Solid State Phys.*, 30, 227, 1975.
14. **Duffy, M. G., Boudreaux, D. S., and Polk, D. E.,** Systematic generation of random networks, *J. Non-Cryst. Solids*, 15, 435, 1974.
15. **Boudreaux, D. S. and Gregor, J. M.,** Structure simulation of metal-metalloid glasses. III, *J. Appl. Phys.*, 48, 5057, 1977; **Boudreaux, D. S.,** Structure simulation of metal-metalloid glasses. II, in *Amorphous Magnetism II*, Levy, R. A. and Hasegawa, R., Eds., Plenum Press, New York, 1977, 463.
16. **Waseda, Y., Okazaki, H., and Masumoto, T.,** The structure of amorphous iron-phosphorus alloys, in *Proc. Int. Conf. Structure of Non-Crystalline Materials*, Taylor and Francis, London, 1977, 202.
17. **Bletry, J. and Sadoc, J. F.,** Determination of the three partial interference functions of an amorphous Cobalt-Phosphor ferromagnet by polarized neutron scattering, *J. Phys. F*, 5, L110, 1975.
18. **Sadoc, J. F. and Dixmier, J.,** Structural investigation of amorphous Co-P and Ni-P alloys by combined X-ray and neutron scattering, *Mater. Sci. Eng.*, 23, 187, 1976.
19. **Egami, T., Williams, R. S., and Waseda, Y.,** Structure of amorphous alloys studied by energy dispersive X-ray diffraction method, in *Proc. 3rd Int. Conf. Rapidly Quenched Metals*, Cantor, B., Ed., The Metals Society, London, 1978, 318.
20. **Bhatia, A. B. and Thornton, D. E.,** Structural aspects of the electrical resistivity of binary alloys, *Phys. Rev. B*, 2, 3004, 1970.

21. **Ruppersberg, H., Lee, D., and Wagner, C. N. J.,** Observation of chemical short-range order in an amorphous Ni-Ti alloy, *J. Phys. F,* 10, 1645, 1980.
22. **Wright, J. G.,** Structure of amorphous Co deposits, *J. Phys. F,* 4, L187, 1974.
23. **Mizoguchi, T., Kudo, T., Irisawa, T., Watanabe, N., Niimura, N., Misawa, M., and Suzuki, K.,** Structure of a metal-metal amorphous alloy, Cu-Zr determined by neutron diffraction, in *Proc. 3rd Int. Conf. Rapidly Quenched Metals,* Cantor, B., Ed., The Metals Society, London, 1978, 384.
24. **Briant, C. and Burton, J. J.,** Icosahedral microclusters: a possible structural unit in amorphous alloys, *Phys. Status Solidi B,* 85, 393, 1978.
25. **Suzuki, K., Fukunaga, K., and Masumoto, T.,** Partial structures in amorphous Pd-Si alloy, *Sci. Rep. Res. Inst. Tohoku Univ. Ser. A,* 26, 1, 1976.
26. **Turnbull, D. and Cohen, M. H.,** On the free-volume model of the liquid-glass transition, *J. Chem. Phys.,* 52, 3038, 1970.
27. **Bernal, J. D.,** Geometry of the structure of monatomic liquids, *Nature (London),* 185, 68, 1960.
28. **Bernal, J. D.,** Analysis of the structure of monatomic liquids, *Proc. R. Soc. London Ser. A,* 280, 299, 1964.
29. **Finney, J. L.,** Random packings and the structure of simple liquids. I. The geometry of random close packing, *Proc. R. Soc. London Ser. A,* 319, 479, 1970.
30. **Polk, D. E.,** The structure of glassy metallic alloys, *Acta Metall.,* 20, 485, 1972.
31. **Cargill, G. S., III,** Dense random packing of hard spheres as a structural model for non-crystalline metallic solids, *J. Appl. Phys.,* 41, 2248, 1970.
32. **Whittaker, E. J. W.,** The cavities in a random close-packed structure, *J. Non-Cryst. Solids,* 28, 293, 1978.
33. **Frost, H. J.,** Hole statistics in dense random packing, *Acta Metall.,* to be published.
34. **Yamamoto, R. and Doyama, D.,** The polyhedron and cavity analyses of a structural model of amorphous iron, *J. Phys. F,* 9, 617, 1979.
35. **Nagel, S. R.,** The Polk model and the structure of metallic glasses, *Scr. Metall.,* 12, 43, 1978.
36. **Koizumi, H. and Ninomiya, T.,** A dislocation model of amorphous metals, *J. Phys. Soc. Jpn.,* to be published.
37. **Morris, R. C.,** Disclination-dislocation model of metallic glass structure, *J. Appl. Phys.,* 50, 3250, 1979.
38. **Rivier, N.,** Disclination lines in glasses, *Philos. Mag.,* 40, 859, 1979.
39. **McQuarrie, D. A.,** *Statistical Mechanics,* Harper & Row, New York, 1976, 269.
40. **Weeks, J. D.,** One and two component models for the structure of metallic glasses. An integral equation approach, *Philos. Mag.,* 35, 1345, 1977.
41. **Stillinger, F. H. and Weber, T. A.,** Amorphous state studies with the Guassian core model, *J. Chem. Phys.,* 70, 4879, 1979.
42. **Woodcock, L. V., Angell, C. A., and Cheeseman, P.,** Molecular dynamics studies of the vitreous state. Simple ionic systems silica, *J. Chem. Phys.,* 65, 1565, 1976.
43. **Gaskell, P. H.,** A new structural model for transition metal-metalloid glasses, *Nature (London),* 276, 484, 1978; **Gaskell, P. H.,** A new structural model for amorphous transition metal-phosphides, in *Proc. 3rd Int. Conf. Rapidly Quenched Metals,* Cantor, B., Ed., The Metals Society, London, 1978, 277.
44. **Hoare, M. and Pal, P.,** Statistics and stability of small assemblies of atoms, *J. Cryst. Growth,* 17, 77, 1972.
45. **Chen, H. S. and Park, B. K.,** Role of chemical bonding in metallic glasses, *Acta Metall.,* 21, 395, 1973.
46. The argument is summarized concisely in a recent paper: **Allen, J. W., Wright, A. C., and Connell, G. A. N.,** Hybridization, short-range order and entropy in metallic glass formation, *J. Non-Cryst. Solids,* to be published.
47. **Adams, D. J. and Matheson, A. J.,** Computation of dense random packings of hard spheres, *J. Chem. Phys.,* 56, 1989, 1972.
48. **Bennett, C. H.,** Serially deposited amorphous aggregates of hard spheres, *J. Appl. Phys.,* 43, 2727, 1972.
49. **Sadoc, J. F., Dixmier, J., and Guinier, A.,** Theoretical calculation of dense random packings of equal and non-equal sized hard spheres. Applications to Amorphous metallic alloys, *J. Non-Cryst. Solids,* 12, 46, 1973.
50. **Finney, J. L.,** Modelling the structures of amorphous metals and alloys, *Nature (London),* 266, 309, 1977.
51. **Ichikawa, T.,** The assembly of hard spheres as a structure model of amorphous iron, *Phys. Status Solidi A,* 29, 293, 1975.
52. **Cargill, G. S., III,** Structural models for amorphous alloys, *At. Energy Rev.,* to be published.
53. **von Heimendahl, L.,** Metallic glasses as relaxed Bernal structures, *J. Phys. F,* 5, L141, 1975.
54. **Koskenmaki, D. C.,** Model of the structure of amorphous metals, *Mater. Sci. Eng.,* 23, 207, 1976.
55. **Boudreaux, D. S. and Gregor, J. M.,** Structure simulation of metal metalloid glasses. I, *J. Appl. Phys.,* 48, 152, 1977.
56. **Reeves, C. M.,** The method of conjugate gradients, *Comput. J.,* 7, 149, 1964.

57. **Machlin, E. S.,** Modified pair potentials and alloy phase stability, *CALPHAD,* 1, 361, 1977.
58. **Kobayashi, S., Maeda, K., and Takeuchi, S.,** Computer simulation of atomistic structure of binary amorphous alloys, *J. Phys. Soc. Jpn.,* to be published.
59. **Cargill, G. S., III,** Structural investigation of non-crystalline nickel-phosphorus alloys, *J. Appl. Phys.,* 41, 12, 1970.
60. **Logan, J.,** The Radial Distribution Function of Amorphous Iron-Phosphorus Alloys, Rep. CALT-822-65, California Institute of Technology, Pasadena, 1974, unpublished; **Logan, J.,** The low temperature electrical resistivity of amorphous Iron-Phosphorus alloys, *Phys. Status Solidi A,* 32, 361, 1975.
61. **O'Handley, R. C. and Boudreaux, D. S.,** Magnetic properties of transition metal-metalloid glasses: a charge transfer model, *Phys. Status Solidi A,* 45, 607, 1978.
62. **Takacs, L.,** Electronic structure of transition metal-metalloid crystalline and glassy alloys, *Phys. Status Solidi A,* 56, 371, 1979.
63. **Fukunaga, T., Misawa, M., Fukamichi, K., Masumoto, T., and Suzuki, K.,** Compositional study on short-range structure of Pd-Si and Fe-B amorphous alloys, in *Proc. 3rd Int. Conf. Rapidly Quenched Metals,* Cantor, B., Ed., The Metals Society, London, 1978, 325.
64. **Boudreaux, D. S.,** Theoretical studies on structural models of metallic glass alloys, *Phys. Rev. B,* 18, 4039, 1978.
65. **Waseda, Y., Okazaki, H., and Masumoto, T.,** The Structure of Amorphous Iron-Phosphorus Alloys, Rep. 1667, Research Institute for Iron, Steel and Other Metals, Tohoku University, Sendai, 1978.
66. **Fujiwara, T. and Ishii, Y.,** Structural analysis of models for the amorphous alloy Fe(100-x)-P(x), *J. Phys. F,* 10, 1901, 1980.
67. **Vincze, I. and Boudreaux, D. S.,** unpublished.
68. **Finney, J. L.,** private communication.
69. **Boudreaux, D. S. and Frost, H. J.,** Short-range order in theoretical models of binary metallic glass alloys, *Phys. Rev. B,* 23, 1506, 1981.
70. **Hasegawa, R. and Ray, R.,** Iron-boron metallic glasses, *J. Appl. Phys.,* 49, 4174, 1978; **Ray, R., Hasegawa, R., Chou, C.-P., and Davis, L. A.,** Iron-boron glasses: density, mechanical and thermal behavior, *Scr. Metall.,* 11, 973, 1977.
71. **Cargill, G. S., III,** private communication.
72. **Cowlam, N., Sakata, M., and Davies, H. A.,** Neutron diffraction measurement of the structure factor of a Cu-Ti metallic glass, *J. Phys. F,* 9, L203, 1979.
73. **Ruppersberg, H., Lee, D., and Wagner, C. N. J.,** Observation of chemical short-range order in an amorphous Ni-Ti alloy, *J. Phys. F,* 10, 1645, 1980.
74. **Raoux, D., Sadoc, J. F., Lagarde, P., Sadoc, A., and Fontaine, A.,** Local structure in a Cu(2)Ti Amorphous alloy by EXAFS and X-ray diffraction, in *Proc. Liquid and Amorphous Metals IV Conf.,* Grenoble, July, 1980.
75. **Bennett, C. H. and Finney, J. L.,** Structural changes at the Lennard-Jones glass transition, in *Rapidly Quenched Metals,* Grant, N. J. and Giessen, B. C., Eds., MIT Press, Cambridge, 1975, 231.
76. **Waseda, Y., Okazaki, H., and Masumoto, T.,** Current views on the structure and crystallization of metallic glasses, *J. Mater. Sci.,* 12, 1927, 1977.
77. **Cochrane, R. W., Harris, R., and Zuckermann, M. J.,** The role of structure in the magnetic properties of amorphous alloys, *Phys. Rep.,* 48, 1, 1978.
78. **Machizaud, F., Kuhnast, F. A., and Flechon, P. J.,** Mise en evidence de l'existence d'agregats icosahedriques dans les depots chemiques amorphes de Ni, P, J., *Ann. Chim. (Paris),* 3, 177, 1978.
79. **Maeda, K. and Takeuchi, S.,** Geometrical characterization of computer-constructed metallic amorphous structures, *J. Phys. F,* to be published.
80. **Kasper, J. S.,** *Theory of Alloy Phases,* American Society for Metals, Metals Park, O., 1956, 264.

Chapter 2

THEORY OF MAGNETISM IN NONCRYSTALLINE SOLIDS

T. Kaneyoshi

TABLE OF CONTENTS

I. INTRODUCTION

In recent years, magnetism in structurally amorphous alloys has become the subject of experimental and theoretical interest as a topic in solid state physics. The atomic sites in these solids are inequivalent both structurally and magnetically. Because of this disordered structure, many interesting physical properties not observed in the corresponding crystalline magnets are just now becoming apparent.[1] Our purpose in this chapter is to give a comprehensive view of the theoretical aspects on this now rapidly growing area of magnetism.

II. EXCHANGE INTERACTION AND MAGNETIC ORDERING

The question of whether or not ferromagnetism may exist in noncrystalline substances was theoretically studied first by Gubanov.[2] As ferromagnetism is caused by exchange interaction between neighboring atoms, there is no need to have a rigorous periodicity in the distribution of magnetic atoms. As predicted by Gubanov, a number of amorphous ferromagnets have been found. An important suggestion by Gubanov was that some non-ferromagnetic crystalline materials would become ferromagnetic upon transition to an amorphous state although the interatomic distances are larger in the amorphous than in the crystalline system. This argument is based on the fact that the existence of ferromagnetism in amorphous states should be determined by the relative relation between the range of exchange interaction and the configuration of magnetic atoms; the Curie temperature, T_c, of an amorphous magnet is, on the molecular field approximation, given by

$$3\ k_B T_c = S(S+1) \int J\ (R_{ij}) g(R_{ij}) d^3 R_{ij} \tag{1}$$

where S is the spin value and $g(R_{ij})$ is the atomic density of type j atoms a distance R_{ij} away from a type i atom. If we assume the exchange interaction J(R) has the form given by the Bethe-Slater curve like Figure 1, with an increase in R, J(R) tends to become positive. In the amorphous state the mean distance between atoms is, in general, greater than in the crystalline state. Thus it is easily understood that some materials which are not ferromagnetic in the crystalline state, can be ferromagnetic in the amorphous state. On the other hand, for large values of R, the absolute value of J(R) decreases, which for ferromagnetic materials can lead to the lowering of the Curie temperature. After the suggestion of Gubanov, the possibility of a material being ferromagnetic in the amorphous state but not in its crystalline state was investigated by some authors.[3-4] Recently, some examples of the above possibility were found; amorphous FeF_2 becomes ferromagnetic with a paramagnetic Curie temperature $T_c = 22 \pm 1$ K in contrast to the Néel temperature $T_N = 78$ K found for crystalline antiferromagnetic FeF_2.[5] Also, amorphous Gd-noble metal alloys show a spontaneous magnetization below a Curie temperature, although the crystalline alloy with the same composition is antiferromagnetic.[6] In this way, the character of magnetic ordering can depend entirely on the structurally amorphous nature of a system.

Now, how should we take the form of exchange interaction as a function of the distance between atoms? One way may be to take the Bethe-Slater curve of direct exchange interaction in crystalline magnets. At the present stage, lack of detailed knowledge of the electronic structure in amorphous magnets is a major stumbling block in understanding the definite form of the exchange interaction. However, due to lack of long-range atomic ordering in the amorphous state, we may generally conclude that a description by localized spins with short-range interactions is better for the amorphous magnetic systems than for the corresponding crystalline system because of a strong localization tendency of electrons in amorphous systems. In fact, amorphous rare earth alloys show large residual resistivity amounting

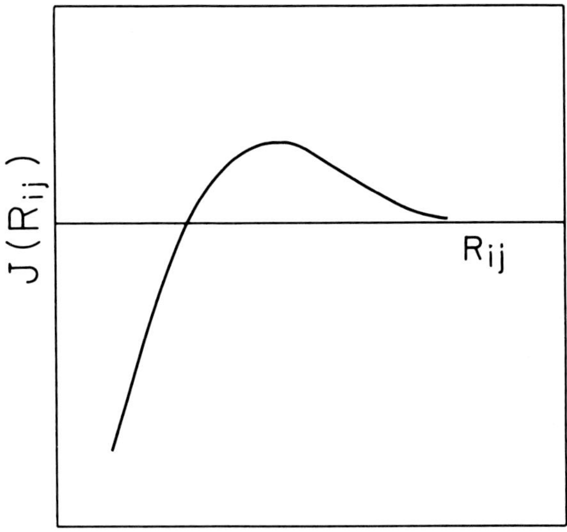

FIGURE 1. The Bethe-Slater curve for exchange interaction.

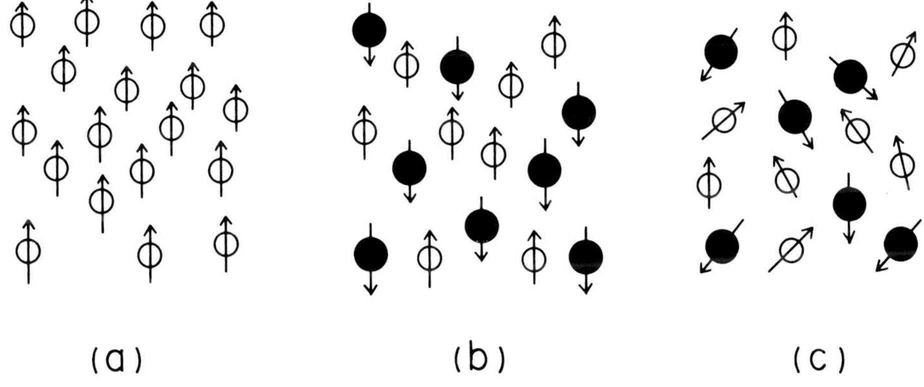

FIGURE 2. The arrangements of magnetic moments normally found in amorphous magnets; (a) aligned ferromagnet, (b) ferrimagnet, and (c) sperimagnet (or canted ferrimagnet).

to a few hundreds of $\mu\Omega$ cm, which means that the mean free path of the conduction electrons is of the order of the interatomic distances. The standard RKKY expression for indirect exchange interaction based on plane wave states may no longer hold for amorphous materials. Therefore, the standard RKKY expression should be replaced by another form including some effects of the short mean-free-path of the conduction electrons.[7] On the other hand, Grest and Nagel[8-9] conclude that to explain the magnetic properties of metallic glasses it is necessary to include superexchange interactions between next-nearest neighbor magnetic atoms being separated by a metalloid atom in addition to the direct and indirect exchange interactions.

The atomic magnetic moments in amorphous magnets usually take one of the arrangements shown in Figure 2.[10] Pure amorphous transition metals and amorphous transition metal-metalloid alloys are reasonably well described as aligned ferromagnets as in Figure 2(a). The transition metal-heavy rare earth alloy systems are usually ferrimagnets, such as in Figure 2(b), with the rare earth moments (black circles) pointing oppositely to the transition metal moments (white circles) forming a two-subnet work structure. Coey[10] terms the spin configuration of Figure 2(c) as a sperimagnet; and if the white circles are all replaced by black circles, it becomes a speromagnet or an asperomagnet. The same configuration also has been called a canted ferrimagnet.[11] From this figure, we can immediately see that, in contrast to the case of crystalline magnets, antiferromagnetism may not be a suitable ordering phase in amorphous magnets. For example, the superexchange is sensitive to structural aspects such as bond angle and distance so that long range antiferromagnetic order may be absent in amorphous insulating magnets. On the other hand, since the exchange mechanism of transition metal alloys can be described in terms of band magnetism, the long-range antiferromagnetic order could be present in their amorphous states. Nevertheless, spin glass-like ordering with random spin orientations is generally observed instead of antiferromagnetism in amorphous magnets, since the local field at each site can easily take a random direction because of the topologically disordered structure. This subject will be discussed in more detail in Section V.

III. EFFECTIVE FIELD THEORY

The first step in the interpretation of the magnetic properties of a solid is usually the application of an effective field theory. The theory can be relied on for an appropriate description of the major aspects of the phenomena being studied. In Section A, effective field theories of amorphous magnets and spin glasses are discussed from a unified point of view. In connection with the theory of spin glasses, we present in Section B an effective field theory of amorphous "random" antiferromagnets discussed by Simpson.[12]

A. Ferromagnetic Case

For a spin 1/2 system, the expectation of a spin S_i is given by

$$\frac{<S_i>}{S_i} = \sigma_i = \tanh\left(\frac{1}{2}\beta\Sigma_j J_{ij}\sigma_j\xi_j\right), \quad \beta = \frac{1}{k_B T} \tag{2}$$

The factor ξ_i is a random variable which takes the value 1 or 0 according to whether or not the site i is occupied by a magnetic atom. Equation 2 is an expression for a particular configuration of magnetic atoms, and hence it is necessary to take an average $<\ldots>_r$ over all their possible configurations. For this purpose, we introduce a variable defined by

$$m_i = \frac{<\xi_i\sigma_i>_r}{<\xi_i>_r} \tag{3}$$

which means the averaged moment of σ_i. We transform Equation 3 as follows[13]

$$m_i = \frac{1}{<\xi_i>_r}\int d\omega \tanh\left(\frac{1}{2}\beta\omega\right)\left[-\frac{1}{\pi}\operatorname{Im}<\xi_i G(\omega+i\epsilon)>_i\right] \tag{4}$$

with

$$G(\omega+i\epsilon) = <(\omega + i\epsilon - \sum_j J_{ij}\sigma_j\xi_j)^{-1}>_{ir} \tag{5}$$

where $<\ldots>_{ir}$ indicates a conditional average over all possible distributions of magnetic atoms except the i-th site. $<\ldots>_i$ denotes the configurational average of the i-th atom. In order to perform the ensemble average of Equation 5, we introduce random functions X_j and x_i defined by

$$x_j = \sigma_j \xi_j, \qquad X_i = \sum_j J_{ij} (x_j - \langle x_j \rangle_{ir}) \tag{6}$$

Then, we must expand Equation 5 with X_i and perform the conditional random average:

$$G(\omega + i\epsilon) = \sum_{\nu=0}^{\infty} \frac{\langle (X_i)^\nu \rangle_{ir}}{(\omega + i\epsilon - W_i)^{\nu+1}} \tag{7}$$

where W_i is a conditionally averaged mean field defined by

$$W_i = \sum_j J_{ij} \langle x_j \rangle_{ir} \tag{8}$$

To further proceed with the present approach, let us introduce two approximations as follows:

Approximation I (Reference 15):

$$\langle (X_i)^{2\nu} \rangle_{ir} = [\langle (X_i)^2 \rangle_{ir}]^\nu$$

$$\langle (X_i)^{2\nu+1} \rangle_{ir} = 0 \tag{9}$$

In this case Equation 4 reduces to

$$m_i = \frac{1}{2\langle \xi_i \rangle_r} \left[\langle \xi_i \tanh \left\{ \frac{1}{2} \beta (W_i + \Delta_i) \right\} \rangle_i + \langle \xi_i \tanh \left\{ \frac{1}{2} \beta (W_i - \Delta_i) \right\} \rangle_i \right] \tag{10}$$

with

$$\Delta_i = \sqrt{\langle (X_i)^2 \rangle_{ir}} \tag{11}$$

Approximation II (Gaussian approximation):

$$\langle (X_i)^{2\nu} \rangle_{ir} = \frac{2\nu!}{(2!)^\nu} \frac{1}{\nu!} \{ \langle (X_i)^2 \rangle_{ir} \}^\nu$$

$$\langle (X_i)^{2\nu+1} \rangle_{ir} = 0 \tag{12}$$

Equation 4 is then given by

$$m_i = (2\pi)^{-1/2} \int_{-\infty}^{\infty} e^{-\frac{y^2}{2}} \frac{1}{\langle \xi_i \rangle_r} \langle \xi_i \tanh \left\{ \frac{1}{2} \beta (W_i + y\Delta_i) \right\} \rangle_i dy \tag{13}$$

At this point, if we neglect the positional correlation between magnetic atoms, or if we take the approximations,

$$\langle \ldots\ldots \rangle_{ir} \simeq \langle \ldots\ldots \rangle_r$$

$$\langle x_\ell x_{\ell'} \rangle_r \simeq \langle x_\ell \rangle_r \langle x_{\ell'} \rangle_r \quad \text{for } \ell \neq \ell' \tag{14}$$

Equations 10 and 13 reduce to

$$m = \frac{1}{2}\left[\tanh\left\{\frac{1}{2}\beta(W_0 + \Delta)\right\} + \tanh\left\{\frac{1}{2}\beta(W_0 - \Delta)\right\}\right] \quad (15)$$

for Approximation I, or

$$m = (2\pi)^{-\frac{1}{2}}\int_{-\infty}^{\infty} e^{-\frac{y^2}{2}} \tanh\left\{\frac{1}{2}\beta(W_0 + y\Delta)\right\} dy \quad (16)$$

for Approximation II, with

$$W_0 = mc\,J_0 \qquad \text{and} \qquad \Delta = J_1\,C^{1/2}q^{1/2} \quad (17)$$

where we define $J_0 = \sum_i J_{ij}$, $J_1^2 = \sum_j (J_{ij})^2$ and $<\xi_i>_r = c<<1$. We also introduce the parameter $q = q_i$ as follows

$$q_i = \frac{<\xi_i\sigma_i^2>}{<\xi_i>_r} \quad (18)$$

In order to obtain self-consistent equations for Equation 15 or Equation 16, we must study Equation 18. Within the framework of the approximation in Equation 14, we have

$$q = 1 - \frac{1}{2}\left[\text{sech}^2\left\{\frac{1}{2}\beta(W_0 + \Delta)\right\} + \text{sech}^2\left\{\frac{1}{2}\beta(W_0 - \Delta)\right\}\right] \quad (19)$$

for Approximation I, or

$$q = 1 - (2\pi)^{-\frac{1}{2}}\int_{-\infty}^{\infty} e^{-\frac{y^2}{2}} \text{sech}^2\left\{\frac{1}{2}\beta(W_0 + y\Delta)\right\} dy \quad (20)$$

for Approximation II. For each approximation, if we put $\Delta = 0$, Equations 15 and 16 reduce to

$$m = \tanh\left(\frac{W_0 m}{2k_B T}\right) \quad (21)$$

which is a well-known result.

For the possible solutions corresponding to these approximations with $J_0 > 0$, we can find three choices as follows:

(a) $m \neq 0, q \neq 0$; ferromagnetic phase
(b) $m = 0, q \neq 0$; "spin glass" phase
(c) $m = 0, q = 0$; paramagnetic phase

Using $k_BT/C^{1/2}J_1$ and $J_0C^{1/2}/J_1$ as coordinate axes, phase diagrams are illustrated in Figure 3. Figure 3.1 is the phase diagram corresponding to the usual Weiss approximation. That is to say, in order to obtain the "spin glass" phase, it is important to take account of the "structural fluctuation" Δ_i from the averaged mean field W_0 for each spin. The transition from phase (a) to phase (b) takes place when $W_0 \simeq \Delta$. For the situation in which $W_0 < \Delta$

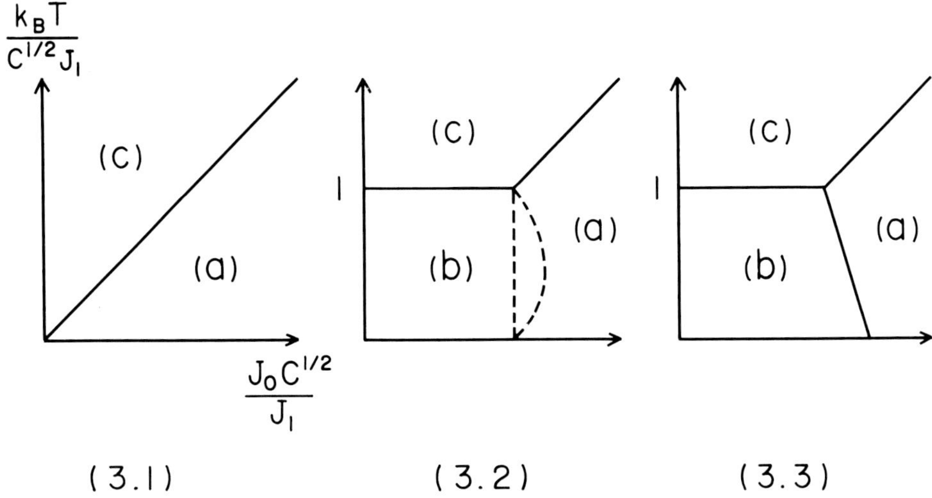

FIGURE 3. The phase diagrams of a spin system, calculated within the M.F.A.; (3.1) Weiss approximation, (3.2) two-δ function approximation, and (3.3) Gaussian approximation.

[phase (b)], therefore, the spin ordering may be spin glass like. The transition temperature T_g from phase (b) to phase (c) is given by

$$T_g = \frac{J_1 c^{1/2}}{2k_B} \qquad (22)$$

for each approximation. Experimentally, it is known that the ordering temperature is a function of the impurity concentration ($\propto c^{0.5\text{-}0.75}$) for spin glass alloys.[14] Except for the scalings of coordinate axes, Figure 3 is identical to that of Southern[15] using the random bond model. In the region of Figure 3.2 bounded by the broken curves, the solution is unstable and the transition from phase (a) to phase (b) is first order. If the interaction is of an RKKY type, the molecular field at a given point is given by a sum of N independent, random variables $J_{ij}\sigma_j$, because of the oscillatory nature of the RKKY interaction. For $N \rightarrow \infty$, the central limit theorem may be valid and the limiting distribution may be a Gaussian (Approximation II). As discussed in the previous section, however, the rapid decay of the polarization in an RKKY sense in amorphous magnets may severely restrict the range of the indirect exchange between magnetic atoms. In this case, there is no reason for the Gaussian approximation to be more appropriate in the "spin glass" phase than the other approximation. In solid solution magnetic alloys, we may conclude that Approximation I is more appropriate for the "spin glass" phase, since the distribution of internal fields may be given by the two δ-functions. In fact, Adachi et al.[16] found that the transition from phase (a) to phase (b) in solid solution magnetic alloys is similar to a first order one.

Now, let us consider amorphous ferromagnets. In this case, the order parameter is given by the averaged moment m, so that the approximations $\langle x_j \rangle_{ir} = m \langle \xi_j \rangle_{ir}$ and $\langle x_\ell x_{\ell'} \rangle_{ir} = m^2 \langle \xi_\ell \xi_{\ell'} \rangle_{ir}$ may be allowed in Equation 11. Then, W_i is given by

$$W_i = m \int J(R_{ij}) g(R_{ij}) dR_{ij} = m \overline{W}_0$$

In order to evaluate the factor $\langle \xi_\ell \xi_{\ell'} \rangle_{ir}$, we need to know the three sites spatial correlation function $f_3(R_i, R_\ell, R_{\ell'})$. At the present time, function f_3 for an amorphous solid is unknown.[17] When the relation $\Delta_i = m\Delta_0$ is substituted into Equation 15, we have

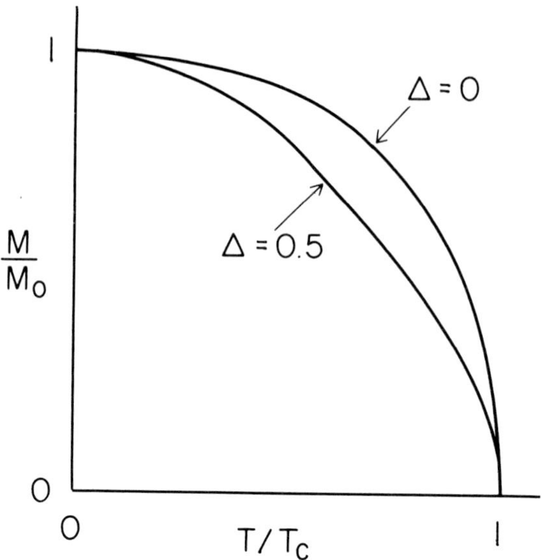

FIGURE 4. The reduced magnetization curve for an amorphous ferromagnet.

$$m = \frac{1}{2}\left[\tanh\left\{\frac{m}{2k_BT}(W_0 + \Delta_0)\right\} + \tanh\left\{\frac{m}{2k_BT}(W_0 - \Delta_0)\right\}\right] \quad (23)$$

which is identical to the result obtained by Handrich.[18] The reduced magnetization curve obtained from Equation 23 is drawn in Figure 4. In this approximation, Curie-Weiss temperature T_c is given by $T_c = W_0/2k_B$, independent of the structural fluctuation Δ. The quantity Δ, however, affects the magnetization behavior in such a way that an increase of Δ leads to a depression of magnetization over the entire temperature range for $T<T_c$, which is generally observed in amorphous ferromagnets.[19]

The form of Equation 23 implies an approximate distribution of internal fields (or Equation 7) as given by Figure 5(a). The approximation is valid for a small structural fluctuation Δ. Increasing the structural fluctuation of local fields results in decrease of the intensity and broadening of the width of the two δ-function distributions of local fields. Accompanied by the decrease of the intensity, a new distribution of local fields appears at a mean field value W_i (or Figure 5(b)) and motional narrowing-like phenomena on the structural fluctuation may be observed for the distribution of local fields. In the case of large structural fluctuations, therefore, a broad distribution of local fields centered at a mean field value W_i may result. Thus, the Gaussian approximation for the structural fluctuation (Figure 5(c)) may correspond to a special case of the upper limit of its fluctuation. Some experimentalists[19] have used Equation 23 in order to analyze their data. At the present time, however, there is little justifiable basis for fitting data such as Figure 4 to Equation 23. In fact, the distribution of internal fields obtained from Mössbauer experiments as shown in Section VI is very different from that shown in Figure 5(a).

B. Antiferromagnetic Case

In this part, we present an effective field theory of amorphous "random" antiferromagnets discussed by Simpson.[12]

For a system of dipole moments with only two quantized orientations, the dipole moment of the i-th atom is given by

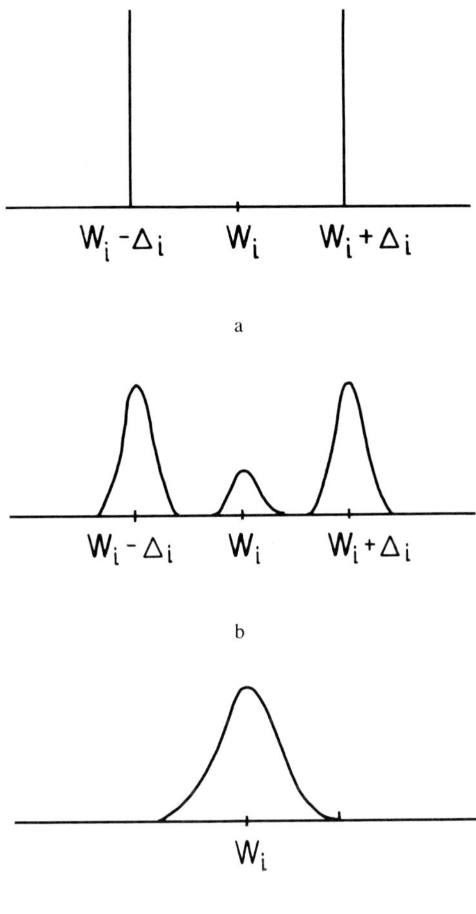

FIGURE 5. The distribution of internal field W_i, (a) approximation I, for which motional narrowing-like phenomena on the structural fluctuation may be observed (b), (c) Gaussian approximation.

$$P_i = P_0 \tanh \frac{P_0}{k_B T} (H_a + H_i^e) \tag{24}$$

with

$$H_i^e = \sum_j J_{ij} P_j \tag{25}$$

where P_0 is the quantized dipole moment at $T = 0$, and H_a the applied external field. Equation 24 is an expression for a particular configuration of magnetic atoms. As usual, it is necessary to take an average over all their possible configurations.

For this purpose, Simpson used the following approximations:

1. The quantity J_{ij} is short ranged because its origin comes from a direct-or superexchange mechanism.
2. The spread in the distribution of J_{ij} is very narrow. Thus, Equation 25 can be written as

$$H_i = \sum_j J_{ij} P_j \simeq \bar{J} \sum_{Z'} P_j \tag{26}$$

where $\bar{J}$ is the average value of J_{ij} and Z' the number of near neighbors.

3. The number of Z' is large because the random close-packed structure has a coordination number of $Z = 12{\sim}14$.

Because of these reasons, and since $P_j = \pm P_0$, Simpson assumes that the distribution function of the internal fields H_i^e is a Gaussian. Thus, in contrast with the spin glass problem discussed by Edwards and Anderson,[20] the origin of the Gaussian distribution function for the present internal field problem is very different. As discussed in Section A, the Gaussian distribution for the internal fields may correspond to a special case of large structural fluctuations. In such a case, it is an open question whether the antiferromagnetic phase is more stable than the spin glass phase at low temperatures. This problem will be discussed in Section V. For this reason, we refer the Simpson's theory to the case of an amorphous ''random'' antiferromagnet instead of a ''true'' amorphous antiferromagnet in which some partial correlation among the spin orientations of near neighbors may exist.

As an order parameter, one may use the following standard deviation σ,

$$\begin{aligned} \sigma &= \overline{((H_i^e)^2)}^{1/2} \simeq \alpha |\bar{J}| (\bar{Z})^{1/2} (\overline{P_j^2})^{1/2} \\ &= \alpha P_0 |J| (Z)^{1/2} \overline{\left(\left(\tanh \frac{P_0}{k_B T} H_j^e\right)^2\right)}^{1/2} \end{aligned} \tag{27}$$

where $\bar{Z}$ is the average number of near neighbors. The parameter α is a constant equal to or greater than unity depending on both the form and the width of the distribution of J_{ij}.

Defining another Gaussian function

$$a = \frac{P_0}{k_B T} H_j^e \tag{28}$$

with a standard deviation σ_a, Equation 27 reduces to

$$\frac{\sigma_a}{\sigma_0} = \overline{(\tanh^2 a)}^{1/2} \tag{29}$$

with

$$\sigma_0 = \alpha |\bar{J}| P_0 (\bar{Z})^{1/2}$$

The standard deviation of a (Equation 28), thus, is given by

$$\frac{\sigma}{\sigma_0} = \frac{k_B T}{P_0 \sigma_0} \sigma_a \tag{30}$$

From Equations 29 and 30, we can obtain the value of σ as a function of T by usual procedure of the molecular field theory. The transition temperature T_N is then given by the condition that Equation 29 has the same slope as Equation 30 at $\sigma_a = 0$. Thus, the transition temperature T_N of a ''random'' antiferromagnet is given by

$$T_N = \frac{P_0 \sigma_0}{k_B} = \frac{\alpha |\bar{J}| P_0^2 (\bar{Z})^{1/2}}{k_B} = \alpha |\bar{J}| \tilde{C} (\bar{Z})^{1/2} \tag{31}$$

This result indicates that the T_N of an amorphous random antiferromagnet will be appreciably reduced in comparison with that of its crystalline counterpart given by $T_N^0 = |J_0|\tilde{C}Z_0$. However, the value of T_N can not be reduced to zero because of $|\bar{J}| \neq 0$, which is clearly in contrast with the experimental fact.[21]

In an amorphous "random" antiferromagnet, Simpson argues that the susceptibilities parallel and perpendicular to the direction of the dipole moment of a particular atom must exist since the dipole moments will be oriented in an applied field direction. Then, the perpendicular atomic susceptibility $\chi_{i\perp}$ of the i-th atom is given by

$$\chi_{i\perp} = \frac{M_i}{H_a} = \beta_i(1 + \sum_j J_{ij}\chi_j) \tag{32}$$

with

$$\beta_i = \frac{P_0 \tanh \dfrac{P_0 H_i^e}{k_B T}}{H_i^e} \tag{33}$$

Performing the random average of Equation 32, we have

$$\chi_\perp = \frac{1}{\dfrac{1}{\bar{\beta}} + \bar{Z}\,\bar{J}} \tag{34}$$

On the other hand, the parallel atomic susceptibility of the i-th atom $\chi_{i\|}$ is given by

$$\chi_{i\|} = \beta_i' \,(1 + \sum_j J_{ij}\chi_j) \tag{35}$$

with

$$\beta_i' = \frac{\tilde{c}}{T}\left(1 - \tanh^2 \frac{P_0 H_i^e}{k_B T}\right) \tag{36}$$

Therefore, the averaged parallel atomic susceptibility $\chi_\|$ reduces to

$$\chi_\| = \frac{1}{\dfrac{1}{\bar{\beta}} + \bar{Z}|\bar{J}|} \tag{37}$$

In comparison with the theory of the spin glass phase discussed in Section A, we can easily obtain the relationship between σ (or Equation 27) and q (or Equation 18). Thus we have

$$q = \left(\frac{\sigma}{\sigma_0}\right)^2 \tag{38}$$

Then, Equation 36 reduces to

$$\bar{\beta}' = \frac{\tilde{c}}{T}\,(1 - q)$$

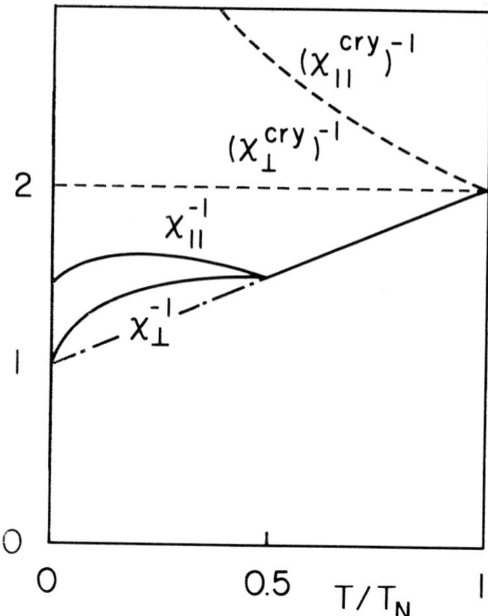

FIGURE 6. The variations of reciprocal perpendicular and parallel atomic susceptibilities with temperature are depicted for a random antiferromagnet.

so that Equation 37 is given by

$$\chi_{||} = \frac{\tilde{c}(1 - q)}{T + \tilde{c}\,\bar{Z}|\bar{J}|(1 - q)} \tag{39}$$

This equation is identical to that for the spin glass phase.[22]

In Figure 6, variations of reciprocal perpendicular and parallel atomic susceptibilities with temperature are depicted. In contrast with the case of the spin glass phase, Simpson argues that the susceptibility of an amorphous "random" antiferromagnet should be equal to the perpendicular susceptibility, since the perpendicular susceptibility $\chi_{\perp}$ is always greater than the parallel susceptibility $\chi_{\parallel}$. However, we would like to suggest that the magnetization process in a random Ising antiferromagnet for applied field parallel to spin directions may be similar to that of metamagnetism, but for perpendicularly applied field it may be similar to that of spin glasses.

It should be pointed out that no considerations either of the quantum mechanical effect or of structural effects of random closed packing (R.C.P.) of magnetic atoms have been considered in the discussion of this theory. An important missing point of this theory is the neglect of misfit found in a "true" amorphous antiferromagnet. As is discussed in Section V, these effects may severely restrict the existence of an amorphous antiferromagnet.

IV. SPIN WAVES

Spin wave resonance was first observed in an amorphous Ni ferromagnet.[23] At that time, whether spin waves exist stably or not under the structural condition of short-range order was interesting from the theoretical point of view, since in such a system the so-called "wave vector" k is not a good quantum number and no simply defined Brillouin zone exists.[30]

Now, there are a number of experimental investigations on spin wave excitations in amorphous ferromagnets. Several interesting questions have been raised by the experimental results. Especially in neutron scattering measurements of spin waves in the amorphous ferromagnet Co_4P, Mook et al.[24] have observed a roton-like structure with a minimum near the wave vector Q corresponding to the first peak in the structure factor a(k). There are some theoretical controversies about the interpretation of this experimental result. Alben[25] has performed some Monte Carlo calculations for spin waves in an amorphous ferromagnet. He explains the roton-like behavior in terms of static structural correlations rather than dynamic effects. In this way, there might be a close relationship between the readily measurable magnetic properties such as the spin wave spectrum and the more elusive structural information.

In amorphous ferromagnets, the Green's function for spin waves is given by[3]

$$EG_{ij}(E) = \delta_{ij} + 2S\sum_m J_{mi}G_{ij}(E) - 2S\sum_m J_{im}G_{mj}(E) \qquad (40)$$

At the present time, there is no firm theoretical basis for the density expansion of spin waves in amorphous ferromagnets. For the case of a dense random packing of hard spheres of magnetic atoms, however, it may be more appropriate to expand Equation 40 with some kind of "virtual crystal" approximation or a quasicrystalline approximation (Q.C.A.), since the system has the coordination number $Z \cong 12$, like f.c.c lattices. The Q.C.A. just corresponds to replacing the well-known expression for the spin wave energy of an ideal crystal

$$\epsilon(\vec{k}) = \sum_j J_{ij}\left(1 - e^{-i\vec{k}\cdot\vec{R}_{ij}}\right) \qquad (41)$$

by the expression

$$\epsilon_{Q\cdot A}(\vec{k}) = \int J(R_{ij})g(R_{ij})\left(1 - e^{-i\vec{k}\cdot\vec{R}_{ij}}\right)d^3R_{ij} \qquad (42)$$

In the previous section, we have proposed that the magnetic properties of amorphous magnets should be determined inevitably by the relative relation between the range of exchange interactions and the configuration of magnetic atoms. This situation should be also reflected in the dispersion relation of spin wave excitations. In order to indicate clearly the relative relation between $J(R_{ij})$ and $g(R_{ij})$ which may be reflected in the spin wave dispersion relation, let us take a simple model for the exchange interaction and the pair correlation function as

$$J(R_{ij}) = \beta R_{ij}^2 e^{-\alpha R_{ij}^2}$$

and

$$g(R) = \begin{cases} -\gamma(R - R_P)^2 + C & \text{for} \quad R_0 \lesssim R < d_0 \\ (2\pi)^2 \dfrac{\cos(bR)}{(bR)^2} + 1 & \text{for} \quad d_0 < R \end{cases}$$

with $b = 2\pi/d_0$. Here, as has been used by Kok and Anderson,[26] we take the constants γ and C which are determined by the boundary condition that g(R) is continuous at $R = d_0$; the peak position R_P is chosen to be $0.995\ d_0$, so that $g(R) = 0.005$ at $R_0 = 0.9\ d_0$,

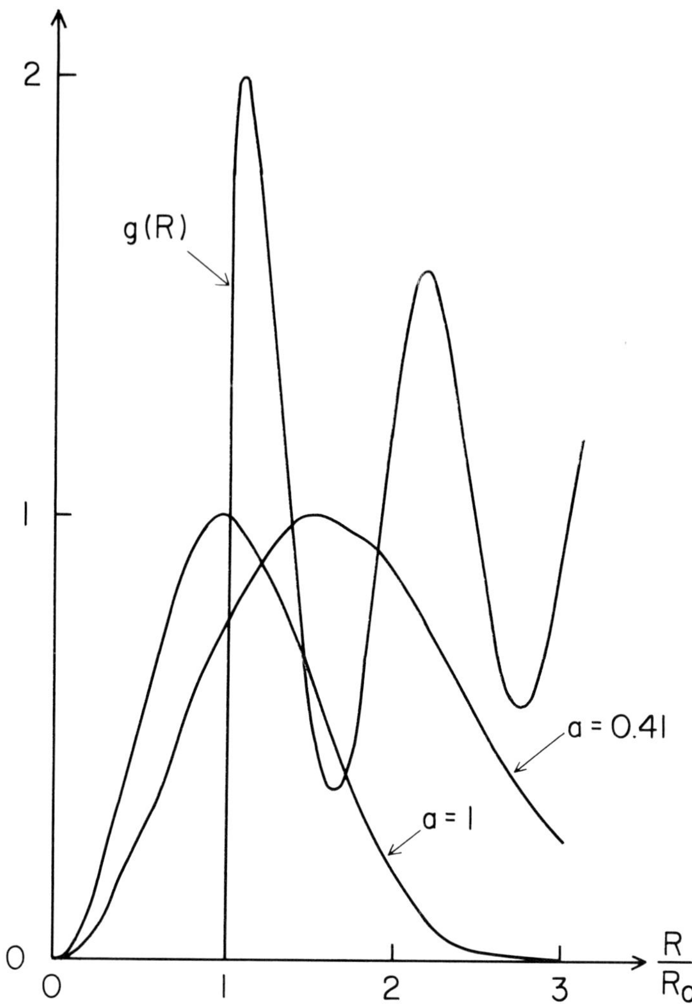

FIGURE 7. The relative relations between a model radial distribution function and the spatial dependence of exchange interaction with the damping factors a as shown.

producing a sharp narrow first peak as in the crystalline case (Figure 7). In Figure 7, two typical cases for the spatial dependence of the exchange interaction with $a = \alpha R_0^2$ are also depicted. For the two values of a = 0.411 and a = 1, the numerical results of Equation 42 are shown in Figure 8. For comparison, we plotted the dispersions of $\epsilon_{Q.A.}$ (k) scaled by their maximum values.

From these results, we can understand the important fact that the dispersion of spin waves may have entirely different structures depending on the spatial dependence of exchange interaction; the dispersion for a = 1 shows that the roton-like minimum may be observed at $kR_P \simeq 2\pi$ in accordance with the neutron scattering experiment of Mook et al., although the calculated roton-like minimum in the Q.C.A. is less sharp and less pronounced than in Mook's data. On the contrary, the result for a = 0.411 has a rather different structure from that for a = 1. An interesting fact is that the roton-like minimum may not be observed in the case of long-range interaction, although it is desirable for the appearance of ferromagnetism. The result for a = 1 is similar to that of Alben's computer calculations. Only the contribution from the first neighbors of dense random packing of hard spheres seems to be important for the appearance of the roton-like minimum in the Q.C.A.

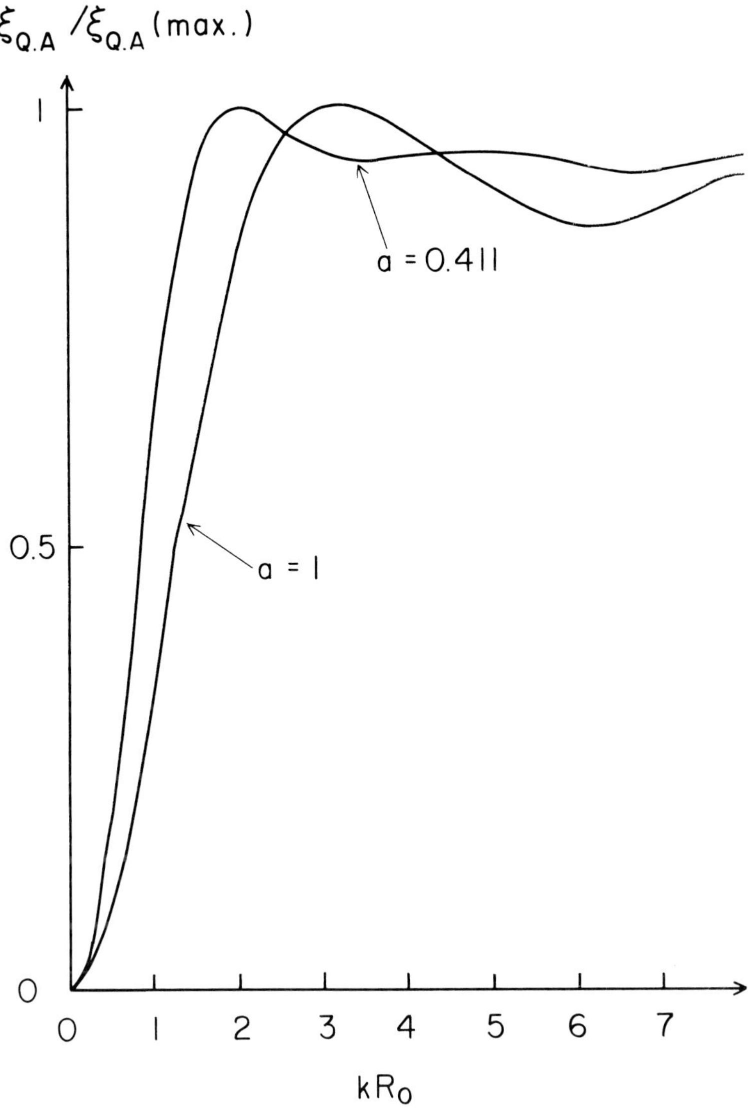

FIGURE 8. The spin-wave dispersion relations for the quasicrystalline approximation.

In neutron inelastic scattering measurements of amorphous ferromagnets, some interesting peaks have been observed in addition to those of spin wave excitations; especially for low energies of neutron scattering an additional central elastic peak is observed at nearly zero energy ($E \cong 0$), the intensity of which increases as momentum transfer k of neutron increases.[27] Right at the first peak of the structure factor a(k), a small middle peak being independent of energy is also observed.[24] In relation to these measurements, Handrich[28] has recently suggested that within the theory of Reference 3 a new type of elementary excitation (he called it the amorphon) exists, which can be described only by mixed states as a consequence of the random structure, and also that the amorphons may explain the quasi-elastic central peak at $E \cong 0$ and $k \cong 0$ and the middle "quasi-elastic peak" right at the first peak of a(k).

Now, the dispersion of long wavelength spin wave states in amorphous ferromagnets is given by

$$E = 2SND_a k^2 \tag{43}$$

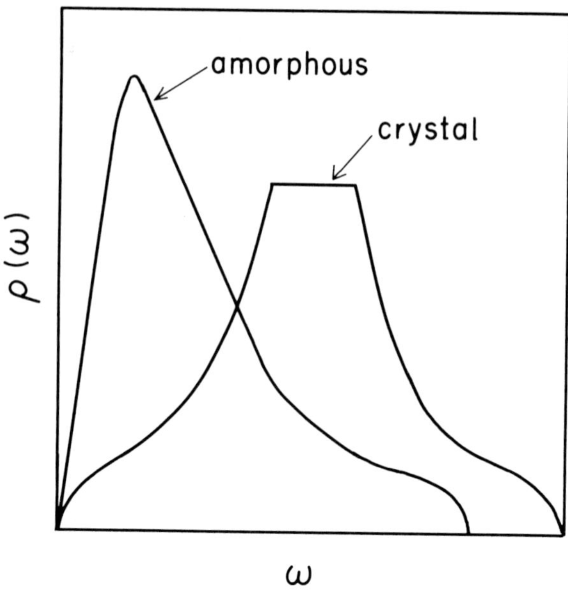

FIGURE 9. A schematic representation for the spin-wave density of states in an amorphous ferromagnet.

where the stiffness constant D_a depends in a more complicated way both on the exchange interaction and the configuration of magnetic atoms than that derived from Equation 42.[29] In general, the stiffness constant D_a of amorphous ferromagnets is less than that of its corresponding crystalline ferromagnet (D_c), which means that spin waves in an amorphous ferromagnet have lower energies of excitation than those in the crystalline case. Spin waves can be easily excited in an amorphous ferromagnet, so that the density of states $\rho_{am}(\omega)$ is expected to have a form as shown in Figure 9. This result has been obtained by Tahir-Kheli in the random bond model.[30] These considerations mean that as the temperature is increased from absolute zero, the magnetization falls more rapidly than for the crystalline case and the Curie temperature T_c is lower than that of the crystalline case. Simpson[31] gives an interesting explanation for this phenomenon. That is, in Equation 41, if J_{ij} is strong, the angle $\Delta\theta_{ij}$ between the two interacting spins becomes correspondingly small. Conversely, if the interaction is weak, the angle between the spins tends to be large. This inverse correlation between J_{ij} and $\Delta\theta_{ij}$ allows us a crude estimate for the effect of structural disorder on the spin wave dispersion relationship by simply averaging Equation 41 over certain distributions of J and $\Delta\theta$.

Let $J_{ij} = J_0(1+\alpha_{ij})$ and $\Delta\theta_{ij} = \langle ka\rangle(1+\beta_{ij})$, where α_{ij} and β_{ij} take a range of values. The quantity a is the lattice constant. Then the spin wave frequency of the "amorphous" linear chain, ω_a, is approximately given by, for the nearest-neighbor interaction,

$$\omega_a \simeq \left\langle \frac{J_0(1+\alpha_{ij})}{h}\,(1-\cos\langle ka\rangle(1+\beta_{ij}))\right\rangle_r \tag{44}$$

For small values of the wave vector k, Equation 44 reduces, upon making a crude assumption that $\beta_{ij} \simeq -\alpha_{ij}$, to

$$\omega_a \simeq \frac{J_0}{2h}\langle ka\rangle^2(1-\sigma) \tag{45}$$

where σ is the mean square of α_{ij}. In this way, Simpson argues that the influence of structural fluctuation on the long-wavelength spin wave states is to reduce their frequency. However, his explanation is based on the so-called "lattice model" of amorphous magnets; the magnetic atoms are arranged in a periodic lattice, and the only random variable is the fluctuation of exchange integral from its mean interaction. As is easily understood from Equation 42 in comparison with Equation 41, whether the stiffness constant D_a is smaller than D_c or not also depends upon the relative relationship between the exchange interaction and the configuration of magnetic atoms, even if we take the Q.C.A. for the spin wave dispersion relation.

In order to understand whether the long-wavelength spin-wave states in amorphous ferromagnets can be clearly defined or not, we must estimate the imaginary part of the dispersion relationship. Roth[32] has suggested that the imaginary part should be of the order of k^5. On the other hand, Kaneyoshi[29] obtained an imaginary part of the order of k^7. In each case the long-wavelength spin-wave states can be clearly defined in a dense random packing amorphous ferromagnet. At the present time, spin waves have been clearly observed in several amorphous ferromagnetic alloys by neutron scattering. When spin waves are the dominant low-energy excitation, the behavior of magnetization M(T) should be given by

$$M(T) = M(0)[1 - BT_{3/2} - CT_{5/2} + \ldots] \tag{46}$$

In simple spin-wave theory, the coefficient B is related to the spin-wave dispersion coefficient D determined by neutron scattering as

$$B_{SW} = 2.612 \left(\frac{g\mu_B}{M(0)}\right)\left(\frac{k_B}{4\pi D}\right)^{3/2} \tag{47}$$

One of the most puzzling aspects of the neutron scattering results is the following. Although the value B_{SW} calculated from Equation 47 by using D determined by neutron scattering is usually equal to the value B obtained from low-temperature magnetization measurements in crystalline ferromagnets, in some amorphous ferromagnets the calculated value of B_{SW} is smaller than the measured B.[33] That is to say, the magnetization drops off much more rapidly with increasing temperature than one would anticipate from the observed spin wave dispersion relation. This fact would seem to necessiate the existence of another type of excitation. However, Birgeneau et al.[34] recently found that for the amorphous metallic ferromagnets $(Fe_xNi_{1-x})_{75}P_{16}B_6Al_3$ the observed B's are properly accounted for by the measured spin-wave dispersion relation (or Equation 47) in contrast to previous studies in other magnetic glasses.

The $T^{3/2}$ law was first observed in amorphous CoP alloys[35] with the range of the $T^{3/2}$ variation extending more widely than that expected for the corresponding crystalline state. Chien and Hasegawa[36] have explained the dominance of the $T^{3/2}$ term in an unusually large temperature range. The temperature range in which the $T^{3/2}$ term dominates depends on the ratio of the $T^{5/2}$ term and $T^{3/2}$ term in Equation 46, i.e.,

$$\frac{C}{B}T = \frac{C_{5/2}}{B_{3/2}}\left(\frac{T}{T_c}\right)$$

For crystalline ferromagnets, $C_{5/2}/B_{3/2}$ is of the order of 1. For amorphous ferromagnets, on the other hand, $C_{5/2}/B_{3/2}$ has values of only about 0.3. Krey[37] has shown that from the computer calculations for the relaxed dense random packings of hard spheres consisting of $N = 888$ sites, the experimental result of Chien and Hasegawa can be reproduced in the case of the damped short-range interaction. From this result, Krey has concluded that a description by localized spins with short-range interactions should be much better for the ferromagnetic amorphous alloys than for the corresponding crystalline metals.

In relation to the spin waves in amorphous ferromagnets discussed in the first part of this section, we shall comment on the magnetic excitations in a random ferromagnetic ("asperomagnetic") rare earth alloy (e.g., $DyNi_{3.4}$),[38] in which only the rate earth component is magnetic. In an "asperomagnetic" material, the magnetic moments are randomly distributed in a cone of finite collinear ferromagnet. Recently, Bhattachargee et al.[39] have studied the long-wavelength spin-wave excitations in amorphous asperomagnetic materials. The Hamiltonian is given by

$$H = -\sum_{i,j} J_{ij}J_iJ_j - D_i\sum_i J^2_{iZ'_i} \tag{48}$$

where D is the anisotropy constant and J_{ij} the exchange interaction between the i-th and the j-th rare earth moments. Z'_i is the local anisotropy axis at the i-th rare earth site. Then they have obtained the coupled Green's functions for spin-wave excitations as follows

$$\omega G_{\ell m}(\omega) = \delta_{\ell m} - D\{-2JfG_{\ell m}(\omega) + J(1-f)[G_{\ell m}(\omega) + G_{\ell m}(\omega)]\} + 2J\phi\sum_j J_{\ell j}G_{\ell m}(\omega) - J\sum_i J_{i\ell}[(\phi+1)G_{im}(\omega) + (\phi-1)G_{im}(\omega)] \tag{49}$$

and

$$\omega g_{\ell m}(\omega) = -DJ\{2fg_{\ell m}(\omega) - (1-f)[g_{\ell m}(\omega) + G_{\ell m}(\omega)]\} - 2J\phi\sum_j J_{\ell g}g_{\ell m}(\omega) + J\sum_i J_{i\ell}[(\phi+1)g_{im}(\omega) + (\phi-1)G_{im}(\omega)] \tag{50}$$

where $G_{\ell m}(\omega) = \langle\langle a_\ell; a_m^+\rangle\rangle$ and $g_{\ell m}(\omega) = \langle\langle a_\ell^+; a_m^+\rangle\rangle$. The parameters ϕ and f are given by the average values of the directional angles between the moment and the anisotropy axis at each site. It seems that these coupled Green's functions have very different forms in comparison with that of Equation 40. However, these equations can be solved in the same way as that of amorphous ferromagnets (or Equation 40). The spin-wave dispersion of the Q.C.A. in an amorphous asperomagnet is given by[40]

$$\omega^2 = \omega_0^2 + kJ^2\gamma(\epsilon(0) - \epsilon(\vec{k})) + kJ^2\phi[\epsilon(0) - \epsilon(\vec{k})]^2 \tag{51}$$

with

$$\omega_0^2 = 4J^2D(2f-1)[Df - (1-\phi)\epsilon(0)] \tag{52}$$

$$\gamma = D(2f-1) + \phi[Df - (1-\phi)\epsilon(0)] \tag{53}$$

where $\epsilon(\vec{k})$ is defined by

$$\epsilon(\vec{k}) = N\int J_{ij}g(R_{ij})\exp(-\vec{k}\cdot\vec{R}_{ij})dR_{ij} \tag{54}$$

In contrast with the case of amorphous ferromagnets, spin waves in amorphous asperomagnetic alloys may not be clearly defined. For example, when the ground state is "spin glass-like" with $\phi \simeq 0$ and $f \simeq 1$, the spin-wave dispersion for the long-wavelength limit $k \to 0$ is given by

$$\omega^2 = \omega_0^2 + \alpha k^2 + i\beta Dk^3$$

in which the imaginary term arises from the random orientations of the local anisotropy axis. Thus, while spin-wave-like excitations have been observed in amorphous ferromagnets by inelastic neutron scattering experiments, such clear observation might be impossible for a speromagnet.[40]

V. DISCUSSION ON THE EXISTENCE OF AMORPHOUS ANTIFERROMAGNETS

Amorphous magnets have a topologically disordered R.C.P. structures, which corresponds to a f.c.c.-like lattice with nearest-neighbor interactions. Introduction of sublattice models into such structures results in a number of misfit in the system, so that antiferromagnetism may be energetically unfavorable in amorphous magnets. If amorphous antiferromagnetism exists, as discussed in the previous section, the spin waves will have lower energies of excitation than those of the corresponding crystalline case, and may be easily excited even by zero point energy of the system. These work against the stability of antiferromagnetism, since some fraction of the spins can have large precession amplitudes even at low temperatures and might occasionally flop to another direction. Because of the topologically disordered structure, in the regions of low local exchange fields, the zero point energy of these regions may cause a considerable reduction of the sublattice magnetization in amorphous antiferromagnets. This question has been studied by Handrich[41] using the so-called "lattice model" of amorphous magnet.

Spins may also have local anisotropy fields originating from ligand Coulomb fields. In an amorphous magnet, the directions of these fields are random because of the random local environments. If the effect of local anisotropy is larger than the molecular field of each "sublattice", "spin glass" like ordering is energetically more favorable than antiferromagnetic one. In other words, the existence of antiferromagnetism is structurally restricted in amorphous solids.[42]

Simpson[12] argues that if an amorphous antiferromagnet exists, the effective field w_0 of each spin may have a distribution illustrated in Figure 10. Figure 10(a) corresponds to a simple crystalline antiferromagnet with nearest-neighbor interactions. If such a material is distorted, where the deviations of the atomic positions are small and completely random in size and direction, then, the effective field distribution may be given as in Figure 10(b). He observes that if the distortion is further increased and modified to the extent that the material becomes amorphous, the effective field distribution may become as shown in either Figure 10(c) or Figure 10(d). As discussed in the previous section, however, we recall that both Figure 10(c) and Figure 10(d) represent the "spin glass" phase rather than the antiferromagnetic phase, since structural fluctuation Δ in such a system is larger than the averaged effective field $|w_0|$. Thus, we face a restriction for the existence of an amorphous antiferromagnet.

Recently, Kobe[43] has discussed the problems of misfit and the effective field in an amorphous antiferromagnet. He takes an N = 40 disk with Ising spin S = 1/2 in a square box bounded by chosen periodical boundary conditions, and chooses the Hamiltonian as follows

$$H = -\sum_{i<j} J_{ij} S_i^Z \cdot S_j^Z \tag{55}$$

where the exchange interaction takes the form of Figure 11. Then, the spin configuration of the magnetic ground state and the distribution of the effective field w_0, corresponding to Figure 8.b discussed by Simpson were obtained. Kobe defines the misfit parameter m as

$$m = 1 - \frac{|E_0|}{E_{i\alpha}} \tag{56}$$

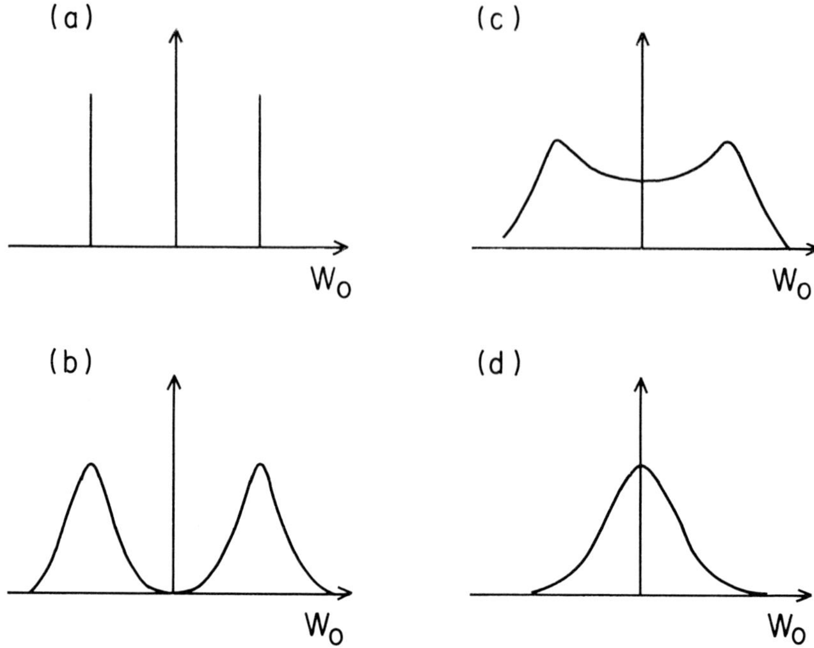

FIGURE 10. The distributions of internal field for (a) simple crystalline antiferromagnetic case, (b) amorphous antiferromagnet, (c) and (d) spin glass.

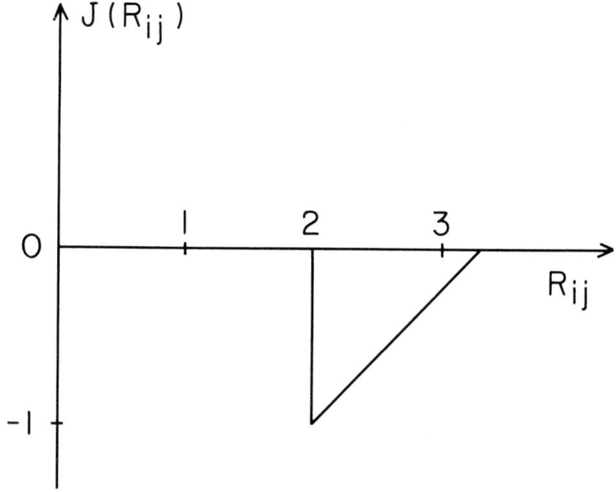

FIGURE 11. A model exchange interaction.

where E_0 is the ground state energy and $E_{i\alpha}$ the energy of an ideal system without any misfit, in which all spins are aligned parallel. The value of m is then given by 0.42 ± 0.05. From these results, however, we can not directly draw conclusions on the existence of long-range ordered amorphous antiferromagnetism.

At the present time, there is no theoretical work on amorphous antiferromagnetism taking account of the R.C.P. structure characteristic to amorphous magnets.

VI. NONUNIFORM DISTRIBUTION OF LOCAL FIELDS

In recent years, a variety of amorphous ferromagnetic materials have been reported to have a resistivity minimum well below the magnetic ordering temperature. According to the accepted Kondo theory, with increasing impurity concentration, the minimum is suppressed and finally washed out as magnetic ordering sets in, since magnetic ordering of the local spins destroys the freedom of spins to flip. Thus the observation has brought us an important theoretical problem which contains the origin of the observed resistivity minimum. Several important experimental aspects of this finding are still the subject of much controversy. This observation has been explained as either a Kondo-type effect or as a structural effect associated with a configurational indeterminancy in the atomic arrangements. For these experimental controversies one is advised to consult the review paper by Tsuei.[44]

In this section we will treat the phenomena as a Kondo-type effect. In order to clarify the coexistence of the resistivity minimum and ferromagnetism, however, another mechanism different from the accepted Kondo theory is required. It was first pointed out by Hasegawa and Dermon[45] that the logarithmic temperature dependence of the resistivity for these systems may originate from an electron-magnon interaction. But, even in these amorphous ferromagnets, if magnetic excitations move through the system coherently, the resistivity minimum may not be observed, since at each magnetic atom the freedom of the spin to flip is suppressed because of the coherent interactions between localized spins.

For this purpose, Madhukar and Hasegawa[46] considered that small clusters of paramagnetic impurities are randomly distributed throughout the material. They proposed that local spin deviations within clusters can provide the freedom for spin-flip scattering of electrons because of the structural disorder. This model is in contrast to the existence of ferromagnetism, but may be valid for the case of concentrations of localized spins not high enough for a long-range order to appear.

On the other hand, it has been suggested that in the propagation of a magnon with low ω_q through a specimen there should exist some regions in which the coherent motion of a magnetic excitation is broken.[47] Within the region, it is necessary for a magnetic excitation of a particular site to be transferred incoherently to its neighbor sites on account of the exchange interaction between them. The spins in it (including the case of only one spin) are weakly coupled to each other and to the surrounding spins, and therefore relatively little energy is required for a spin flip. Thus, the coexistence of a resistivity minimum and ferromagnetism in an amorphous magnet should be closely connected to the amorphous nature of the system. When an amorphous ferromagnet transforms to the crystalline one by annealing, the logarithmic temperature dependence of the resistivity disappears, as shown by Yoshiie et al.[48]

From a somewhat different point of view, Slechta[49] has discussed the coexistence phenomenon. When the long range magnetic order prevails, each spin is in an effective molecular field H_{eff} which plays the role of an external magnetic field. Then, a spin can act as a Kondo scatterer if

$$\mu H_{eff} < k_B T \tag{57}$$

He has studied numerically whether there exist Kondo-like spins satisfying the restriction (Equation 57) in an amorphous ferromagnet. His results are complementary to those obtained by Kaneyoshi and Honmura.[47]

We have assumed that the internal field in such a system would have different values at each site. The distribution of the internal fields can be obtained from Mössbauer measurements on amorphous magnets.[50] The situation of a ferromagnet is illustrated in Figure 12. In the crystalline case, the hyperfine fields H_{hf} have sharply defined values (Figure 12a). In

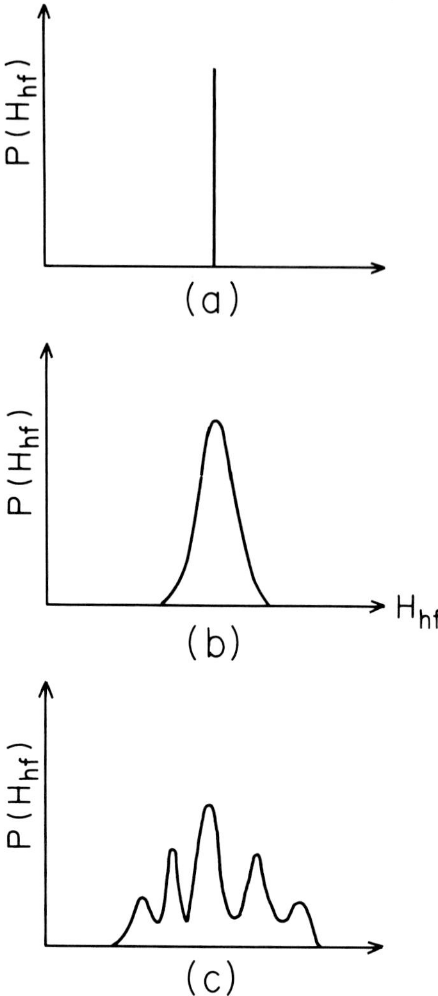

FIGURE 12. The distribution of the effective field H_{hf}. (a) crystalline case, (b) amorphous case with distribution of interatomic spacing, and (c) amorphous case with different coordination shells.

an amorphous solid, H_{hf} has some distributions because of certain degrees of fluctuation in nearest-neighbor distances (Figure 12b). If the number of neighbor atoms interacting via short-range exchange is not constant, there occur several maxima in the distribution of H_{hf} covering a wide range of magnitude due to the presence of different nearest-neighbor coordination shells (Figure 12c). In fact, Sharon and Tsuei[51] have shown by Mössbauer studies that a nonnegligible fraction of the magnetic atoms in certain amorphous alloys have near-zero hyperfine field. For amorphous Fe_{1-x} Si_x systems, Marchal et al.[52] have obtained a distribution of H_{hf} as shown in Figure 13. At the present time, it seems that there are some controversies on the analysis of the distribution of H_{hf} of Sharon and Tsuei. In the experiment of Marchal et al., the temperature-dependent "high-field part" of the distribution was assigned to icosahedrally coordinated Fe atoms with statistically varying composition of their environment. The nontemperature-dependent "low-field part" was traced to only weakly coupled iron atoms dispersed over the alloy in the form of small clusters. In this way, these Mössbauer measurements indicate that there are always a certain fraction of the spins only

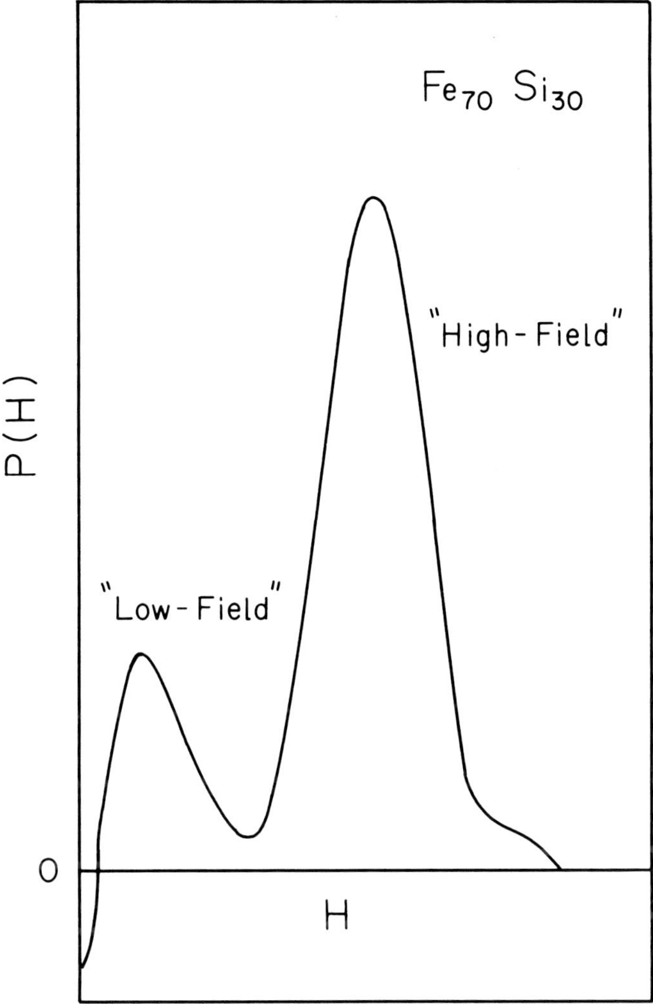

FIGURE 13. The distribution of H_{hf} for amorphous $Fe_{70}Si_{30}$ system after Marchal et al.[52] is schematically depicted.

weakly coupled to the other spins because of the statistical distribution of the moments in amorphous alloys.

From the theoretical point of view, it is interesting to study the nonuniform distribution of the internal fields in amorphous magnets. Even in the molecular field approximation (Equation 2), there is a macroscopic system of algebraic nonlinear equations to be solved. Because of this fact, we have tried to formulate the generalized effective field theory of a disordered magnet, in which we take account of the different fluctuation of local field at each site. However, the only possibility of solving the equations is an application of numerical methods.

Now, let us use the quasi-classical method presented in the book of Tyablikov.[53] The Hamiltonian is

$$H = -\mu \sum_{i} (\vec{H} \cdot \vec{S}_i) - \frac{1}{2} \sum_{i \cdot j} J_{ij} (\vec{S}_i \cdot \vec{S}_j) \quad (58)$$

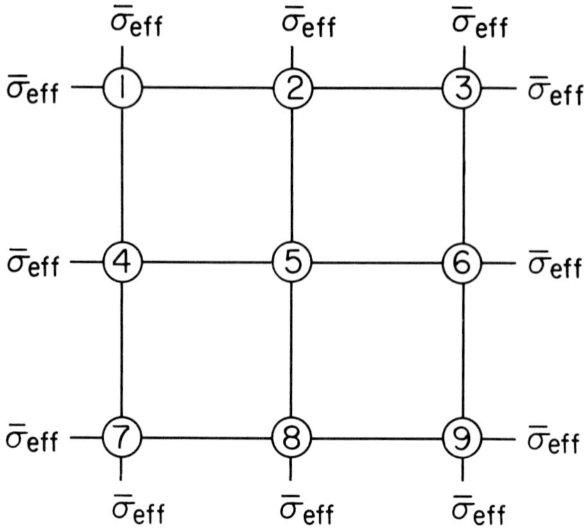

FIGURE 14. The two-dimensional reference cluster of N=9.

where $\vec{S}_i$ is the classical vector given by $\vec{S}_i = \vec{\gamma}_i S_i$. The parameters $\vec{\gamma}_i$ are unit vectors. Then, the Hamiltonian (Equation 58) reduces to

$$\epsilon_0 = -\mu \sum_i (\vec{H} \cdot \gamma_i) \vec{S}_i - \frac{1}{2} \sum_{i \cdot j} J_{ij} (\vec{\gamma}_i \cdot \vec{\gamma}_j) S_i S_j \tag{59}$$

The directions of the vectors $\vec{\gamma}_i$ are determined by minimizing the energy ϵ_0;

$$-\sum_j J_{ij} S_j \gamma_j^\alpha - \mu H^\alpha = Z\lambda_i \gamma_i^\alpha \tag{60}$$

$$\sum_{\alpha=x,y,z} (\gamma_i^\alpha)^2 = 1 \qquad \text{for all i} \tag{61}$$

where λ_i is the Lagrange multiplier. For simplicity Slechta[54] puts H=0 in Equation 60 and takes the nearest-neighbor interaction. By solving Equation 60, the ground state of the system is in general given by the eigenvectors which correspond to the lowest eigenvalue. In the case of a crystalline magnet, we can put $\lambda_i = \lambda$ for all i in Equation 60. In a disordered magnet, however, this simplification cannot be allowed because of the randomness existing in the system.

In Figure 14, the two-dimensional reference cluster of nine spins imbedded in an effective medium studied by Slechta is presented. The random bond clusters are obtained from the reference system by using a random number generator. The procedure of Slechta is that the mutual orientations of $\vec{S}_i$ are given by the minimization of the energy for $T = 0$. These orientations are then maintained throughout $0<T<T_M$, where T_M is obtained from the numerical solution of

$$\sigma_i = \tanh\left(\frac{\alpha_i}{k_B T}\right) \tag{62}$$

with

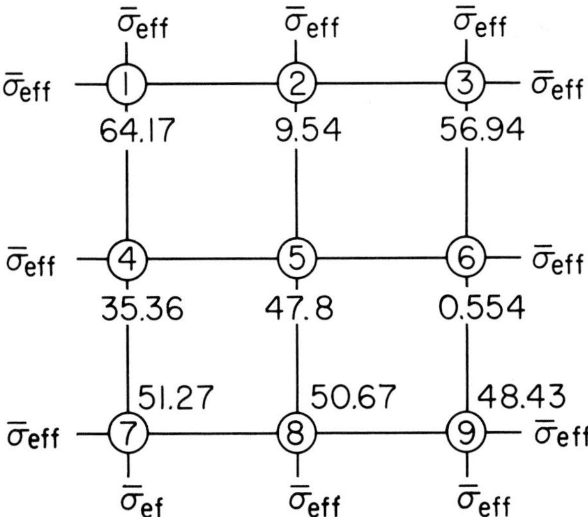

FIGURE 15. The distribution of transition temperature T_i^M for $\Delta J = 16.92$ K.

$$\alpha_i = \frac{1}{2}\sum_j J_{ij}\sigma_j(\vec{\gamma}_i \cdot \vec{\gamma}_j) \tag{63}$$

In Figure 15, the distribution of local temperature T_i^M of magnetic ordering at each site is given assuming a half width ΔJ of the fluctuation of J equal to 16.92 K. This example indicates that the nonuniform distribution of internal fields may exist in an amorphous ferromagnet. In particular, the T_i^M for i = 2, 4, and 6 in Figure 15 are considerably lower than those of the rest. Therefore, above a certain critical temperature these spins behave as if they are nearly magnetically free.

Recently, from the Monte Carlo calculation of effective fields in an amorphous ferromagnet Grest and Nagel[8] have shown that when we include superexchange interactions between the next-nearest-neighbor magnetic atoms being separated by a metalloid atom in addition to the direct and indirect exchange interactions, there is a long tail in the distribution of effective fields which goes through zero and is nonzero at relatively large negative fields. This is similar to the Mössbauer data of Sharon and Tsuei, and suggests that the anomalous resistivity minimum may be understood as a Kondo-type effect.

In this chapter, the local Heisenberg and Ising models have been used for the theoretical description of magnetism in amorphous materials. In conclusion, we believe that magnetism in amorphous states is effectively determined by the relative relation between the nature of exchange interaction and the local configuration of magnetic atoms, although basic atomic-scale properties, like magnetic moment, are governed by "chemical disorder".[11] However, many of the real amorphous magnets are metallic. At the present time, lack of detailed knowledge of the electronic structure in amorphous magnets is a major stumbling block in clear understanding of the magnetism. For the last few years the electronic states in topologically random systems have received much attention. For example, the density of states for pure amorphous transition metals is strongly asymmetric similar to the f.c.c. systems, although due to the structural disorder the broadening of a peak at the upper band edge is observed. Richter and Kobe[55] have discussed how the itinerant ferromagnetism is weakened by the structural disorder.

REFERENCES

1. **Levy, R. A. and Hasegawa, R., Eds.,** *Amorphous Magnetism II*, Plenum Press, New York, 1977.
2. **Gubanov, A. I.,** Quasi-classical theory of amorphous ferromagnets, *Sov. Phys. Solid State*, 2, 468, 1960.
3. **Kaneyoshi, T.,** Contribution to the spin-wave theory of amorphous ferromagnets, *J. Phys. C*, 5, 3504, 1972.
4. **Tyagi, I. S., Kishore, R., and Joshi, S. K.,** Disordered induced ferromagnetism in itinerant electron system, *Solid State Commun.*, 18, 1145, 1976.
5. **Litterst, F. J.,** Susceptibility of non-crystalline ferromagnetic FeF_2, *J. Phys. (Paris) Lett.*, 36, L-197, 1975.
6. **Buschow, K. H. J.,** Magnetic coupling of rare earth moments in amorphous alloys, *Solid State Commun.*, 27, 275, 1978.
7. **Kaneyoshi, T.,** Indirect exchange interaction in a disordered magnet, *J. Phys. F*, 5, 1014, 1975.
8. **Grest, G. S. and Nagel, S. R.,** Effective-field distributions and resistivity minima in amorphous ferromagnets, *Phys. Rev. B*, 19, 3571, 1979.
9. **Kaneyoshi, T.,** Comment on the Grest and Nagel model in amorphous ferromagnetic alloys, *J. Phys. Soc. Jpn.*, 49, 1191, 1980.
10. **Coey, J. M. D.,** Amorphous magnetic order, *J. Appl. Phys.*, 49, 1646, 1978.
11. **Alben, R., Budnick, J. I., and Cargill, G. S., III,** Magnetic structures, in *Metallic Glasses*, American Society for Metals, Metals Park, O., 1978, chap. 12.
12. **Simpson, A. W.,** An effective field model of the amorphous antiferromagnet, *Phys. Status Solidi*, 40, 207, 1970.
13. **Kaneyoshi, T.,** Contribution to effective field theories of disordered magnets, *J. Phys. C*, 8, 3415, 1975.
14. **Cannella, V.,** Low field magnetic susceptibility of dilute alloys: are dilute CuMn and AuMn magnetically ordered?, in *Amorphous Magnetism*, Hooper, H. O. and de Graaf, A. M., Eds., Plenum Press, New York, 1973, 195.
15. **Southern, B. W.,** Effective-field approximations for a disordered magnets, *J. Phys. C*, 9, 4011, 1976.
16. **Adachi, K., Imura, R., Matsui, M., and Sawamoto, H.,** Spin glass state of $Mn_x\ Ni_{1-x}$ Sb; synthesis in high pressure and high temperature atmospheres, *J. Phys. Soc. Jpn.*, 44, 114, 1978.
17. **Kaneyoshi, T.,** A molecular field theory of amorphous ferromagnets, *J. Phys. C*, 6, 3130, 1973.
18. **Handrich, K.,** A simple model for amorphous and liquid ferromagnets, *Phys. Status Solidi B*, 32, K55, 1969.
19. **Tsuei, C. C. and Lilienthal, H.,** Magnetization distribution in an amorphous ferromagnet, *Phys. Rev. B*, B13, 4899, 1976.
20. **Edwards, S. F. and Anderson, P. W.,** Theory of spin gasses, *J. Phys. F*, 5, 965, 1975.
21. **Egami, T., Sacli, O. A., Simpson, A. W., Terry, A. L., and Wedgwood, F. A.,** Amorphous antiferromagnetism in some transition element-phosphorous pentoxide glasses, in *Amorphous Magnetism*, Hooper, H. O. and de Graaf, A. M., Eds., Plenum Press, New York, 1973, 27.
22. **Sherrington, D. and Kirkpatrick, S.,** Solvable model of a spin-glass, *Phys. Rev. Lett.*, 35, 1792, 1976.
23. **Ajiro, A., Tamura, K., and Endo, H.,** Ferromagnetic resonance in amorphous Ni film, *Phys. Lett. A*, A35, 275, 1971.
24. **Mook, A. H., Wakabayashi, N., and Pan, W.,** Magnetic Excitations in the amorphous ferromagnet Co_4P, *Phys. Rev. Lett.*, 34, 1029, 1975.
25. **Alben, R.,** Spin waves in a model for an amorphous ferromagnet, *AIP Conf. Proc.*, 136, 1976.
26. **Kok, W. C. and Anderson, P. W.,** Comments on the paramagnetic Curie temperature in amorphous magnetic alloys, *Philos. Mag.*, 24, 1141, 1971.
27. **Rhyne, J. J., Lynn, J. W., Luborsky, F. E., and Walter, J. L.,** Spin excitations in amorphous transition-metal boron glasses, *J. Appl. Phys.*, 50, 1583, 1979.
28. **Handrich, K.,** A new type of elementary excitations in amorphous solids (amorphons), *Phys. Status Solidi B*, 97, K71, 1980.
29. **Kaneyoshi, T.,** Spin waves in amorphous ferromagnets, *J. Phys. Soc. Jpn.*, 45, 1835, 1978.
30. **Tahir-Kheli, R. A.,** Spatially random Heisenberg spins at very low temperatures I, dilute ferromagnet, *Phys. Rev. B*, B6, 2808, 1972.
31. **Simpson, A. W.,** A simple semi-classical spin wave model for amorphous magnetism, *Wiss. Z. Tech. Hochsch. Dresden*, 3, 1020, 1974.
32. **Roth, L. M.,** Effective medium approximation for ferromagnetic spin waves in amorphous systems and alloys, *AIP Conf. Proc.*, 29, 150, 1976.
33. **McColl, J. R., Murphy, D., Cargill, G. S., III, and Mizoguchi, T.,** Spin wave dispersion and temperature dependence of magnetization in an amorphous Co-P alloy, *AIP Conf. Proc.*, 29, 172, 1976.
34. **Birgeneau, R. J., Tarvin, J. A., Shirane, G., Gyorgy, E. M., Sherwood, R. C., Chen, H. S., and Chien, C. L.,** Spin-wave excitations and low-temperature magnetization in the amorphous metallic ferromagnetic $(Fe_xNi_{1-x})_{75}P_{16}B_6Al_3$, *Phys. Rev. B*, 18, 2192, 1978.

35. **Cargill, G. S., III and Cochrane, R. W.,** Amorphous cobalt-phosphorus alloys; atomic arrangements and magnetic properties, *J. Phys. (Paris)*, 35, C4-269, 1974.
36. **Chien, C. L. and Hasegawa, R.,** $T^{3/2}$ dependence of hyperfine field and spin-wave excitations in ferromagnetic metallic glasses, *Phys. Rev. B*, 16, 2115, 1977.
37. **Krey, U.,** Spin wave excitations and the magnetization curve of amorphous ferromagnets, *Z. Phys. B*, 31, 248, 1978.
38. **Coey, J. M. D., Chappert, J., Revouillat, P. J., and Wang, T. S.,** Magnetic structure of an amorphous rare-earth transition-metal alloys, *Phys. Rev. Lett.*, 36, 1061, 1976.
39. **Bhattachargee, A. K., Coqblin, B., Jullien, R., Plischke, M., Zobin, D., and Zuckermann, M. J.,** Long-wavelength magnetic excitations in randomly anisotropic ferromagnets, *J. Phys. F*, 8, 1793, 1978.
40. **Kaneyoshi, T.,** Spin waves in amorphous ferro- and aspero-magnets, *J. Magn. Magn. Mater.*, 15—18, 1343, 1980.
41. **Handrich, K.,** Comment on the ground state of amorphous antiferromagnets, *Phys. Status Solidi B*, 58, K93, 1973.
42. **Egami, T. and Simpson, A. W.,** On the amorphous antiferromagnet, *Phys. Status Solidi B*, 53, K117, 1972.
43. **Kobe, S.,** Ground state of an Ising antiferromagnet with a dense random packing structure, in *Amorphous Magnetism II*, Levy, R. A. and Hasegawa, R., Eds., Plenum Press, New York, 1977, 529.
44. **Tsuei, C. C.,** Some controversial aspects of electronic and magnetic interactions in the amorphous metallic state, in *Amorphous Magnetism II*, Levy, R. A. and Hasegawa, R., Eds., Plenum Press, New York, 1977, 181.
45. **Hasegawa, R. and Dermon, J. A.,** Electrical resistivity and Curie temperature of amorphous (Fe-Ni)-P-C alloys, *Phys. Lett. A*, 42, 407, 1973.
46. **Madhukar, A. and Hasegawa, R.,** Mechanism for resistivity minimum in amorphous ferromagnets, *Solid State Commun.*, 15, 61, 1974.
47. **Kaneyoshi, T. and Honmura, R.,** On the possibility of the coexistence of the resistivity minimum and ferromagnetism in an amorphous ferromagnet, *Phys. Lett. A*, 46, 1, 1973.
48. **Yoshiie, T., Yamakawa, K., and Fujita, F. E.,** Electrical resistivity minimum in amorphous iron films, *J. Phys. Soc. Jpn.*, 37, 572, 1974.
49. **Slechta, J.,** A criterion of the existence of the anomalous Kondo effect in amorphous magnetic materials, *J. Non-Cryst. Solids*, 18, 137, 1975.
50. **Litterst, F. J. and Kalvius, G. M.,** Investigations of non-crystalline materials and liquid crystals by the Mössbauer effect, in *Proc. Int. Conf. Mössbauer Spectroscopy*, Hrynkiewicz, A. Z. and Sawiciki, J. A., Eds., Cracow, 1975, 189.
51. **Sharon, T. E. and Tsuei, C. C.,** Magnetism in amorphous Fe-Pd-P alloys, *Phys. Rev. B*, B5, 1047, 1972.
52. **Mangin, P., Marchal, G., Piecuch, M., and Janot, C.,** Mössbauer spectra analysis in amorphous system studies, *J. Phys. E*, 9, 1101, 1976.
53. **Tyablikov, S. V.,** *Method in the Quantum Theory of Magnetism*, Plenum Press, New York, 1969.
54. **Slechta, J.,** A study of structurally disordered single element magnetic materials beyond the homogeneous molecular field approximation, *Phys. Status Solidi B*, 67, 595, 1975.
55. **Richter, J. and Kobe, S.,** Density of states and itinerant ferromagnetism in amorphous systems with dense random packed hard sphere structure, *Phys. Status Solidi B*, 85, 223, 1978.

Chapter 3

ELECTRONIC STRUCTURE OF METALLIC GLASSES

W. L. Johnson and M. Tenhover

TABLE OF CONTENTS

I. INTRODUCTION

Since their original discovery by Pol Duwez,[1] metallic glasses have occupied the interest of an increasing number of scientists. The rapid growth of research activity in this field can be attributed in great part to the many diverse and novel properties possessed by these materials. Several specific properties are discussed in the chapters of this volume. These include magnetism, chemistry, mechanical behavior, superconductivity, and electronic transport. The electronic structure of metallic glasses influences all of these properties. At the risk of overstating the case, one might say that a knowledge of both atomic and electronic structure is fundamental to understanding all other properties of metallic glasses.

By comparison with other areas of solid state science, the study of electrons in noncrystalline materials is still in an early stage of development. Understanding of electronic structure in solids prior to 1960 was premised almost exclusively on the Bloch theorem and the accompanying assumption of long-range atomic order. Noncrystalline materials pose a new challenge to both theoretical and experimental science. In this chapter, an attempt is made to survey some of the recent theoretical and experimental work on the electronic structure of metallic glasses. Section II reviews some of the theoretical approaches to calculating electronic structure in disordered metallic alloys. Emphasis is placed on d-band alloys since a large portion of the work on metallic glasses concerns alloys containing transition metals as the major constituent. Section III surveys the current state of experimental research on electronic structure. An effort is made to provide a theoretical basis for interpreting the rather broad spectrum of results within a coherent framework. Section IV covers the superconducting properties of metallic glasses. The importance of electronic structure in determining the systematics of superconductivity is stressed. Finally, in Section V, the problem of electronic transport is reviewed. Several current theoretical models are described and the predictions of the models compared with existing data. It is the express purpose of the authors to relate all of the topics discussed to the basic problem of electronic structure. This theme provides a continuous thread which links the four sections that follow.

II. THEORETICAL BACKGROUND

A. An Overview

The problem of determining the electronic structure of metallic glasses can be compared with the corresponding problem in crystalline metals, crystalline alloys, and intermetallic compounds. In both cases, one can start from either the nearly free electron approach appropriate to simple metals or the tight binding approach appropriate to d-band or f-band metals. The perfect periodicity of the atomic arrangement in crystalline metals and intermetallic compounds allows considerable simplification in calculating electronic structure since the wave functions satisfy Bloch's theorem. In a substitutionally disordered crystalline alloy, this simplification is already lost. Chemical disorder destroys the periodicity. In metallic glasses, the situation is further complicated by the absence of an underlying lattice. Metallic glasses are both chemically and topologically disordered. A complete description of the atomic arrangement in metallic glasses requires the coordinates of every atom in the system. In contrast, for crystalline metals and intermetallic compounds, only the positions of the atoms in a single unit cell are required. Knowledge of the atomic arrangement in metallic glasses is obtained primarily from diffraction experiments in the form of the pair correlation functions $\rho_{ij}(r)$ which each represents the number of j-type atoms per unit volume at a distance r from an i-type atom, averaged over all i-type atoms.[2,3] As r becomes large, $\rho_{ij}(r)$ tends toward the average density of j-type atoms $\bar{\rho}_j = c_j\bar{\rho}$ where c_j is the concentration of j-type atoms and $\bar{\rho}$ is the average atomic density of the glass. Each $\rho_{ij}(r)$ is a one-dimensional map of pair arrangements in which all directional information is lost by averaging. Determination of electronic structure is in turn limited by the incomplete knowledge of atomic arrangements.

In the following subsections, the nearly free electron and tight binding limits of the electronic structure theory are outlined for the case of chemically and topologically disordered systems. A brief description of the most widely used methods of calculating the electronic spectral distribution function D(E) is given. The discussion is not strictly rigorous and the reader is referred to the references given for a more detailed discussion of the theory. Emphasis is placed on those concepts which will be used to interpret the experimental data on metallic glasses in Sections III, IV, and V.

B. Tight Binding and Atomic Orbitals

Since a large number of glass forming alloys contain transition and rare-earth metals as principle constituents, it is appropriate to discuss electronic structure models based on the atomic orbitals approach. We wish to describe d- and f-electron wave functions and the corresponding contributions to the density of states function D(E). The tight binding or LCAO (Linear Combination of Atomic Orbitals) approach is well suited to this task.[4] We focus specifically on the case of d-band alloys. The five atomic d-states of the i-th transition metal atom of the α-type can be written $|i\alpha m\rangle$ where m labels the d-orbitals (i.e., d_{xy}, d_{zy}, etc.). Using this notation, the LCAO wave function for a binary alloy containing two transition metals can be written[4]

$$|\psi_E\rangle = \sum_{im} A_i^m |i\alpha m\rangle \tag{1}$$

where α refers to the type of atom at the i-th site. In the position representation, one can write

$$|i\alpha m\rangle = \phi_\alpha^m (\vec{r} - \vec{r}_i) \tag{2}$$

where $\vec{r}_i$ is a position vector of the atom and ϕ_α^m is the m-th atomic d-orbital of an α-type atom. It should be noted the ϕ_α^m may be chosen to be a solution of an atomic Schrödinger equation with suitable boundary conditions appropriate to a solid as in the Wigner-Seitz method.[5,8] We assume that the one electron potential can be written as a superposition of "atomic-like" potentials

$$V(\vec{r}) = \sum_i V_{i\alpha}(\vec{r} - \vec{r}_i) = \sum_i V_{i\alpha} \tag{3}$$

where $|i\alpha m\rangle$ are now taken as solutions of the one electron atomic-like Schrödinger equation

$$H_{i\alpha}|i\alpha m\rangle = E_\alpha^0 |i\alpha m\rangle \tag{4}$$

where

$$H_{i\alpha} = T + V_{i\alpha}$$

If one ignores the overlap of the $|i\alpha m\rangle$, i.e., if $\langle i\alpha m|j\beta m'\rangle = \delta_{ij}\,\delta_{mm'}\delta_{\alpha\beta}$ then the Schrödinger equation for $|\Psi_E\rangle$ becomes a matrix equation of the form

$$(E_\alpha^0 - \bar{\Gamma}_{i\alpha m} - E)\,A_i^m + \sum_{\substack{j\neq i\\ m'}} \bar{\bar{\Gamma}}_{i\alpha m}^{j\alpha' m'} A_j^{m'} = 0 \tag{5a}$$

The Γ symbols are one and two center integrals of the form

$$\bar{\Gamma}_{i\alpha m} = \langle i\alpha m | \sum_{j=i} V_{j\alpha'} | i\alpha m \rangle \tag{5b}$$

$$\bar{\bar{\Gamma}}_{i\alpha m}^{j\alpha' m'} = \langle i\alpha m | V_{j\alpha'} | j\alpha' m' \rangle \tag{5c}$$

The solution of the secular equation corresponding to Equation 5 yields the eigenvalues E and coefficients A_i^m. In principle, the density of states can be constructed in this manner. A more direct method of obtaining D(E) employs an expansion in terms of moments developed by Cyrot[6] and others. The moments of D(E) are defined as[6,7]

$$\mu_n = \int E^n D(E) dE \tag{6}$$

and can be expressed as

$$\mu_n = \mathrm{Tr} \langle H^n \rangle \tag{7}$$

where Tr refers to the trace of the matrix $\langle H^n \rangle$ and $\langle\rangle$ refers to an ensemble average. *H* is the one-electron Hamiltonian of the solid.

In what follows we shall concentrate on the simplest features of D(E) in metallic glasses. In particular, we want to analyze the simplest features of D(E) for metallic glasses containing one or two transition metals. We thus focus on the case of a binary alloy. Examples of such metallic glasses to be discussed in the following sections are ETM_{1-x} LTM_x alloys where ETM is an early transition metal (e.g., Ti, Zr, Y, Nb) and LTM is a late transition metal (e.g., Mn, Fe, Co, Ni, Cu, Rh, Pd etc.). Experimental photoemission studies enable one to profile D(E) for $E < E_F$ (E_F = Fermi energy). We wish to compare such spectra with theoretical models in an effort to understand the systematics of D(E).

The Hamiltonian matrix *H* can be symbolically written

$$H = E^0 + \bar{\Gamma} + \bar{\bar{\Gamma}} \tag{8}$$

and Equation 5 as a matrix equation involving the vector $\vec{A}$ as

$$H\vec{A} = (E^0 + \bar{\Gamma} + \bar{\bar{\Gamma}})\vec{A} = E\vec{A} \tag{9}$$

Physically, $\bar{\Gamma}$ represents the average shift of the free atom d-levels on forming the solid while $\bar{\bar{\Gamma}}$ represents the mixing of atomic levels on different sites i and j to form a band. In the language of energy bands, $\bar{\Gamma}$ gives the center of gravity for the d-levels each type of orbital in the solid while $\bar{\bar{\Gamma}}$ is related to the width of the subbands formed from each level.

Hodges, Watson, and Ehrenreich[8] (HWE) have discussed the change in mean energy of d-orbitals on going from a free atom to a solid in terms of the renormalized atom approach. This change is well approximated for a pure metal by renormalizing the neutral atom valence wave functions so as to give exact charge neutrality within a Wigner-Seitz sphere of volume equal to the mean atomic volume in the metal. The renormalized energy of the d-level, called E_R, was shown to agree well with the average d-band energy in the solid obtained by an accurate band structure calculation.[8] Thus this method provides a good means of estimating Γ in metals. Table 1 lists the values of E_R for 3d and 4d transition metals. This will be used later for reference. For a pure metal of α-type atoms, the center of gravity of the d-band is thus given to a good approximation as[4,5]

Table 1
RENORMALIZED ATOM D-LEVELS TAKEN FROM HODGES, WATSON, AND EHRENREICH[8]

Element	Renormalized state	E_{RA}(eV)
Ti	3d	−0.6
V	3d	−0.2
Cr	3d	−0.2
Mn	3d	−1.1
Fe	3d	−2.2
Co	3d	−3.1
Ni	3d	−4.3
Cu	3d	−5.6
Y	4d	−1.5
Zr	4d	−1.0
Nb	4d	−0.8
Mo	4d	−1.0
Ru	4d	−2.6
Rh	4d	−3.9
Pd	4d	−6.2
Ag	4d	−8.9

$$\langle E_\alpha \rangle \approx E_\alpha^0 + \Gamma_\alpha = E_R \qquad (10)$$

where E_R is the renormalized atom energy and $\bar{\Gamma}_\alpha$ is an average of $\bar{\Gamma}_{i\alpha m}$ over all i and m.

For a binary alloy, we have two free atom levels E_α^0 and E_β^0. In such an alloy, one might still expect the renormalized atom approach to give a good estimate of $\langle E_\alpha \rangle$ and $\langle E_\beta \rangle$ if the renormalization allows for two Wigner-Seitz cell volumes appropriate to each atom. On the other hand, one has to allow for the possibility of charge transfer effects. That is, the total charge within the α-type and β-type Wigner-Seitz cells may not be equal. The charge transfer may be of either a d-like or s-like character. A typical transition metal atom has roughly a $d^{N-1}s$ valence electron configuration where N is the total number of valence electrons. The charge transfer effect can be understood most simply by analogy to a simple diatomic heteronuclear molecule.[5] Suppose we were to calculate the bonding orbitals formed when two transition metal atoms of different species are brought together. Assume for simplicity a single nondegenerate d-orbital on each atom. Let the free atom energies be E_α^0 and E_β^0 which differ by an amount $\Delta = E_\beta^0 - E_\alpha^0$. Let the overlap (two center) integral be $V_\beta = \langle d\alpha | v_\beta | d\beta \rangle$ where $|d\alpha\rangle$ and $|d\beta\rangle$ are atomic orbitals centered on the α-type and β-type atom. In this simple problem we obtain bonding and antibonding states at respective energies

$$E_b = \tfrac{1}{2}[E_\alpha^0 + E_\beta^0] \quad [V^2 + \Delta^2]^{1/2}$$

$$E_a = \tfrac{1}{2}[E_\alpha^0 + E_\beta^0] + [V^2 + \Delta^2]^{1/2} \qquad (11)$$

The bonding state is doubly occupied in the molecular ground state and the bond polarity is given by[5]

$$\alpha_p = \frac{\Delta}{\sqrt{\Delta^2 + V^2}} \qquad (12)$$

as can be seen by a simple comparison of the probability of the bonding electrons being found on the α or β sites.

For the case of a binary alloy, we must consider all of the occupied and unoccupied d-states. Those which are of a bonding type will tend to build up charge on the atom whose free level E^0 is lowest and will be preferentially occupied in the solid. The bond polarity for d-electrons should again roughly be proportional to

$$\alpha_p \propto \frac{\Delta}{\sqrt{\Delta^2 + V_{\alpha\beta}^2}} \tag{13}$$

where $\Delta = \Delta E_R$ is the difference in renormalized atom levels $(E_R^\beta = E_R^\alpha) = \Delta E_R$ and $V_{\alpha\beta}$ is a typical two center integral for nearest-neighbor $\alpha - \beta$ atomic pairs. The difference in total charge on α and β sites is determined by composition, short-range order (numbers and type of neighbors around a given atom) and α_p. The charge transfer effect tends to raise the energy of a state with lower E_R and lower the energy of a state with higher E_R. Its net effect is to establish a common chemical potential $\mu \approx E_F$ in the alloy. The charge transfer effect increases with the ratio$|\Delta E_{RA}/V_{\alpha\beta}|$ as seen in Equation 13. Taking $E_R^\beta > E_R^\alpha$, for example, we would expect to find the α and β subbands of the alloy centered at

$$\langle E_\alpha \rangle = E_R^\alpha + E_\alpha^{c.t.}$$

$$\langle E_\beta \rangle = E_R^\beta + E_\beta^{c.t.} \tag{14}$$

where $E_{\alpha(\beta)}^{c.t.} > 0$ represents the change in d-level energy due to charge transfer or d-bond polarity. Given a fixed $V_{\alpha\beta}$, we would expect $E_{\alpha(\beta)}^{c.t.}$ to increase with increasing $\Delta E_R = E_R^\beta - E_R^\alpha$. This is indeed observed experimentally as will be seen in Section III of the article.

Having determined the mean level positions $\langle E_\alpha \rangle$ and $\langle E_\beta \rangle$ amounts to determining the d-band centers, while the effect of $\bar{\bar{\Gamma}}$ is to broaden the d-levels of the alloy into bands. Defining $\Delta E^* = \langle E_\beta \rangle - \langle E_\alpha \rangle$, we would now expect to find two resolved (α and β) subbands when the typical matrix element is $\bar{\bar{\Gamma}} \ll \Delta E^*$. When $\bar{\bar{\Gamma}} \gg \Delta E^*$, we would expect the α and β subbands to merge into a single band. These are, respectively, the split band, and virtual crystal (rigid band) limits of the band structure.[8,9] Various methods of treating the $\bar{\bar{\Gamma}}$ problem

$$H'\vec{A}_\alpha = [\langle E_\alpha \rangle + \bar{\bar{\Gamma}}]\,\vec{A}_\alpha = E\vec{A}_\alpha \tag{15}$$

have been discussed by Velicky, Kirkpatrick, and Ehrenreich (VKE)[10] The problem has several simple limiting cases. These are the split band $|\Delta E^*/\bar{\bar{\Gamma}}| \gg 1$, virtual crystal $|\Delta E^*/\bar{\bar{\Gamma}}| \ll 1$, and dilute alloy limits. If a binary alloy is represented by $TM_{1-x}^\alpha\ TM_x^\beta$, then the latter limit refers to the case where either $x \ll 1$ or $(1-x) \ll 1$. Figure 1 shows results of a coherent potential approximation for a model density of states D(E) for several cases. The results are taken from Reference 10. In the figure, δ is related to $\Delta E^*/|\bar{\bar{\Gamma}}| \sim \delta$. The cases shown ($\delta = 0.25, 0.75$) illustrate the evolution of D(E) from a virtual crystal to split band limit for various values of the concentration x. In their model, VKE take nondegenerate d-orbitals with $\Gamma_{i\alpha}^{j\beta} = \Gamma_{i\alpha}^{j\alpha} = \Gamma_{i\beta}^{j\beta}$ (i and j are restricted to first neighbors). That is, they use a single value for two center integrals of all nearest neighbor atoms irrespective of species. VKE note several important points relating to moments μ_n of D(E). If the partial moments μ_n^α and μ_n^β are defined as

$$\mu_n^\alpha = \mathrm{Tr}_\alpha \langle H^n \rangle = \sum_{im} \langle i\alpha m|H^n|i\alpha m\rangle$$

$$\mu_n^\beta = \mathrm{Tr}_\beta \langle H^n \rangle = \sum_{jm'} \langle j\beta m'|H^n|j\beta m'\rangle \tag{16}$$

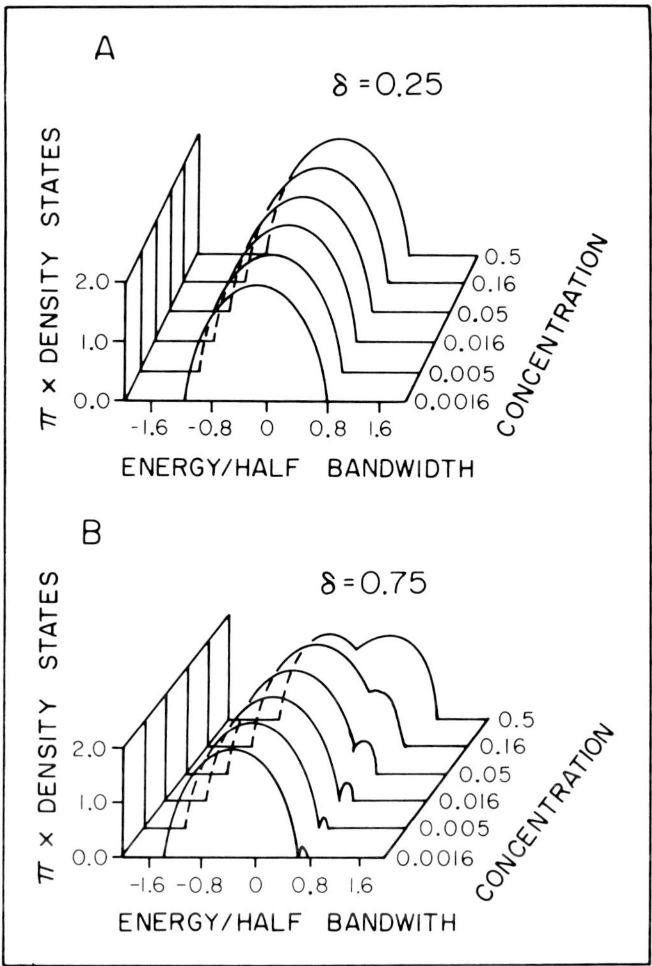

FIGURE 1. Electronic density of states in the C.P.A. approximation calculated for $\delta = 0.25$ and $\delta = 0.75$. The figure illustrates the evolution of D(E) from the "rigid band" limit to the "split band" limit. (From Velicky, B., Kirkpatrick, S., and Ehrenreich, H., *Phys. Rev.*, 175, 747, 1968. With permission.)

where the first (second) sum runs over only sites i (j) occupied by α-type (β-type) atoms, one can discuss the features of the separate subbands. One finds for example that μ_1^α and μ_1^β, the shift in subband centers due to band formation, are always such as to cause the subbands to separate in energy. This is the familiar level-repulsion effect in the theory of heteronuclear molecules.[5] In the split band limit, one finds subbands whose widths and heights are proportional to the square root of the concentration of the respective species.[10] If W^α and W^β are the subband widths, then

$$W^\alpha \propto (1 - x)^{1/2}$$

$$W^\beta \propto (x)^{1/2} \tag{17}$$

These predictions are indeed roughly followed by experimental data on binary metallic glasses of the form ETM_{1-x} LTM_x as will be seen in Section II. The higher moments μ_n, μ_n^α, and

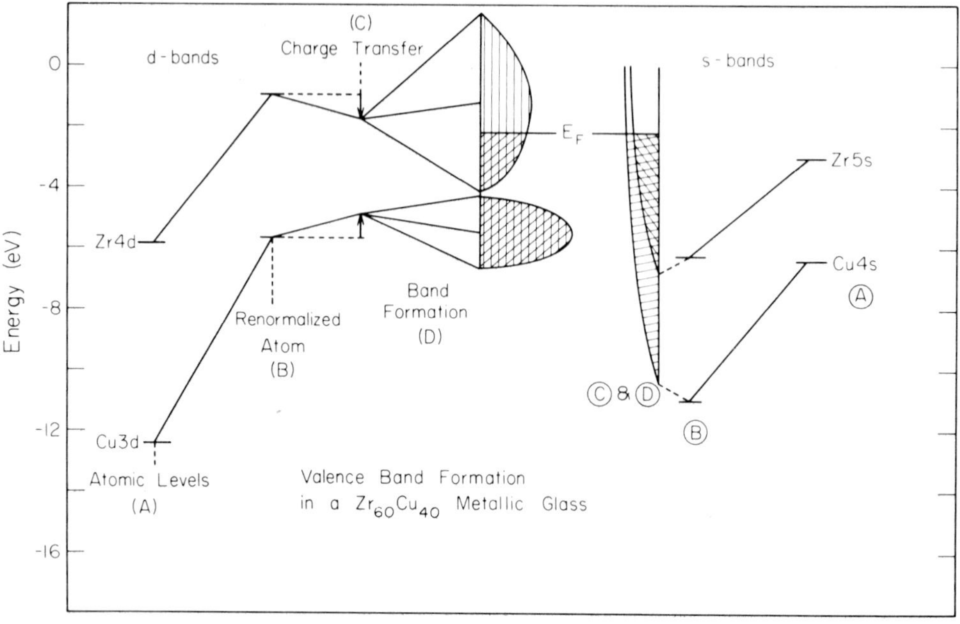

FIGURE 2. A schematic illustration of d- and s-band formation in a binary transition metal alloy $Zr_{60}Cu_{40}$. The atomic levels, renormalized atom levels, charge transfer effect, and band formation are illustrated.

μ_n^β $(n > 2)$ can be evaluated for a specific model. These give the detailed shape of D(E) for a given alloy. Specific results will be mentioned in Section III in an attempt to understand the experimental data.

In Figure 2, a semiquantitative summary of the above discussion of d-band formation is presented for the case of a glassy $Zr_{0.6}Cu_{0.4}$ alloy. Shown are the free atom levels, renormalized atom levels, charge transfer effect, and band formation. Also shown is a corresponding diagram for the s-levels. The s-levels form a free electron-like conduction band and can be correctly placed with respect to E_F only by a detailed band calculation. The energies depicted for E_R^{Zr}, E_R^{Cu}, E_{Zr}^0, and E_{cu}^0 are taken from Reference 10. The other energies involved are chosen arbitrarily for the purpose of illustration. In Section II, the details of such a diagram are discussed.

C. Nearly Free Electrons; s- and p-Bands

A few metallic glasses containing only simple metals have been obtained by the method of melt quenching. These include such alloys as $Mg_{70}Zn_{30}$,[11] $Ca_{60}Al_{40}$.[12] The pseudopotential method described by, for example, Harrison,[13] and other authors[14] can be directly applied to calculate the electronic density of states D(E) for such liquid alloys once the partial pair correlation functions $\rho_{ij}(r)$ and corresponding structure factors $S_{ij}(\vec{q})$ are known. The reader is referred to the references for further discussion of the method.

In the present article, we will focus on s and p band contributions to D(E) in transition metal alloys. This contribution was for the most part neglected in Section II.B. Metallic glasses containing only transition metals can be generally characterized by an electronic configuration $d^{N-1}s$ for the transition metal atoms. For transition metal-metalloid glasses, (e.g., $Pd_{80}Si_{20}$, $Fe_{80}B_{20}$, $(Mo_{0.6}Ru_{0.4})_{80}B_{20}$) the metalloid element (Si,B) has a typical valence configurations of s^2p^{N-2} or sp^{N-1} where N is the metalloid valence. Thus, such alloys will, in general, contain metalloid s and p contributions to D(E). In this case, the band structure will be quite complex. We first examine the case of alloys containing only transition metals.

The s-band of the transition metals is rather broad ($\sim$ 10 eV) spanning a range of energies typically larger than and including the d-band region. Figure 2 illustrates schematically the s-band density of states $D_s(E)$ for a $Zr_{0.6}Cu_{0.4}$ alloy. The band should be roughly free-electron-like with $D_s(E) \propto E^{1/2}$. However, in the energy range where $D_s(E)$ overlaps $D_d(E)$, one expects a significant hybridization of atomic s and d wave functions. The eigenfunctions have a mixed character in this region. Harrison has discussed this hybridization in detail and shown that its effect is to modify the pseudopotential by an additional term[5,18]

$$V_{sd} = \sum_{d} \frac{\langle \vec{K} | \Delta(\vec{r}) | d \rangle \langle d | \Delta(\vec{r}) | \vec{K} \rangle}{E_{\vec{k}} - E_d} \tag{18}$$

where $\vec{K}$ represents a free electron (s-band) plane wave state and $\Delta(\vec{r})$ is the hybridization potential. This term should be added to the usual pseudopotential V_0 of the conduction electrons to give an effective pseudopotential

$$V^{eff} = V_0 + V_{sd} \tag{19}$$

This term has the net effect of scattering plane wave electrons into d-band states. Mott has used this scattering mechanism[15] to explain the resistivity of transition metals as will be discussed in Section V.

When metalloid elements are added to a transition metal alloy, one must consider the interaction of the metalloid s and p states with the transition metal s and d states. Again, the s electrons can be regarded collectively as forming a nearly free electron conduction band although it is not an easy matter to establish the location of the bottom of this free electron conduction band with respect to the d-band levels. It is also of importance to be able to locate the mean position of the metalloid p levels with respect to the transition metal d-levels and E_F. This can be roughly accomplished by taking the center of gravity of the d-band to be given by the renormalized atom levels already discussed in Section II.B. In the simplest approximation, one expects to find the metalloid p-levels roughly near the respective free atom levels.[5,16] Charge transfer and renormalization effect will of course tend to modify the mean metalloid p-band energies. Table 2 gives a list of atomic s and p valence levels for the most commonly occurring metalloid atoms in metallic glasses. The data are taken from Hermann and Skillman.[16] These data will be used in the following sections to aid in interpreting photoemission data on metallic glasses which contain metalloid elements.

D. Intra-Atomic Correlations; Formation of Localized Magnetic Moments; Spin Fluctuations

The d-band model discussed in Section II.B completely neglects the effect of electron exchange and correlation and is strictly a one-electron approach to the d-band structure. Friedel,[4] for example, discusses the effect of neglecting correlations. The most important effects neglected are the intra-atomic Coulomb and exchange effects. For a free transition atom, Hund's rules specify that the lowest energy electronic configuration is obtained by maximizing the total spin S of the d-electrons. In a solid, this tendency is in direct competition with the band energy involved in transferring electrons from low energy states with both spin states filled to higher energy states to obtain the parallel spin configuration. The difference in the energy for a parallel spin configuration of two d-electrons and an antiparallel configuration is defined as the intra-atomic exchange integral $U_{mm'}$ and is given by

$$U_{mm'} = \langle im(1), im'(2) | e^2/r_{12} | im(2), im'(1) \rangle \tag{20}$$

where $|im(1), im'(2)\rangle$ is a two electron state with electrons (1) and (2) in atomic states $|im\rangle$ and $|im'\rangle$ and r_{12} is the separation of the electrons. When $U_{mm'}$ becomes large with

Table 2
ATOMIC S AND P VALENCE LEVELS FOR METALLOID ELEMENTS COMMONLY FOUND IN METALLIC GLASSES

Element	State	Energy (eV)
Be	2s	−8.2
	2p	−4.1
B	2s	−12.5
	2p	−6.6
C	2s	−17.5
	2p	−9.0
Al	3s	−10.1
	3p	−4.9
Si	3s	−13.6
	3p	−6.5
P	3s	−17.1
	3p	−8.3
S	3s	−20.8
	3p	−10.3
Ge	4s	−14.4
	4p	−6.4
As	4s	−17.3
	4p	−7.9

respect to the band energy $\bar{\bar{\Gamma}}$, one might expect Hund's rule to apply. Thus for $U \gg \bar{\bar{\Gamma}}$, one would expect magnetic moments (localized spin states) and magnetism to result. The reader is referred to References 17, 18, and 19 for a more detailed discussion of localized moments in transition metals.

One particularly interesting question is whether or not a given transition metal will form a magnetic moment when introduced as a dilute impurity in an otherwise nonmagnetic matrix.[17,20] Anderson[17] has provided a eloquent model which answers this question for a nondegenerate impurity d-orbital. Moriya[19] has examined the case of a fivefold degenerate impurity d-orbital. The following are roughly the criteria for formation of a localized moment on a dilute impurity:

1. $|U_i| \gtrsim W_i$ where U_i is the intra-atomic Coulomb energy for d-electrons of opposite spin on the impurity atom and W_i is the width of the impurity d-subband as in Section II.B. (U_i should actually be summed over all occupied impurity d-states m.)
2. $|\langle E_i \rangle - E_F| \lesssim U_i$ where $\langle E_i \rangle$ is the mean energy of the impurity d-band.

Thus, the impurity d-band must lie close to E_F (by comparison with U_i) and U_i must dominate the impurity bandwith energy W_i.

The transition from the case of no localized moment to moment formation is not completely sharp. For intermediate cases, spin fluctuations occur. The localized moment forms in a transient manner with a finite lifetime.

The same theory can roughly be applied to more concentrated alloys where the impurity level broadens to form a subband characteristic of a concentrated alloy (see Figure 1). We will use these ideas to discuss formation of localized moments and spin fluctuations in metallic glasses containing nonmagnetic 4d metals together with 3d metals (e.g., $Zr_{70}Fe_{30}$)

in Section IV. The formation of localized moments and spin fluctuations are particularly important in understanding superconductivity in such alloys.

III. EXPERIMENTAL RESULTS AND COMPARISON WITH THEORY

In this section we will review and discuss the progress made in experimentally determining the electronic structure of metallic glasses. Most of the work, and most of the interest in this area has been on metallic glasses containing transition metals. We will restrict the discussion here to this particular type of metallic glass. This is not a serious restriction, since many of the interesting effects connected with the electronic structure of metallic glasses can be found within this group.

Metallic glasses represent a new material for the study of electronic structure. There are several unique features which make them especially interesting for this work. The most obvious feature is the absence of long range order. This, at first glance, might be expected to lead to strikingly different electronic structures as compared to the corresponding crystalline phases. The atomic and electronic structures are intimately related, so that a detailed knowledge of the electronic structure might provide additional insight concerning the atomic arrangement of glasses. Metallic glasses can usually be formed over some composition range. This permits investigation of the properties with chemical content without the complications encountered in crystalline materials, namely the formation of definite compounds. From the experimental point of view, the ability to make metallic glasses has permitted a variety of experiments which cannot be readily performed on liquids or thin film samples of amorphous materials.

The ultimate goal of experiments on electronic structure is to relate it to the properties of the materials. Knowledge of the electronic structure is crucial to an understanding of properties such as superconductivity and magnetism in these materials. Of special interest is the connection between electronic structure and formation of glasses.

The starting point for discussing the electronic properties of crystalline metals is the construction of the Fermi surface. The shape and size of which is intimately involved in the transport coefficients as well as the optical and equilibrium properties. One of the most important tests of a band structure calculation is how well it can reproduce the details of the Fermi surface. In crystalline materials there are a variety of techniques which can provide information about the Fermi surface (such as cyclotron resonance and the de Haas-van Alphen effect). These measurements usually require single crystals of very pure substances and frequently must be done at low temperature. These kind of experiments are out of the question for amorphous materials, and practically speaking they are usually quite difficult to perform in crystalline alloys.

If the scattering of electrons in a material is weak, then the electron wavevector $\vec{K}$ is approximately a good quantum number and the Fermi surface is definable and nearly spherical (nearly free electron case). As discussed in the previous section, we might expect simple metal glasses (e.g., MgZn and CaAl) to fall in this category. The transition metal glasses we wish to discuss here are in contrast found to be in the strong scattering regime where $\vec{K}$ is not a good quantum number and the concept of a Fermi surface is not even approximately valid.[21]

The natural concept to carry over from the physics of crystalline materials is the idea of a density of states function D(E). For these materials the density of states function carries the maximum amount of information available to us on the electronic structure. Most experiments on the electronic structure, by necessity, have concentrated on determining D(E) or just its value at the Fermi surface $D(E_F)$. Useful as this function is, it is a poor substitute for knowledge of the entire Fermi surface and band structure. Because of this, our experimental knowledge of electronic structure cannot be as satisfying as that in crystalline materials.

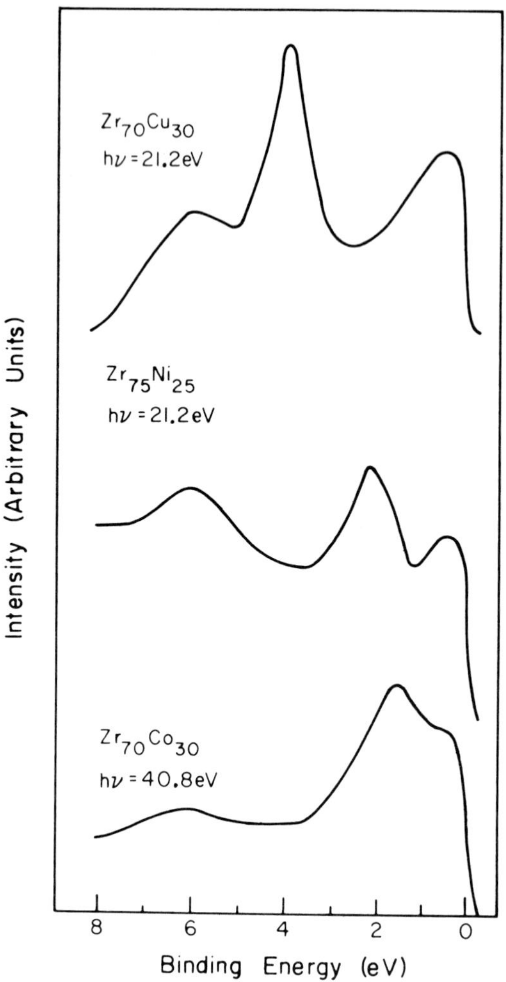

FIGURE 3. The UPS spectra of several Zr-base metallic glasses taken from References 22 and 23.

As we mentioned above, most of the experiments in this area have dealt with determining D(E) and D(E_F). The techniques used naturally divide themselves into those which determine D(E) and those whose goal it is to measure D(E_F).

The most direct methods for obtaining information on D(E) involve the various photoemission techniques, such as Ultraviolet Photoemission Spectroscopy (UPS) and X-ray Photoemission Spectroscopy (XPS). In addition to these, some Soft X-ray Spectroscopy (SXS) and Auger Electron Spectroscopy (AES) measurements have been made on metallic glasses. The basic idea behind all these techniques is similar. A beam of nearly monochromatic electrons or photons is focused on the sample and the ejected electrons or photons are energy analyzed. In Figure 3 we show the UPS spectra from References 22 and 23 obtained for a few Zr-based metallic glasses. These measurements were made using a He-I($h\nu$ = 21.2 eV) or He-II($h\nu$ = 40.4 eV) ultraviolet source and energy analyzing the emitted electrons. The solid lines in Figure 3 are the results of measuring the intensity of the emitted electrons as a function of energy in which the energy is taken with respect to the Fermi energy (E_F = 0). These measurements are not selective, that is to say that the spectra are a superposition of contributions from both components (Zr and Cu, Ni, or Co). In addition to this, the contribution from each type of atom comes from the various orbitals near E_F. In $Zr_{70}Cu_{30}$, for example, the UPS curve might have contributions from Zr(4d, 5p, and 5s) and Cu(3d, 4p, and 4s) orbitals. Complicating the analysis further, the intensity we measure is not directly convertible to D(E) since the contributions from different types of orbitals are weighted by different transition probabilities as well as final state densities.[25] For this reason the intensities are usually given in arbitrary units. Riley et al.[24] have used the fact that the transition probabilities are functions of energy to deconvolute the experimental spectra of $Pd_{81}Si_{19}$ into partial density of states functions. Here the relevant levels are the Pd (4d)and Si (3p and 3s). Combining XPS, UPS, and AES and using the fact that the transition probabilities for d, p, and s electrons are different for each of these, they were able to schematically separate the different contributions. They found that the Pd (4d) band was below E_F and that the main contribution to $D(E_F)$ is from p-like levels associated with Si. We shall return to this later when we discuss the $Pd_{1-x}Si_x$ system in more detail.

For many electronic properties an important parameter is the value of the density of states function at the Fermi energy $D(E_F)$. There are a few techniques which can determine $D(E_F)$ for metallic glasses. We shall discuss three of them.

The most direct measurement of $D(E_F)$ is usually obtained from heat capacity measurements. We can write the total electronic energy (U) of the glass as:

$$U = \int_0^\infty E\, D(E)\, f(E) dE \tag{21}$$

where f(E) is the Fermi distribution function

$$f(E) = \left[\frac{1}{e^{(E - E_F)/K_BT} + 1} \right] \tag{22}$$

and the electronic heat capacity is

$$C_{el} = \frac{dU}{dT} = \int_0^\infty D(E) \left(\frac{df}{dT} \right) dE \tag{23}$$

The electrons in an energy range k_BT of E_F will make the largest contribution to C_{el}, so that if D(E) is not rapidly changing on the scale of k_BT we can set $D(E) = D(E_F)$.

$$C_{el} = D(E_F) \int_0^\infty E \left(\frac{df}{dT} \right) dE \tag{24}$$

The integral can be evaluated using a Sommerfield expansion to give to leading order

$$C_{el} = \frac{\pi^2}{3} D(E_F) K_B^2 T \tag{25}$$

From a practical standpoint the most important assumption made in obtaining this formula is that $D(E_F)$ is not rapidly changing near E_F. This should not be a serious problem for metallic glasses. Measurements of heat capacity (C_p) are therefore expected to give reliable values for $D(E_F)$. Problems do occur when there are contributions to C_p which cannot be separated from the electronic term. This was encountered in the glass $Fe_{20}Ni_{60}P_{14}B_6$ where a magnetic contribution occurs at low temperature preventing a determination of the electronic term.[26] Also, to obtain accurate values of $D(E_F)$, the effects of electron-phonon renormalization must[27] be accounted for. This is discussed in more detail in Section IV.

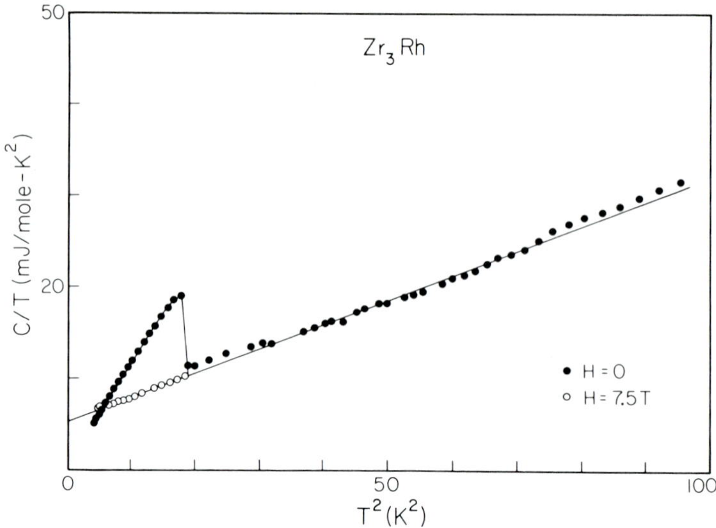

FIGURE 4. The results of specific heat measurements on the metallic glass Zr_3Rh. (From Garoche, P. and Johnson, W. L., *Solid State Commun.*, 39, 403, 1981. With permission.)

In Figure 4 we have plotted the results of a C_p experiment on the metallic glass Zr_3Rh.[28] This glass is a superconductor and below the transition temperature (T_c) we see the heat capacity anomaly associated with this. Applying a magnetic field of 7.5 T reduces T_c to below 2 K and thus permits measurements at low temperature in the normal state. This is an important point because the lower in temperature the experiment can be performed, the more accurate the determination of the electronic term. Mizutani et al.[29] have shown that the low temperature heat capacity of metallic glasses in the normal state can be analyzed by the following expression

$$C_p = (\gamma + \gamma')\,T + \beta T^3 + \delta T^5 \tag{26}$$

where γ ($= 1/3\pi^2 D(E_F)k_B^2$) represents the normal electronic contribution, γ' is the non-electronic contribution generally observed in insulating glasses,[28] $\left(\beta = \frac{12\pi^4}{5} Nk_B \frac{1}{\theta_D}\right)^3$ (θ_D-Debye temperature) is the harmonic and δ the anharmonic lattice contributions. Below T_c the normal electronic term (γ) disappears, allowing a determination of γ'. Garoche and Johnson[28] found that for Zr_3Rh,γ' was 50 times smaller than γ. Earlier results have been reported for $Zr_{70}Pd_{30}$.[31] To a very good approximation, we can ignore γ' in analyzing the C_p data for the purpose of obtaining $D(E_F)$.

Plotting C_p/T vs. T^2 (Figure 4) gives a straight line allowing one to determine γ (intercept) and β (slope). Deviations from a straight line can be caused by anharmonic lattice contributions (T^5 and T^7 terms) which seem important for this glass at $T^2 = 50\ K^2$.

Another method which has been used for finding $D(E_F)$ in metallic glasses comes from the theory of dirty limit superconductors.[32] From the microscopic theory of superconductivity it is possible to derive a formula relating γ, the normal state resistivity (ρ_N) and the slope of the upper critical field (H_{c2}).[33] In the case of an isotropic dirty limit superconductor,[34] assuming that the transport properties can be described by a Boltzmann equation, we have

$$D(E_F) = \frac{\pi^2 \hbar}{8e^2 K_B \phi_0 \rho_N} \left[\frac{dH_{c2}}{dT} \bigg|_{T_c} \right] \tag{27}$$

Table 3
PHOTOEMISSION EXPERIMENTS ON A VARIETY OF METALLIC GLASSES

Alloy	Technique	Ref.
Pd–Si	UPS	36
	UPS, XPS, AES	24
	UPS, XPS	39
Fe–B	XPS	42
Nb–Ni	UPS, XPS	51, 53
Fe–P–B	UPS, XPS	54
Ta–Ni	UPS	51
Ti–Co	UPS	51
Zr–Cu	XPS, UPS	22, 50
Zr–Rh	UPS	51
$Mo_{40}Ru_{40}B_{20}$	UPS, XPS	56

where ϕ_0 (= hc/2e) is the fundamental flux quantum. Metallic glasses are expected to be isotropic and in the extreme dirty limit,[35] the validity of the above relation is therefore dependent primarily on the ability of the Boltzmann transport equation to correctly describe its properties. In crystalline superconductors the agreement between $D(E_F)$ from C_p measurements and from the above formula is generally within 20%. As will be shown in Section IV, comparable agreement is obtained in metallic glasses.

A third way of finding $D(E_F)$ for metallic glasses is measurements of the magnetization. In the free electron case we know that we can write for the temperature independent contribution to the susceptibility (Pauli paramagnetism)

$$\chi_{Pauli} = 2\mu_B^2 \, D(E_F) \tag{28}$$

where μ_B is the Bohr magneton. This relation is valid only for the case of a free electron gas. There are other contributions to the temperature independent susceptibility which have to be calculated or simply ignored in order to use χ_{Pauli} to find $D(E_F)$ in general.[35] This method is expected to be the least reliable of the three.

A fairly large amount of work has been done using the various photoemission techniques on metallic glasses. A detailed list is given in Table 3. Likewise, information and references for the experiments determining $D(E_F)$ are given in Tables 4 and 5. For convenience we have converted the results of the various determinations of $D(E_F)$ to equivalent values of γ. We distinguish the γ's by γ_c, γ_H, and γ_m for heat capacity, critical field, and magnetic susceptibility measurements, respectively. A single type of measurement on a material is usually not adequate to understand its properties, especially with materials as complicated as these. For this reason we choose here to discuss in detail a few systems in which a large amount of work has concentrated.

The $Pd_{1-x}Si_x$ glasses are of particular interest since we have available to us the results of photoemission, C_p, and magnetic susceptibility measurements. Waclawski et al.[36] have studied the compositional dependence of the UPS spectra in the $Pd_{1-x}Si_x$ system. They found that the Pd 4d band was nearly full in the glass and that its binding energy increased monotonically with Si content (see Figure 5). The number of valence electrons as well as E_F increases as we add Si, forcing the Pd 4d levels to move to a higher binding energy. Based on this we would expect that $D(E_F)$ would decrease as the Pd 4d band sinks to higher binding energies. In support of this expectation, Müller et al.[37] found that the magnetic

Table 4
γ_c AND $\gamma_{H_{c2}}$ VALUES FOR SEVERAL METALLIC GLASSES

Metallic glass	γ_c (mJ/mole-K^2)	$\gamma_{H_{c2}}$ (mJ/mole-K^2)	Ref.
$La_{80}Ga_{20}$	6.1	6.0	116, 117
$Pd_{74.5}Cu_9Si_{16.5}$	1.17	—	29
$Fe_{80}P_{14}B_6$	6.3	—	26
$Ni_{80}P_{14}B_6$	2.4	—	26
$Zr_{70}Pd_{30}$	4.7	4.5[a]	31
$Zr_{65}Cu_{35}$	3.9	4.1[a]	118
$Zr_{75}Rh_{25}$	5.4	5.3[a]	28
$Mo_{40}Ru_{40}P_{20}$	2.5	2.8[a]	35

[a] This work.

Table 5
ELECTRONIC HEAT CAPACITY PARAMETER DETERMINED FROM SPECIFIC HEAT (γ_c), CRITICAL FIELD AND RESISTIVITY MEASUREMENTS (γ_H), AND PAULI PARAMAGNETIC SUSCEPTIBILITY MEASUREMENTS (γ_m) FOR SEVERAL MO–RU BASE METALLIC GLASSES

Metallic glass	γ_c(mJ/mole-K^2)	γ_H (mJ/mole-K^2)	γ_m (mJ/mole-K^2)	Ref.
$(Mo_{1-x}Ru_x)_{80}P_{20}$				
x = 0.2	—	—	4.4 ± 0.35	35
0.3	—	3.2 ± 0.5[a]	—	35
0.4	—	3.0 ± 0.5[a]	3.2 ± 0.2	35
0.5	2.8 ± 0.3[b]	2.8 ± 0.4[a]	—	35
0.6	—	2.6 ± 0.4[a]	2.5 ± 0.2	35
0.7	—	2.1 ± 0.3[a]	—	35
0.8	—	—	1.7 ± 0.1	35
$(Mo_{0.6}Ru_{0.4})_{100-x}B_x$				
x = 10	—	3.45 ± 0.6	—	80d
12	—	3.85 ± 0.7	—	80d
14	3.4 ± 0.3	3.76 ± 0.6	—	80d
16	3.1 ± 0.3	3.40 ± 0.6	—	80d
18	2.6 ± 0.3	2.83 ± 0.5	—	80d
20	2.0 ± 0.3	2.01 ± 0.4	—	80d
22	2.1 ± 0.4	1.84 ± 0.4	—	80d
24	—	1.62 ± 0.3	—	80d
$(Mo_{0.6}Ru_{0.4})_{100-x}Si_x$				
x = 18	—	2.21 ± 0.3	—	57
20	—	2.34 ± 0.3	—	57
22	—	2.11 ± 0.3	—	57
24	—	2.34 ± 0.3	—	57
26	—	2.65 ± 0.3	—	57
28	—	2.50 ± 0.3	—	57
30	—	2.76 ± 0.4	—	57

[a] This work.
[b] R. D. Shull and W. L. Johnson, unpublished results.

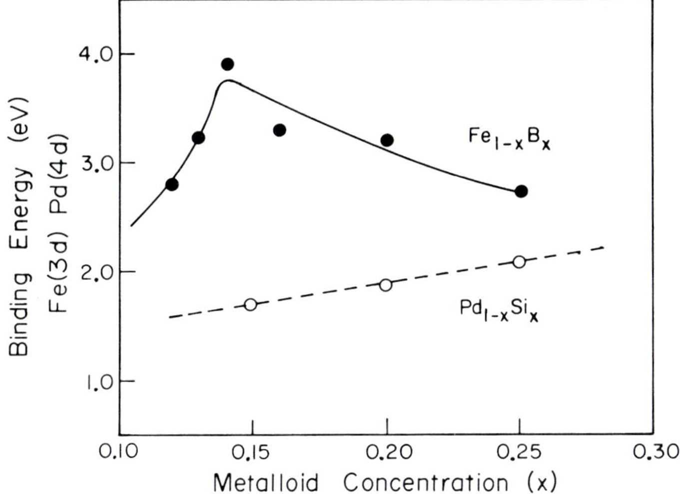

FIGURE 5. The binding energy of Fe (3d) and Pd (4d) levels taken from the photoemission data of References 42 and 46.

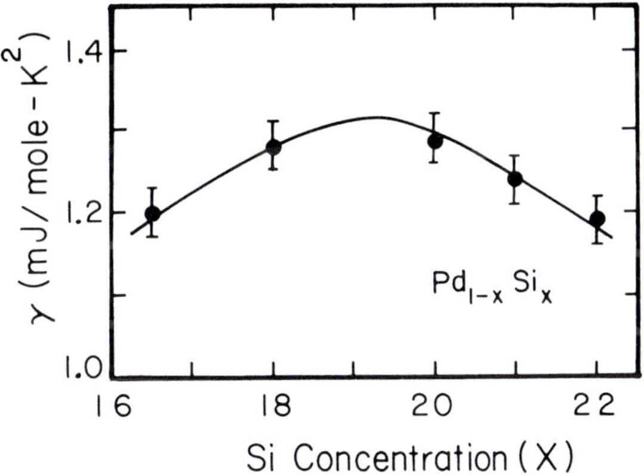

FIGURE 6. The electronic heat capacity parameter, γ_c, for $Pd_{1-x}Si_x$ metallic glasses. (Data from Mizutani, U., Hartwig, K. T., Massalski, T. B., and Hopper, R. W., *Phys. Rev. Lett.*, 41, 661, 1978. With permission.)

susceptibility (χ) of liquid $Pd_{1-x}Si_x$ alloys drastically decreased as x increases. They claim from these measurements that the electrons at E_F were principally d-like and that the drop in χ with x results from a shift of E_F to regions of lower density of states. In addition to this, a number of authors have concluded that the main contribution to $D(E_F)$ for $Pd_{1-x}Si_x$ glasses is from the Pd 4d band.[38-40] In Figure 6 we show the results of C_p measurements by Mizutani et al.[29] in $Pd_{1-x}Si_x$ glasses. We see that γ_c (and therefore $D(E_F)$) reaches a maximum at about 19% Si. In their analysis, Mizutani et al.[29] concluded that contributions to γ_c from effects other than $D(E_F)$ are relatively small and probably monotonic with composition. We may therefore assume that the γ_c (Figure 6) reflects a change in the density of states in the conduction band. Mizutani et al.[29] attributed the relatively large values of γ_c to a tail of the Pd 4d band at E_F. Indeed, the value of γ_c is roughly twice as large as one would expect for a filled d band material (e.g., Ag).

As we mentioned earlier, Riley et al.[24] tried to separate the photoemission spectra of a

$Pd_{81}Si_{19}$ glass into partial density of states contributions. In doing this they concluded that the Pd 4d band did not contribute to $D(E_F)$ but that $D(E_F)$ was mainly from the 3p levels of Si as a result of a Pd 4d-Si 3p hybridization. They actually saw a local maximum in the 3p levels of Si which might lead to the curve in Figure 6. One further piece of information which has some bearing on this question is the Hall Effect measurements of Mizutani et al.[41] The Hall coefficient gives information on the conduction electron concentration. They found that for $Pd_{1-x}Si_x$ glasses the electron concentration was nearly constant below $x = 0.18$, increased rapidly for $x = 0.18$ to $x = 0.20$, and increased slowly for $x \geqslant 0.20$. The rapid increase they see is in the same range as the peak observed in γ.

Combining the results of these measurements we are faced with something of a paradox. The UPS and magnetic susceptibility measurements implied that $D(E_F)$ should decrease with x, while the Hall Effect measurements imply an increase with x, and the C_p measurements found a peak in $D(E_F)$. The key to understanding these results is to realize that the different experiments measure different aspects of the electronic structure. The UPS and magnetic susceptibility measurements are most sensitive to the d electrons. The Hall coefficient measurements yield conduction electron concentration values which would be expected in these alloys to give more weight to the s and p electrons. And finally, the C_p measurements directly determine the sum of all contributions to $D(E_F)$. With this in mind, we interpret the results of the various experiments as indicating that the d band contribution to $D(E_F)$ decreases with x, while the s and p contributions increase with x. We suggest that the interplay of these two effects is the origin of the peak in γ_c. Using our analysis of the previous section we find that the renormalized atom 4d levels for Pd and 3p atomic levels of Si are at -6.2 and -6.5 eV, respectively, so that we expect the 3p levels of Si to be near or at E_F and we expect considerable hybridization. The correct description of the states at E_F for $Pd_{1-x}Si_x$ glasses is that they are the result of a Pd 4d-Si 3p hybridization and that as x increases the d-character decreases while the p-character increases.

Another system which has attracted considerable attention is the $Fe_{1-x}B_x$ glasses. From the photoemission work of Matsuura et al.[42] we know that the Fe 3d band is near E_F and that as expected, it is more than half occupied. In contrast to the $Pd_{1-x}Si_x$ system, the Fe 3d levels in $Fe_{1-x}B_x$ show a distinct peak in binding energy near $x = 0.14$ (see Figure 5). An immediate consequence of this is that the contribution to $D(E_F)$ from the Fe 3d band is smallest for $Fe_{0.86}B_{0.14}$. Similar anomalies have been found in density,[43] coordination number,[44] magnetic moment,[45] glass transition temperatures,[46] and modes of crystallization[47] as a function of B content in these glasses. The origin of these anomalies is unknown, explanations have ranged from a structural stabilization[48] to a simple alloy mixing effect.[49] The XPS results of Matsuura et al.[42] show that the Fe 3d levels are most sensitive to this anomaly, which suggests that there might also be a magnetic contribution to this behavior. Matsuura et al.[42] also noticed a small hump centered at about -10 eV (with respect to E_F) in the valence band spectra of the $Fe_{1-x}B_x$ glasses. Using the values of the 3d Fe level and B levels from Tables 1 and 2 of Section II we expect that this small hump is a contribution from the B 2p levels. The interesting feature of this hump is that its intensity and binding energy seems to be independent of B content. Matsuura et al.[42] speculated that the hump results from forming a bonding state between B 2p and Fe 3d electrons. They explained the lack of variation in its intensity and binding energy by suggesting that B occupies "interstitial-like" and "substitutional-like" sites in the glass. The bonding state leading to the hump is formed from Fe and "interstitial-like" B atoms. It is further assumed that the concentration of the "interstitial-like" B atoms is less than 14% B, so that the intensity and binding energy are saturated in $Fe_{1-x}B_x$ alloys with $x \geqslant 0.14$. The "substitutional-like" B atoms are then assumed to contribute and perhaps deform the Fe 3d band. Unfortunately, one does not have the experimental information necessary to confirm the description provided by Matsuura et al.[42] for these glasses. From Section II (Equation 18) one doesn't expect much hybridization

Table 6
SUBBAND SPLITTING OF ZR-BASED GLASSES

Metallic glass	ΔE^{RA} (eV)	ΔE^{exp}
$Zr_{76}Fe_{24}$	1.2	0.8
$Zr_{70}Co_{30}$	2.1	1.0
$Zr_{76}Ni_{24}$	3.3	1.7
$Zr_{70}Cu_{30}$	4.6	3.2
$Zr_{70}Pd_{30}$	5.2	3.0

between the B 2p and Fe 3d levels so that their identification of the hump as the Fe–B bonding state could be questioned. Heat capacity measurements as a function of composition and nuclear resonance data to give more information on the B environment and coordination number are especially needed.

We now turn to a discussion of Zr-based and Mo–Ru-based metallic glasses. Many of these glasses exhibit superconductivity, which makes them of special interest to us, since this allows us to find $\gamma_{H_{c2}}$. In Tables 4 and 5 we list the values of γ derived from the three experiments we described earlier. We see that for all cases the γ derived from the various techniques are in surprisingly good agreement. This is an important point for our discussion of the systems $(Mo_{0.6}Ru_{0.4})_{1-x}B_x$, $(Mo_{0.6}Ru_{0.4})_{1-x}Si_x$, and $(Mo_xRu_{1-x})_{80}P_{20}$ where only a limited amount of C_p measurements are available.

By far, the most extensive UPS and XPS studies of metallic glasses have been on the Zr-based alloys.[22,23,50,51] In Figure 3 we showed the UPS spectra for a few Zr glasses. Oelhafen et al. have performed UPS measurements as a function of composition for the following systems: Zr–Cu,[22] Zr–Pd,[22] Zr–Ni,[51] Zr–Co,[51] and Zr–Fe.[51] They found from these studies that the contribution to D(E) at E_F was mainly from Zr 4d orbitals and the d-states of the late transition metal (LTM) were at higher binding energies. This observation has also been confirmed by SXS experiments on Zr–Pd glasses.[52] For $Zr_{70}Cu_{30}$ (Figure 3) for example, the peak at 4 eV is primarily from the Cu 3d levels and the peak near E_F is associated with the Zr 4d levels. Also from their compositional studies, Oelhafen et al. found that the width of the peak associated with the LTM increases with increasing concentration of the LTM. Finally, as Figure 3 shows, the distance between the two subbands decreases as the distance between Zr and LTM in the periodic table decreases. In the terminology of Section II, $Zr_{70}Cu_{30}$ is an example of the split band case while $Zr_{70}Co_{30}$ approaches the virtual crystal limit.

Let us define ΔE^{exp} as the difference in energy of the center of the LTM peak and the maximum of the Zr levels. We do this to provide some measure of the subband splitting for our comparisons between alloys and with the predictions of Section II. ΔE^{exp} does not correspond to the difference between the centers of the two subbands since it is not possible to determine the position of the center of the Zr 4d band from these experiments. The Zr 4d subband is less than half full here so that its center lies above E_F.

In Section II we listed the factors which are responsible for the positions and shapes of the features observed in D(E). The first step in locating the levels is to use the renormalized atom levels from Table 1. We list the splitting predicted from these levels (ΔE_R) in Table 6. The ΔE_R values give a reasonable estimate for the ΔE^{exp} in magnitude and predict the decrease in ΔE^{exp} in going from Zr–Cu to Zr–Fe. Note that the lower values of ΔE^{exp} compared to ΔE_R are partially due to the difficulty in determining the center of the Zr subband we mentioned above. The next factor that needs to be considered is the effect of charge transfer or d-band polarity. As pointed out in Section II this has the effect of bringing the two

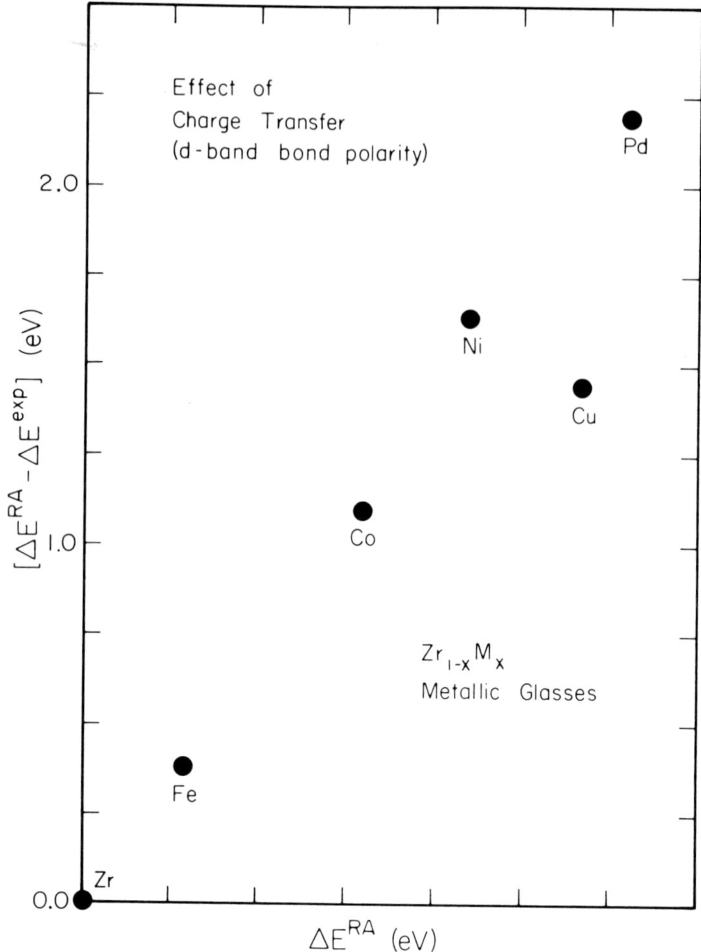

FIGURE 7. Illustration of the effect of charge transfer on band splitting in Zr-base metallic glasses. See text for details.

subbands closer together. A final effect is the level repulsion due to formation of a band. This causes the two subbands to separate, increasing the level splitting and partially cancelling the attraction resulting from charge transfer. These steps are summarized for a $Zr_{60}Cu_{40}$ glass in Figure 2 of Section II. To get an idea of the various contributions we plot the difference in energy between the observed levels (ΔE^{exp}) and the difference in the renormalized atom values (ΔE_R) vs. ΔE_R for each LTM in Figure 7. We see that deviation from our predicted values of the subband splitting (i.e., renormalized atom values) increases with increasing ΔE_R. We mentioned in Section II that we would expect the energy shift due to charge transfer or d-band polarity to increase with increasing subband splitting. This suggest that the attraction of the subbands due to charge transfer slightly dominates the level repulsion from band formation. The reasonable agreement between ΔE^{exp} and ΔE_R shows that the two effects nearly cancel for these glasses. It is difficult to separate the various contributions experimentally. Some authors[53-55] have stated that based on the core level shifts that charge transfer does not or does occur in certain alloys. The XPS and UPS results of Amamou and Krill[50] on $Cu_{60}Zr_{40}$ are especially interesting on this question. They found that the binding energies of the Zr 4p and Cu 3p orbitals in the glass shifted compared to the pure metals while the Zr 3d levels did not shift. These results illustrate the complexity of analyzing core level shifts in terms of charge transfer. The charge transfer described in Section II dealt primarily with d electrons, however, there is likely to be a compensating transfer of s and p electrons. These two effects may make some of the core levels behave differently than others as perhaps Amamou and Krill see in $Cu_{60}Zr_{40}$.

Another feature of the photoemission spectra of interest is the width of the subbands. The LTM subbands in the Zr glasses are considerably smaller than in their respective parent metals. This is a simple alloying effect which can be neatly described by the methods of VKE mentioned in Section II. The observed increase in width of the LTM subband with increasing LTM content approximately follows Equation 17.

As we mentioned in Section II, calculating the detailed shape of D(E) requires finding the higher moments μ_N. The UPS and XPS spectra of the Zr glasses are remarkably featureless, so it is difficult to discuss the effect of the higher moments on their experimentally derived D(E). We shall return to this point when discussing the Mo–Ru-based glasses.

Oelhafen et al.[51] have proposed that there is a correlation between the electronic structure and glass-forming ability in these materials. They found that the larger the LTM d-band binding energy shift upon alloying, the better the glass-forming ability. They also point out that this binding energy shift must be related to the heats of formation of the alloys and that all binary transition metal alloys that are good glass formers have large exothermic heats of formation. It seems likely that the large negative heat of formation is an important, but certainly not the only factor relevant to glass formation.

With the extensive results we have available to us on photoemission in these alloys, it is regrettable that more experiments to determine γ_c have not been made. In Table 4 we list γ's for $Zr_{65}Cu_{35}$, $Zr_{70}Pd_{30}$, and $Zr_{75}Rh_{25}$. It is convenient at this point to compare $Zr_{70}Pd_{30}$ and $Zr_{75}Rh_{25}$. Ignoring compositional differences, we see (from γ_c) that D(E) increases in going from Pd to Rh. From Table 1 we see that the Rh subband is expected to be about 2.3 eV closer to the Zr subband (and hence E_F) than the Pd subband. The larger γ_c we find for the Rh alloy is a result of having the Rh contribute more to $D(E_F)$ than the Pd subband. This increase in $D(E_F)$ is also accompanied by an increase in T_c for Rh compared to Pd.

The behavior of T_c in the $Zr_{70}X_{30}$ (X = 3d metal) glasses gives some extra insight into the electronic structure of these glasses. From the experimental data in Table 6 or from the renormalized atom levels in Table 1 we expect that the contribution to $D(E_F)$ from the LTM 3d band to increase as we move to the left in the 3d row of the periodic table. From Figure 3 we see that $Zr_{70}Cu_{30}$ should be described as a split band case in which the Cu 3d subband makes little contribution to $D(E_F)$. Moving to Ni, Co, and Fe we gradually approach the virtual crystal limit. Magnetic susceptibility measurements by Amamou and Krill[54] have found that χ_{Pauli} increases sharply in going from $Zr_{70}Ni_{30}$ ($\chi_{Pauli} = 0.67 \times 10^{-6}$ emu/g) to $Zr_{70}Co_{30}$ ($\chi_{Pauli} = 2.2 \times 10^{-6}$ emu/g) indicating either an increasing $D(E_F)$ or enhancement due to many body effects (i.e., the Stoner factor). As we will discuss in Section IV increasing $D(E_F)$ is expected to increase the superconducting transition temperature. In Figure 8 we plot the behavior of T_c vs. X for a number of binary and ternary glasses. For the binary glasses, starting with Cu and moving to Ni and Co we see the expected increase in T_c. Beyond Co, T_c drops rapidly and stays below 1.2 K until reaching V. This is an interesting situation since $D(E_F)$ should increase as we move to Fe, Mn, and Cr. Using the renormalized atom values we would expect the Mn 3d and the Zr 4d subbands to line up and result in a large $D(E_F)$.

The explanation of this behavior is that as we move to the left of Co we progressively tend to satisfy the conditions for forming a localized magnetic moment on the 3d element. The requirements for this to happen were listed in Section II and this question will be discussed in more detail in Section IV. Of interest here is that the behavior of T_c is a result of having the 3d subband first approach E_F increasing T_c and then overlap it so strongly that it fulfills the requirements for forming a localized magnetic moment which acts as a pair

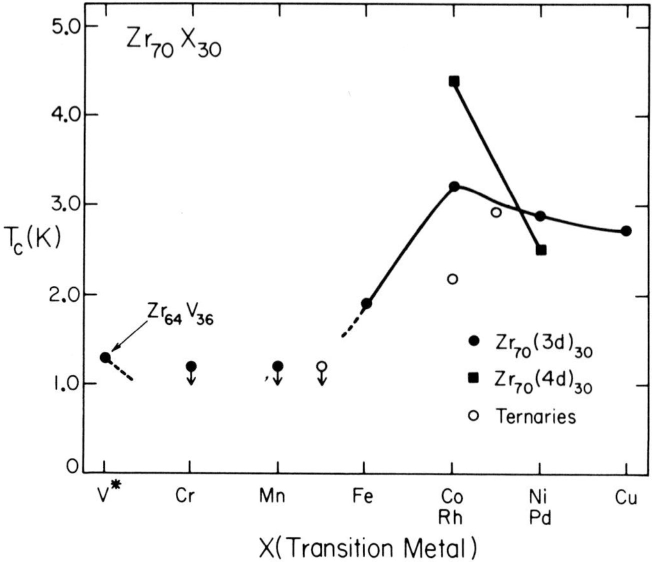

FIGURE 8. The superconducting transition temperature T_c for $Zr_{70}X_{30}$ glasses. The arrows indicate a T_c value below 1.2 K (the lowest experimentally accessible temperature).

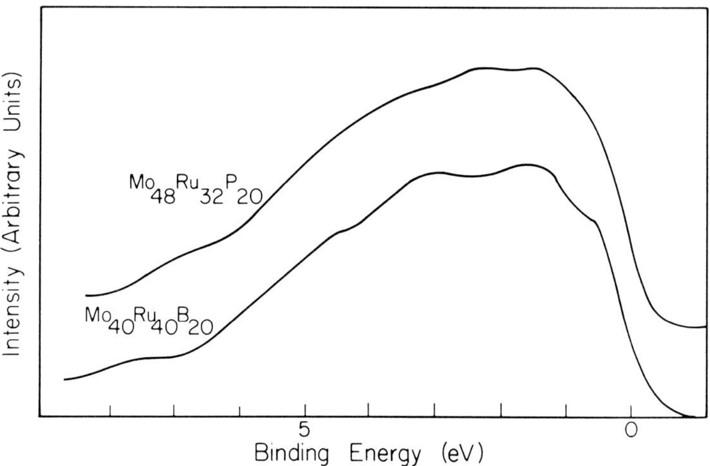

FIGURE 9. The UPS spectra of two Mo–Ru metallic glasses. Data taken from Reference 56.

breaker for superconductivity. Also on Figure 8 we have put some ternary glasses. Note that the T_c is lower than would be expected from the averages of the nearby binaries. That is, the T_c of $Zr_{70}Ni_{15}Fe_{15}$ for example, is much smaller than that of $Zr_{70}Co_{30}$. Replacing some of the Fe by Ni in $Zr_{70}Fe_{30}$ does not alter the fact that the Fe levels are near E_F and that they are capable of forming a magnetic moment. This is a clear example of the breakdown of the rigid band or virtual crystal model.

In the discussion above we have ignored the effects of the higher moments of D(E) on the properties of the materials. We mentioned before that the UPS spectra of the Zr glasses are somewhat featureless. We now wish to discuss the Mo–Ru-based glasses in which some photoemission data and a considerable amount of $\gamma_{H_{c2}}$ information is available.

In Figure 9 we show the UPS spectra of two Mo–Ru glasses from Reference 56. These

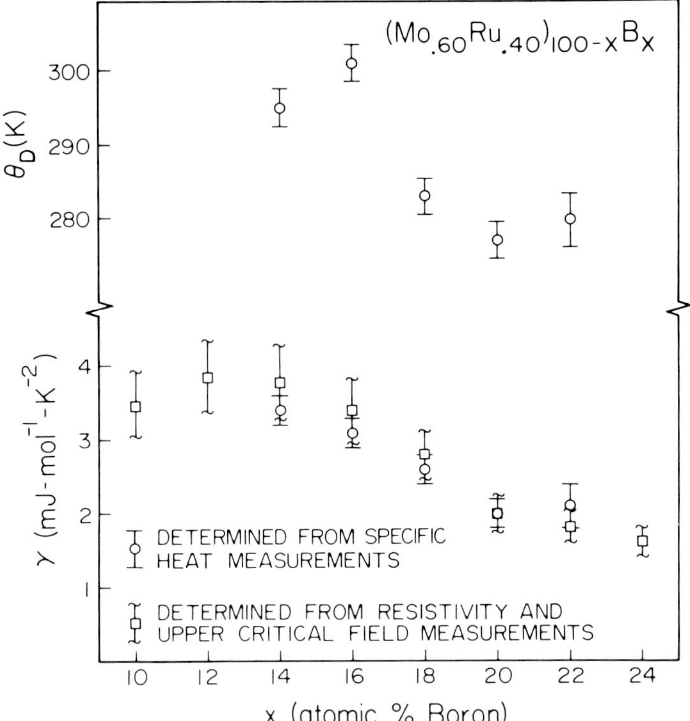

FIGURE 10. Composition dependence of the electronic specific heat coefficient γ and Debye temperature θ_D for $(Mo_{0.60}Ru_{0.4})_{100-x}B_x$ metallic glasses. Data taken from Reference 80d.

glasses are good examples of the virtual crystal limit described in Section II. In contrast to the UPS spectra of $Zr_{70}Cu_{30}$ in which one saw two distinct fairly narrow subbands, these glasses show one very broad band. An interesting feature of the spectra is the fine structure on the MoRuB spectrum, which does not appear on the MoRuP spectrum. Amamou and Johnson[56] suggested that the electronic structure of the pure elements are conserved to a greater extent in the B alloys than in the P alloys. This fine structure should be associated with the contributions from the higher moments of the D(E) function. Using Table 1, we can again understand the UPS results in terms of the renormalized atom values. We would expect the Mo and Ru subbands to overlap strongly in these glasses. This coupled with the fact that the widths of the Mo and Ru d-bands are larger than the Zr band results in the single very broad feature in D(E). Both the P and B levels (Table 2) are expected to be at higher binding energy compared to the Mo and Ru subbands, so that the main contribution to $D(E_F)$ is coming from Mo and Ru 4d levels.

In Tables 4 and 5 we list the γ values for glasses of the form $(Mo_{1-x}Ru_x)_{80}P_{20}$. We again note that γ_c γ_m, and $\gamma_{H_{c2}}$ are in reasonably good agreement. We expect that the γ_m's, because of the assumption made in relating them to $D(E_F)$, are the least reliable values of γ. The good agreement between $\gamma_{H_{c2}}$ and γ_m gives us some confidence about the assumptions made by Reference 15. As the Mo content increases in these glasses we see $D(E_F)$ and T_c increase. The shape of D(E) is not expected to change much with changing Mo to Ru ratio. The situation corresponds roughly to the case of $\delta = 0.25$ in Figure 1 of Section II. For such small values of δ, one expects the rigid band model to be appropriate for describing D(E). Adding Mo simply has the effect of moving E_F more towards the middle of the 4d band thereby increasing $D(E_F)$.

In Figure 10 and Table 5 we have the values of γ for the system $(Mo_{0.6}Ru_{0.4})_{1-x}B_x$.[80d] For

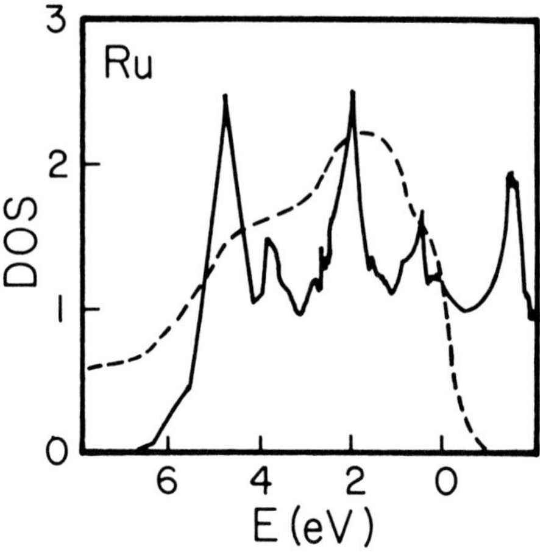

FIGURE 11. Calculated D(E) for hcp Ru taken from Reference 58 (solid line) and UPS spectrum from Reference 56 (dashed line).

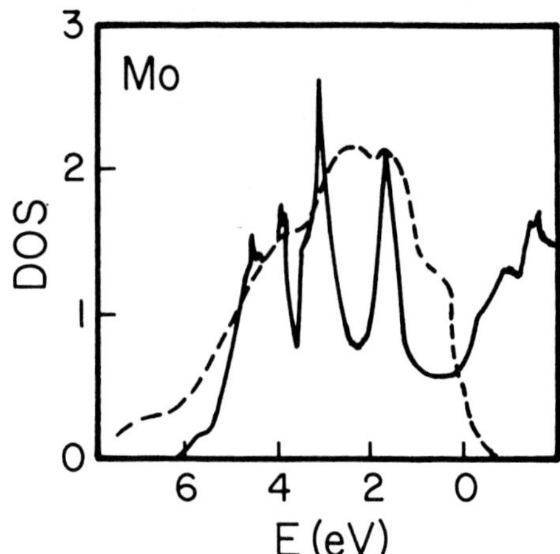

FIGURE 12. Calculated D(E) for bcc Mo taken from Reference 58 (solid line) and UPS spectrum from Reference 56 (dashed line).

this system γ_c values are available and agree well with the $\gamma_{H_{c2}}$ values. The γ_c data also suggest a rapid decrease with B content. The behavior of the B and Si glasses are strikingly different. The γ of the B glasses decreases steadily with B content while the γ of the Si glasses seems to be nearly independent of composition. Referring to the expected level positions in Table 2 for B and Si we would expect that they would be too far below E_F to make a direct contribution to γ. In addition to this, the large change we see in γ for the B glasses cannot be caused by B levels moving in and out of E_F. The origin of the different behavior of γ for these glasses is unknown. A likely explanation is that there is a structural change in the B glasses as a function of B content.[57,80d] This could lead to some structure in $D(E_F)$ vs. B content via the higher moments of D(E). More work is needed to understand this behavior.

Earlier in this section it was mentioned that measurements of the electronic structure might give information on the atomic structure of glasses. Information on atomic structure consists mainly of partial pair correlation functions ($\rho_{ij}(r)$ (see Section II)) obtained from diffraction experiments (X-ray, electron, and neutron). These functions give average distances and coordination numbers. The possibility of obtaining additional information from the electronic structure to supplement the diffraction experiments will be discussed.

In Figures 11 and 12 we show the results of a "muffin tin" calculation of D(E) for bcc Mo and hcp Ru (solid lines) are shown.[58] The obvious differences in the calculated spectra are a direct consequence of the different local point symmetries inherent in the bcc and hcp structures. These are the kind of differences one might hope to see between the photoemission spectra of metallic glasses. Also in Figures 11 and 12 the UPS spectra from Reference 56 of bcc Mo and hcp Ru (dashed lines) are shown. The clear differences one saw in the calculated D(E) have largely disappeared, due primarily to instrumental resolution and possibly surface effects. The electronic structure of metallic glasses is expected to be more complicated than that of pure metals. Metallic glasses always contain two or more elements and the constituent atoms experience a variety of different local environments. In a few special cases, it has been claimed the basis of diffraction experiments,[59-61] EXAFS,[62] NMR,[63] and Mössbauer effect experiments,[64-66] that there is some degree of chemical ordering in certain metallic glasses. In a $Pd_{78}Ge_{22}$ glass, for example, Hayes et al.[62] claimed that there were no Ge–Ge first nearest neighbors. The UPS or XPS spectrum of a $Pd_{78}Ge_{22}$ glass probably would not be sensitive to this chemical ordering since it would still contain contributions from Pd and Ge atoms having a variey of coordination numbers and local symmetries (topological disorder). The fact that there are no Ge–Ge neighbors would be expected to have only a small effect on the measured D(E).

Gaspard[67] has theoretically examined the effects of chemical short range order on the electronic structure of metallic glasses using a tight binding description on the d bands. He found that the changes in the D(E) in going from a completely disordered alloy to an alloy with a strong tendency for chemical ordering was probably not within the resolution of photoemission experiments.

A number of authors[23,38,42] have compared the UPS or XPS spectra of glasses and the corresponding crystalline phases. For all cases considered the spectra have been very similar. This has led several authors to conclude that the short-range order of the glass is similar to that of the crystalline phase. Based on the work of Gaspard and our comments above, this conclusion seems premature. One exception to this has been the work of Pollak et al.[68] on A-15 Nb_3Ge and amorphous Nb_3Ge. They found that there was a clear change of the Nb–Ge bonding from covalent to metallic upon the amorphous to crystalline transition.

Gaspard's work and our comments are certainly not the final word on this matter. In trying to understand the MoRuB glasses it was mentioned that a change in the structure of the glass is possibly responsible for the anomalous behavior of γ in this system. This is an open question which can only be answered by further investigation. In examining the data on the electronic structure of metallic glasses, one realizes that an understanding of these materials can only come from detailed measurements of a wide variety of properties. The combination of UPS and heat capacity experiments is perhaps at present the most powerful means of obtaining direct information about the electronic structure of these novel materials.

IV. SUPERCONDUCTIVITY

Superconductivity in amorphous metals and alloys has been discussed in several recent review articles.[69-71] These review articles cover both superconducting simple metals and their alloys[69,70] as well as transition metals and alloys.[70,71] As in the previous sections, the emphasis will be on metallic glasses as opposed to vapor-quenched amorphous alloys. As such, we concentrate on transition metal alloys since most superconducting metallic glasses contain transition metals as a major constituent. The reader is referred to the above-mentioned articles for a broader discussion. First, we examine the microscopic origin of superconductivity and attempt to relate this to the electronic structure of metallic glasses. Second, we discuss some of the special superconducting properties of metallic glasses which relate to their disordered atomic scale structure.

A. Microscopic Origin of Superconductivity in d-Band Alloys

Any effort to understand the microscopic origin of superconductivity in a class of materials must involve a description of the lattice vibrational spectrum, electronic structure, and the interaction of electrons with lattice vibrations. These three properties are interconnected in an inseparable way. That is, each depends on the other two. A rather detailed discussion of this problem for the case of nearly free electron metals has been given in Reference 70. In the present article, emphasis will be placed on d band superconductors. As such, most of the discussion will be based on a tight binding description aimed at metallic glasses based on transition metals.

We begin with McMillan's approximate solution[72] to the Eliashberg equations.[73] McMillan gives T_c in terms of microscopic parameters as follows[72]

$$T_c = \frac{\langle\omega\rangle}{1.45} \exp\left[\frac{-1.04\,\lambda}{\lambda - \mu^*(1 + 0.62\lambda)}\right] \tag{29}$$

where

λ = electron-phonon coupling
μ^* = effective Coulomb coupling constant for electron repulsion
$<\omega>$ = typical average phonon frequency

The electron-phonon coupling constant is expressed in terms of microscopic parameters as:

$$\lambda = \frac{D(E_F)<I^2>}{M<\omega^2>_M} \tag{30}$$

with $<I^2>$ = square of electronic matrix element of the change in crystal potential as an atom is moved. This is averaged over initial and final electron states lying on the Fermi surface. Also, we have

M = ionic mass

$<\omega^2>_M$ = average squared phonon frequency as defined by McMillan in Reference 72 (31)

The interplay of the parameters $D(E_F)$, $<I^2>$, and $<\omega^2>_M$ is responsible for determining λ and consequently T_c. McMillan's equation is a solution of the Eliashberg equations for the special case of a material having the phonon density of states $F(\omega)$ of bcc Nb. Amorphous materials may have a rather different $F(\omega)$ and McMillan's equation should strictly speaking be modified to account for this. The basic features of the equation should, in spite of this, be very similar.[74] Thus, we shall use Equations 29 and 30 as a basis for analyzing the systematics of T_c in metallic glasses.

Several authors have attempted to analyze λ for d-band metals and alloys in an effort to develop a self-consistent picture of how the various parameters in Equation 30 vary among materials. One of the first notable efforts of this type was made by Friedel and his colleagues.[75,76] Using a tight-binding description of $D(E_F)$ together with the rigid-ion method of calculating $I_{KK'}$, they showed that it is possible to express the product $\eta = D(E_F)<I^2>$ in terms of D(E) and E_F as

$$\eta = -2q^2 \int_0^{E_f} E\, D(E)dE \tag{32}$$

where E is measured with respect to the centroid of the d-band and q_0 is the Slater coefficient describing the exponential decay of the d-orbitals in the limit of large distance from the nucleus. In fact, the integral in Equation 32 is just the contribution to the cohesive energy of the metal, E_c^d, due to d-band formation if one ignores the one-center integrals ($\bar{\Gamma}$) referred to earlier in Sections II and III. We can thus write $\eta \approx q_0^2 E_c^d$. Now E_c^d varies slowly with d-band occupation passing through a broad maximum for a roughly half filled d-shell[4,77] in which all bonding states are occupied and antibonding states empty. This simple relationship predicts that λ is influenced strongly by $<\omega^2>_M$ since $\lambda = \eta/M<\omega^2>_M$ and η varies slowly across a d-band series. Varma and Dynes[78] combined Friedel's analysis with the notion of screening phonons through d-electron polarizability. Using expressions developed by Gomersall and Gyorffy[79] to describe the screening effect, they derive an approximate relationship between $<I^2>$ and $<\omega^2>_M$ for d-band alloys. This relationship can be simply stated as

Table 7
SPECIFIC HEAT DATA AND CALCULATED SUPERCONDUCTING PARAMETERS (SEE TEXT) FOR SEVERAL SUPERCONDUCTING METALLIC GLASSES BASED ON 4D-TRANSITION METALS

Alloy	γ_c (mJ/mole-K^2)	θ_D(K)	T_c(K)	Ref.	λ	$D(E_F)^a$ states/eV-atom-spin)	(Varma Dynes parameter δ) (eV)
$Zr_{75}Rh_{25}$	5.4	191	4.4	28	0.71	1.34	0.53
$Zr_{70}Pd_{30}$	4.7	180	2.4	80a	0.60	1.25	0.48
$Zr_{65}Cu_{35}$	3.9	220	1.7	80b	0.53	1.08	0.49
$Zr_{40}Cu_{60}$	2.8	230	$\lesssim$0.3	80b	$\lesssim$0.39	~0.85	~0.46
$Nb_{75}Ge_{25}$	3.0	222	3.6	80c	0.63	0.78	0.81
$(Mo_{0.6}Ru_{0.4})_{84}B_{16}$	3.1	301	6.4	80d	0.69	0.78	0.88

[a] Renormalized value of $D(E_F)$ obtained from $D^* (E_F) = [3\gamma/2\pi^2 k_B{}^2(1 + \lambda)]$. (See Reference 72 for details.)

$$\left[\frac{<I^2>}{M<\omega^2>}\right] \approx W(1 \pm S) \qquad (33)$$

where W is roughly the width of the d-band and S is the overlap integral for atomic d-orbitals on neighboring atoms. In the language of Section II, $S = <i\alpha m|j\beta m'>$. The plus sign refers to E_F lying within the bonding orbitals (lower half of the d-band) and the minus sign to E_F lying within the antibonding (upper half of the d-band). The "Varma-Dynes" parameter $\delta = <I^2>/M<\omega^2>$ should be roughly constant for transition metals and alloys of a given series to within the factor $(1 \pm S)$. Thus $\lambda = D(E_F)\delta$ should be mainly governed by $D(E_F)$.

To test the above relationships for the case of amorphous alloys requires a knowledge of λ and $D(E_F)$. These can be estimated from specific heat data.[72] Taking $<\omega> \approx \theta_D$ one can obtain λ from Equation 29. $D(E_F)$ is obtained directly from the linear contribution, $C_{el} = \gamma T$, to the normal state low temperature heat capacity due to electrons. Taking into account the renormalization of γ by the electron-phonon interaction gives

$$D(E_F) = \frac{3\gamma}{\pi^2 K_B^2 (1 + \lambda)} \qquad (34)$$

where $D(E_F)$ is the total density of states at E_F (two spin states). The "Varma-Dynes" relation then states that $\delta = \lambda/D(E_F) = W(1 \pm S)$ for materials where a single type of atomic orbital dominates $D(E_F)$. Table 7 lists T_c, θ_D, and γ for a number of metallic glasses for which specific heat data is available. To test the above relation, we consider metallic glasses for which $D(E_F)$ is dominated by a 4d transition metal. From Section III, (photo-emission data), we know that the 4d band of Zr dominates $D(E_F)$ for the alloys $Zr_{1-x}LTM_x$ where LTM = Pd, Rh, and Cu. In $(Mo_{1-x} Ru_x)$ glasses, the 4d band of Mo and Ru dominate $D(E_F)$ while in Nb_3Ge[68] $D(E_F)$ is dominated by the Nb 4d-band. These alloys should thus provide a good test of the Varma-Dynes relation. In Figure 13, we plot λ (calculated as above) as a function of $D(E_F)$ (from Equation 29) for the above metallic glasses together with corresponding data for crystalline Nb, Zr, Mo, T_c and bcc alloys thereof.[72,78] The data lie on two well-defined lines according to whether E_F lies in the upper or lower half of the 4d-band. Varma-Dynes[78] have already shown that crystalline transition metals and alloys follow this behavior. Figure 13 shows that metallic glasses obey exactly the same relation. In other words, λ is a simple function of $D(E_F)$ for all materials in which $D(E_F)$ is dominated by 4d orbitals. We can now see that the principal difference between amorphous alloys and

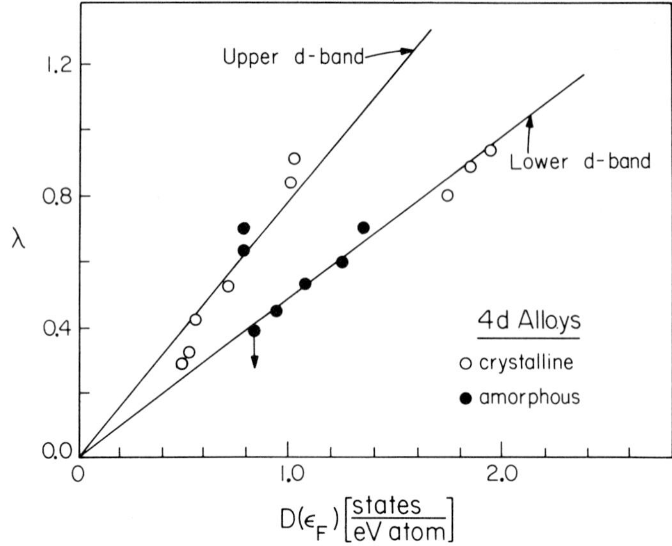

FIGURE 13. The electron-phonon coupling parameter λ plotted as a function of $D(E_F)$ for crystalline and amorphous alloys based on 4d transition metals.

their crystalline counterparts is in the structure of the spectral function D(E) which determines both E_F and $D(E_F)$.

In Section II we saw that D(E) itself is determined mainly by the short range atomic order of metallic glasses. We now see that short-range order becomes the most significant factor determining λ and thus T_c for metallic glasses. A few general remarks are in order at this point. Earlier, we mentioned the difficulties which arise in the theory of T_c due to the necessity of self-consistently determining $D(E_F)$, $<I^2>$, and $<\omega^2>$. The basic assumption made in any calculation of vibrational properties of a solid is the Born-Oppenheimer or adiabatic approximation.[81] The electronic structure is first calculated for a given arrangement of ions. The electronic energy E^e is found to vary with ionic arrangement. This energy is then added to direct ion interaction energies to give a total effective ionic energy. The important fact is that the electronic structure is regarded as the primary property of the solid from which ionic properties (i.e., $<\omega^2>$) follow. The Born-Oppenheimer approximation provides the key to decoupling the self-consistency problem. We find that the electronic structure D(E) is the natural fundamental starting point for a theory of superconductivity. Once we know D(E), we can proceed to deduce λ and T_c for a given class of alloys. Amorphous materials are in summary rather nonunique in that they appear to follow the same general behavior as crystalline alloys based on the same metals.

B. Critical Field Behavior (Effects of Disorder)

An important property of glassy superconductors which again derives from electronic properties is the critical field. The short electronic mean free path, ℓ_e, in metallic glasses ensures that they are type II superconductors[82] with small coherence length ξ. When $\ell_e \ll \xi$, one can write the coherence length in the "dirty limit" as[32]

$$\xi_0 = \xi_d = (0.855)\sqrt{\ell_e \xi_0}\left[1 - \frac{T}{T_c}\right]^{-1/2} \tag{35}$$

where

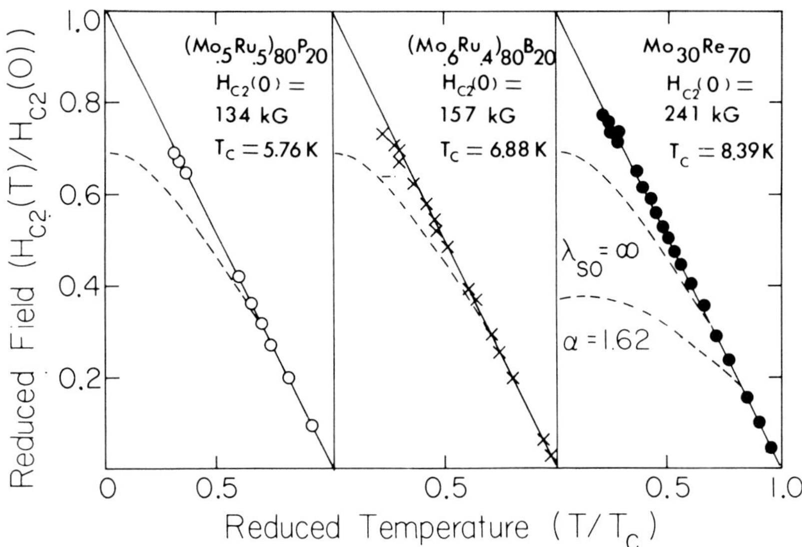

FIGURE 14. The reduced upper critical field $[H_{c2}/DT]_{T_c}$ as a function of (T/T_c) for several amorphous alloys. $H_{c2}(0)$ is defined by linear extrapolation. The values of $H_{c2}(0)$ and T_c are given in the figure.

$$\xi_o = \frac{\hbar V_f}{\pi\Delta(T)}$$

= clean limit coherence length and Δ (T) = energy gap (BCS). Under these circumstances, one defines an upper critical field $H_{c2}(T)$ which is the largest field (at temperature T) for which mixed state superconductivity exists. Extensive theoretical efforts have been made to calculate $H_{c2}(T)$ from the microscopic theory. In the limit where $(\xi/\ell_e) \gg 1$, asymptotically exact results are obtained.[83,84] The effects of paramagnetic limiting[85] and spin-orbit coupling[86] have also been accounted for in the microscopic theories.

Figure 14 shows a series of critical field curves $H_{c2}(t)$ for several metallic glasses where $t = T/T_c$. The theoretical curves shown correspond to a case where the spin orbit coupling goes to infinity. This gives the theoretical curve which is most nearly linear and best describes the data. The data follow a more linear dependence than the theory can predict. This apparent breakdown of the theory for the temperature dependence of $H_{c2}(t)$ in amorphous alloys has been the subject of recent discussion.[87] Carter et al.[88] have attributed the breakdown to microscopic inhomogeneities in amorphous alloys and have used a simple model for such inhomogeneities to describe the experimental data. The model represents the inhomogeneities by a distribution of electronic diffusivities on a local spatial scale. The present authors[87] have suggested that the breakdown of the theory could be related to the effect of strong electron scattering which accompanies the approach to electron localization. Further investigation of this interesting problem is clearly required.

The gradient of the upper critical field in a type II superconductor, $\left[\frac{dH_{c2}}{dT}\right]_{T_c}$ in the limit as $T \rightarrow T_c$ can be theoretically related to microscopic material parameters following Equation 27. This equation can be expressed in convenient units as Equation 33

$$\left[\frac{dH_{c2}}{dT}\right]_{T_c} = \left[-4.48 \times 10^4 \gamma\rho_N\right] \quad (36)$$

where ρ_N is the electrical resistivity and γ is the electronic heat capacity parameter. This formula was already mentioned in Section III. A careful critical survey of experimental data on metallic glasses reveals this formula to be in good agreement with experiment in cases where specific heat, resistivity, and critical field measurements are all available. Figure 15 shows γ_c (from specific heat) vs. $\gamma_{H_{c2}}$ (from Equation 36, $\left[\frac{dH_{c2}}{dT}\right]$ and ρ_N). The prediction of Equation 36 is shown by the solid line. Excellent agreement appears to prevail over a broad range of materials. Equation 36 thus provides a fairly reliable method of estimating γ (or $D(E_F)$) given only ρ and dH_{c2}/dT. This fact was already used in Section III.

Summarizing, we have shown how electronic structure and superconductivity are intimately connected. Both T_c and H_{c2} can be directly related to electronic parameters such as $D(E_F)$ and ρ_N. It is clear that an understanding of the systematics of superconductivity must be preceded by knowledge of the electronic structure of metallic glasses.

C. Spin Fluctuations, Localized Moments, and Pair Breaking

In Sections II and III, we alluded to the phenomena of localized magnetic moment formation in metallic glasses. At the end of Section II, the conditions for formation of a localized moment were described in the context of Anderson's theory.[17] This problem is particularly important for superconductivity. Magnetic moments tend to form whenever the intra-atomic Coulomb interaction, U, of the d-electrons dominates the two center integrals, $\bar{\bar{\Gamma}}$. Stated differently, one must have $U \gtrsim W$. The magnitude of U is determined by the atomic-like properties of the ϕ_α^m d-orbitals discussed in Section II. For a d-band with little structure in $D_d(\epsilon)$, we have roughly $W \propto [D_\alpha(\epsilon)]^{-1}$. For a given transition metal atom (α) in a metallic glass, we expect a localized magnetic moment to form when the $\bar{\bar{\Gamma}}$ integrals between this atom and its neighbors are smaller than some critical value. This in turn suggests localized moment formation should correlate with a large contribution of the given atom to $D_d(\epsilon)$. The second requirement for localization as stated in Section II is $|\langle E_i\rangle - E_F| \lesssim U_i$ where E_i is the average energy of the atomic level. Combined with the first requirement, this suggests that localized moment formation on α-type atoms will correlate with a large partial contribution of α-type d-levels to $D(E_F)$. We have already shown that both T_c and H_{c2} of superconductors tend to increase with $D(E_F)$. We now observe, on the other hand, that if W is too small ($D(E_F)$ too large), localized moments may form. As a precursor to moment formation, one expects spin fluctuations. Either of these will have a pronounced effect on superconductivity. Localized moments (and spin fluctuations) act as a pair-breaking mechanism. Gorkov[89] first discussed the depression of T_c by dilute concentrations of moment carrying atoms in a superconductor. The rate of depression of T_c with increasing moment concentration c is given implicitly by[89,90]

$$\ln(T_c/T_{co}) = \psi(\tfrac{1}{2}) - \psi(\tfrac{1}{2} + 0.14\alpha\, T_{co}/\alpha_{cr}T_c) \quad (37)$$

where T_{co} is the T_c in the absence of magnetic moments and α is a pair-breaking parameter defined as

$$\alpha = \hbar^{-1} c\, D(E_F)\, J^2\, S(S+1)$$

with J = exchange interaction parameter. The critical parameter α_{cr} is given by $\alpha_{cr} = K_B T_{co}/4\hbar\gamma$ where $\ell n\gamma$ is Euler's constant. Spin fluctuations can be viewed as transient moments having a lifetime τ. The lifetime of a Cooper pair in a superconductor is $\tau_S \approx \hbar/\Delta$. When $\tau > \tau_s$, the spin fluctuations behave much like stable localized moments and produce a pair-breaking effect.

The experimental data on superconductivity of Zr-base metallic glasses discussed in Sec-

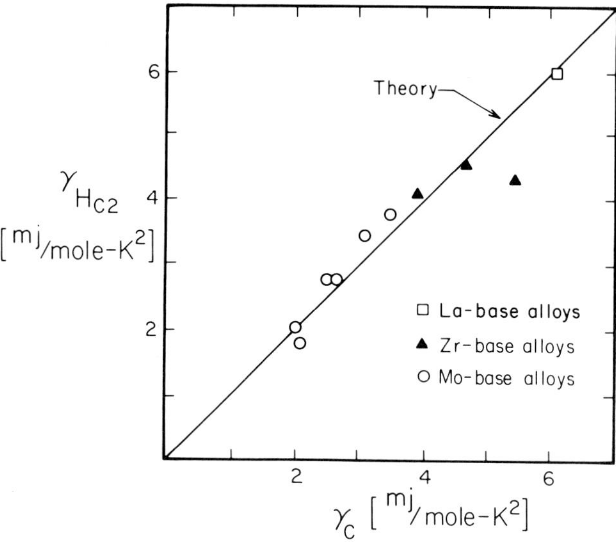

FIGURE 15. A plot of γ_c vs. $\gamma_{H_{c2}}$ (as determined from Equation 36). Data are taken from Tables 4 and 5. The solid line represents the relation expressed by Equation 36 as described in the text.

tion III are a good illustration of these effects. Figure 8 shows the variation of T_c in Zr_{1-x} LTM_x metallic glasses with the group number and row of the LTM constituent. For the 4d elements LTM = Rh, Pd we find T_c increasing from $Zr_{70}Pd_{30}$ to $Zr_{75}Rh_{25}$. This is due as seen from Table 7 and Figure 13 to an increase in $D(E_F)$. The splitting, ΔE^{exp}, of the Zr-4d and Rh-4d levels is less than that of the Zr-4d and Pd-4d levels as discussed in Section III. The leads to a larger contribution of Rh to $D(E_F)$ and an overall larger γ (see Table 7). A similar effect should be expected for LTM = Mn, Fe, Co, Ni, Cu with a gradual increase of δ and decrease of $D(E_F)$ as we progress from Mn to Cu. Note that all of these glasses form at roughly the same range of compositions $0.25 \leqslant x \leqslant 0.40$. As expected, we find that T_c increases on moving from $Zr_{70}Cu_{30}$ to $Zr_{70}Ni_{30}$ to $Zr_{70}Co_{30}$ (see Figure 8). This trend is broken by $Zr_{70}Fe_{30}$. Continuing, we find spin glass behavior and defined magnetic moments in $Zr_{70}Mn_{30}$ which as a consequence is not superconducting at all. This behavior can be simply understood. The intra-atomic integral U_i is roughly the same for all of the 3d transition metals above. For Ni and Cu, we have a split band (large ΔE^{exp}) and the 3d partial density of states $D_{3d}(\epsilon)$ is centered well below E_F. Thus $|\langle E_i \rangle - E_F| \gtrsim U_i$ and the second condition for moment formation is violated. This is also true to a lesser extent for $Zr_{70}Co_{30}$ (see photoemission data in Section III). All of these exhibit the expected trend of increasing T_c with increasing $D(E_F)$. In contrast, the Fe-3d and Zr-4d bands clearly merge in $Zr_{70}Fe_{30}$ as seen in the UPS spectra of Reference 51. Spin fluctuations and ultimately stable moment formation in $Zr_{70}Mn_{30}$ are thus observed. Superconductivity is destroyed by the accompanying pair-breaking effect. Thus, we have explained the variation of T_c in these alloys.

A final comment is worthwhile making. The high degree of disorder in the local atomic environments of amorphous alloys permits a distribution of "local" electronic parameters (i.e., $\bar{\Gamma}$, $\bar{\bar{\Gamma}}$) among the atoms of a given atomic species. It is possible as a result that even when the average electronic structure as measured by the total D(E) appears unfavorable for moment formation, there may nevertheless exist a small number of atomic sites having lower $\bar{\bar{\Gamma}}$ values where moment formation or spin fluctuations are possible. Only a small concentration of such sites will strongly influence superconductivity and result in an observable depression of T_c. The disordered structure of amorphous alloys will thus tend to enhance the tendency for the appearance of localized moments. The addition of 3d transition metals to superconducting metallic glasses should as a consequence tend to frequently result in pair-breaking effects.

V. TRANSPORT

The study of the transport properties of noncrystalline materials has been a very active area of research. Liquid metals were the first examples of these materials to be examined in detail. With the advent of the pseudopotential concept[91] a theory of the resistivity (ρ) of simple metal liquids was developed by Ziman.[92] This theory was one of the first steps in the development of a transport theory for noncrystalline materials. Güntherodt and Kunzi[93] have discussed the similarities between liquid metals and metallic glasses in a recent review. They point out that the magnitude as well as the temperature dependence in the glassy and liquid states are very similar. It seems logical, therefore, to use the same kind of formalism to describe the transport properties for these two cases. Metallic glasses have some obvious advantages over liquids in studying transport in noncrystalline materials. Of special interest is the ability to perform low temperature measurements such as heat capacity as well as measurements of resistivity, Hall effect, and thermopower over extended ranges of temperature. In this section we will give a brief discussion of the behavior of the resistivity and thermopower of nonmagnetic metallic glasses. For additional information on transport in magnetic glasses and topics such as Hall effect and thermal conductivity we refer the reader to References 38, 93, and 94.

As we mentioned above, the starting point for a discussion of transport in noncrystalline materials is the Ziman theory of the electrical properties of liquid metals. This theory was originally designed to treat simple metal liquids. In this case, the conduction electrons are assumed to form a degenerate free electron gas and are scattered by the irregular arrangement of metal ions. If the ions scattered independently the mean free path of the electrons (ℓ_e) would be much too short leading to values of ρ much higher than those observed experimentally. Ziman showed that the strong correlation between the positions of the ions in a liquid gives rise to coherence between waves diffracted by adjacent ions. This scattering can be calculated by perturbation theory leading to a finite relaxation time. Using the standard Boltzmann equation for transport we have

$$\rho = \frac{3}{(2e^2 v_F^2 \, D(E_F)\tau)} \tag{38}$$

where v_F is the Fermi velocity and τ is the electron relaxation time given by

$$1/\tau = \int (1 - \cos\theta)\, f(\theta) d\Omega \tag{39}$$

where f(θ) is the probability for scattering through the angle θ into the solid angle dΩ. The scattering probability is calculated in the Born approximation as

$$f(\theta) = \frac{2\pi}{\hbar} |\langle \vec{K} | V | \vec{K}' \rangle|^2 D(E_F) \tag{40}$$

where $\langle\vec{K}|$ and $|\vec{K}'\rangle$ are the initial and final electron states and V is the potential associated with the ions. We assume that the potential at any point is the sum of potentials centered at each ion

$$V_{total} = \sum_i V(\vec{r} - \vec{R}_i) \tag{41}$$

then

$$<\vec{K}|V_{total}|\vec{K}'> = \frac{V(K)}{N}\sum_i e^{i\vec{K}\cdot\vec{R}_i}$$

where R_i is the position of the center of the i-th ion and V(K) is the pseudopotential. Putting this into Equation 40 and noticing that

$$1/N<|e^{i\vec{K}\cdot\vec{R}_i}|^2> = S(K) \qquad (42)$$

is the structure factor of the liquid and we have

$$\rho = \frac{3\pi}{he^2V_f^2N}\int_0 \left[|V(K)|^2S(K)\right]4\left(\frac{K}{2K_f}\right)^3 d\left(\frac{K}{K_f}\right) \qquad (43)$$

Faber and Ziman extended this treatment to the case of liquid alloys.[95] For a binary alloy we simply replace the bracket quantity in Equation 43 by

$$\left[V_1^2(C_1C_2 + C\,S_{12}) + V\,(C_1C_2 + C_2S_{22}) + 2C_1C_2V_1V_2(S_{12} - 1)\right] \qquad (44)$$

where C_1, C_2 are the atomic fractions of the two components, S_{11}, S_{12}, and S_{22} are the partial structure factors, and V_1 and V_2 are the pseudopotentials.

Evans, Greenwood, and Lloyd[96] extended these ideas to liquid transition metals by replacing the weak ion pseudopotentials V(K) appropriate for simple metals by the t-matrix obtained from "muffin tin" self-consistent potentials.

$$t(\vec{K}\,\vec{K}') = -\left(\frac{2\pi\hbar^3}{\Omega_0 m\sqrt{2mE_F}}\right) \times \sum_{\ell} (2\ell + 1)\,\mathrm{Sin}^2(\eta_\ell[E_f])\,e^{i\eta_\ell}P_\ell(\cos\theta) \qquad (45)$$

For transition metals the d-phase shift η_2 dominates at E_F so that

$$\rho = \left(\frac{30\pi^3\hbar^2}{me^2\Omega_0K_f^2E_f}\right)\mathrm{Sin}^2(\eta_2[E_F])\;S(2K_f) \qquad (46)$$

We will refer to this as the EGL formula. Since $S(2K_f)$ is the only factor which changes with temperature in the above formula, ρ(T) is determined by the temperature dependence of S(K). A number of authors[97,98] have discussed its behavior with temperature. Nagel,[98] for example, has shown that with suitable approximations

$$S(K) = 1 + [S_0(K) - 1]\,e^{-2[W(T)-W(0)]} \qquad (47)$$

where S_0 (K) is the zero temperature structure factor, $e^{-2W(T)}$ is the effective Debye-Waller factor, and θ_D is the Debye temperature. Using Equations 46 and 47 we can also find an expression for the thermopower (Q)

$$Q = \left(\frac{-\pi^2K_B{}^2T}{3\,e\,E_f}\right)(3 - 2q - \tfrac{1}{2}r) \qquad (48)$$

where r is a term related to the energy dependence of the t matrix.

Mott[99] has criticized this treatment of the resistivity of liquid transition metals. He points out the changes in ρ on melting $[(\rho_{Liquid} - \rho_{Solid})/\rho_{Solid}]$ for Fe and Co are 0.01 and 0.09, respectively. From this observation it would seem that the resistivity of these two elements do not depend much on S(K). He suggests that s-d transitions are responsible for the mean-free-paths (ℓ_e) for materials in which ℓ_e is on the order of the interatomic spacings. His model for resistivity is similar to that used for crystalline materials involving s-d transitions with sp and d electrons have different mobilities. The sp electrons which are primarily responsible for conduction can be scattered into the empty d states giving a resistivity related to the density of d states at the Fermi level $D_d(E_F)$. The coupling constant for this process is given by[99]

$$q = \frac{S(2K_f)\,|t(2K_f)|^2}{\int_0^{2K_f} |t(K)|^2 S(K) 4(2K_f)^{-4} K^3\, dK}$$

$$\Gamma = \sqrt{N} \int_{\Omega_0} \psi_s^* V \psi_d\, d^3\vec{r} \tag{49}$$

where N is the total number of atoms, Ω_0 is the volume per atom, ψ_s is an s electron wavefunction, and ψ_d can be taken as an atomic d wavefunction. Using the Born approximation

$$\frac{1}{\tau_s} = \frac{2\pi}{\hbar} \left|\frac{\Gamma}{N}\right|^2 N D_d(E_f) \tag{50}$$

therefore,

$$\rho = \frac{m\Omega_0}{Ze^2\tau_s} \tag{51}$$

where Z is the number of s electrons.

Cochrane et al.[100] proposed a model to explain the low temperature resistivity of amorphous metals based on an analogy to the Kondo effect.[101] The two spin orientations of the Kondo model are replaced by tunneling levels such as those proposed by Anderson et al.[30] to explain the anomalous low temperature heat capacity of glasses. Using this model Tsuei[102] has asserted that the total resistivity of amorphous alloys can be expressed as

$$\rho(T) = A + B\rho_N(T) + C\ell n(T^2 + T_\Delta^2) \tag{52}$$

where A and B are constant, $\rho_N(T)$ is the regular electron-phonon term, c is a measure of the abundance of the effective tunneling configurations and T_Δ is similar to T_K in the Kondo problem. From this model, by analogy with the Kondo system, we expect to see some kind of anomaly in Q at T_Δ.

Mooij[103] has shown that there is a remarkable connection between the resistivity and the temperature coefficient of resistivity for a wide variety of materials. We will call this the Mooij correlation and simply put, he found that the larger values of ρ corresponded to smaller values of α. The condition for a negative α is to have $\rho \geqslant 150\ \mu\Omega - \text{cm}$. Mooij considered not only amorphous materials, but also included crystalline and thin film materials. It would seem that the phenomenon of negative values of α for materials having high ρ is a very general one. Several ideas have emerged to understand this phenomenon and we will mention some recent ones below.

Girvin and Jonson[104] have developed a theory of transport in random metal alloys based on electron-phonon dynamics. They find that in the high ρ limit where the weak-scattering approximation breaks down, the adiabatic-phonon assumption also fails and phonons can actually assist the electron mobility leading to an anomalous negative temperature coefficient of resistivity. These ideas have not been applied to a specific metallic glass yet and predictions for the temperature dependence of Q are not available. Similar ideas have been proposed by Imry[105] for the resistivity of highly disordered systems. He used the scaling theory of Anderson localization[106] for materials just above the Anderson transition. The scaling theory of Anderson localization involves a characteristic length ξ which diverges on both the extended and localized sides as the localization edge is approached. For electron mean-free-paths $\ell_e < \xi$ one cannot distinguish between localized or extended electronic states. Imry's model predicts that for $\rho(T=0) \gtrsim 100\ \mu\Omega-cm$, $\rho(T)$ should decrease with increasing temperature. Both the models of Girvin and Jonson and Imry provide a basis for an explanation of the Mooij correlation. The attractive feature of these models is that they are designed to treat the problem of ρ in a very general sense.

The Ziman theory has been remarkably successful in explaining the magnitude of ρ and its temperature dependence in simple metal liquids. For discussions of this, see References 93 and 107. It is difficult to obtain values for ρ using this theory due to uncertainties in S(K) and V(K) particularly in the case of binary alloys where three partial structure functions and two pseudopotentials are needed. In the case of liquid transition metals the situation is somewhat more complex due to uncertainties in the various relevant parameters. Esposito et al.[108] have developed the EGL formula in terms of a consistent application of the single site approximation upon which the Ziman theory is based. Their work is particularly noteworthy in that it permits a valid systematic comparison with experiments. In addition to this, they used the experimentally obtained structure factors for their calculations. They found that the computed ρ is very sensitive to the values of the input parameters. In Table 8 we list some of the results of their calculations. Except for liquid Ni and amorphous NiP we see that the experimental values of ρ are a factor of two or more smaller than the calculated values. We also see that the mean-free-paths for liquid Fe, liquid Co, and glassy CoP are smaller than an interatomic distance. Esposito et al. found that liquid Ni, liquid Cu, and glassy NiP could be brought into agreement with experiment by reasonable shifts of E_F to compensate for the shortcomings of the single site approximation. However, their results on liquid Fe, liquid Co, and glassy CoP showed that a single site transport theory applied in a consistent manner cannot be made to agree with experiment for any reasonable values of E_F. They concluded that the EGL formula can be applied semiquantitatively to weak scattering materials, but should not be used to compute ρ for strong scattering materials. Weak scattering systems are considered those in which ℓ_e is more than a few interatomic spacings.

Fujiwara[109] has made an equivalent calculation of ρ for amorphous and liquid Fe, Co, and Ni using the s-d transition model proposed by Mott.[99] He calculated the d-electron density of states in a relaxed model of amorphous and liquid metals using the recursion method. In Table 9 we list some of the results of Fujiwara's calculations. We see that the agreement between calculated and experimental values is quite good. We also note that the mean-free-paths obtained here are a few times an interatomic distance in contrast to the results of Esposito et al. for liquid Co and Fe (Table 8).

To illustrate some of the various models we mentioned above, we wish now to consider the behavior of ρ and α in the metallic glasses $(Pd_{1-x}Cu_x)_{1-y}P_y$ and $(Pd_{1-x}Ni_x)_{1-y}P_y$ from References 110 and 111. In Figures 16 and 17 we show the effects of changing the Cu and P content in the PdCuP system. In Figures 18 and 19 we see that as the Cu or P content is increased α becomes progressively smaller. Somewhat different behavior is found for the PdNiP glasses. Here, like the other case, as the P content is increased the room temperature ρ increases (Figure 20) and α decreases. However, the values of ρ and α are insensitive to the Ni content of the glass (Figure 21). So, one needs to address the following questions

Table 8
COMPARISON BETWEEN CALCULATED AND EXPERIMENTAL VALUES OF RESISTIVITY FOR VARIOUS MATERIALS[108]

Material	$\ell_{calc.}$ (Å)	$\rho_{calc.}$ (μΩ-cm)	ρ_{exp} (μΩ-cm)
Liquid Fe	0.50	1130	136
Liquid Co	1.49	329	115
Liquid Ni	6.15	74	83
Liquid Cu	12.35	41	21
Amorphous Ni–P	3.69	168	132
Amorphous Co–P	1.68	349	160

Table 9
CALCULATED VALUES OF ℓ AND ρ FROM FUJIWARA[109]

Material	$\ell_{calc.}$ (Å)	$\rho_{calc.}$ (μΩ-cm)
Amorphous Fe	4.79	136
Liquid Fe	5.34	129
Amorphous Co	6.96	91
Liquid Co	7.68	87
Amorphous Ni	8.12	76
Liquid Ni	9.96	67

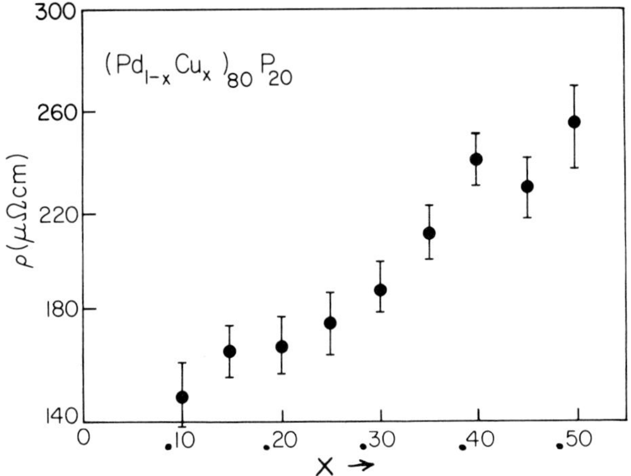

FIGURE 16. The effect of Cu content on the electrical resistivity of $(Pd_{1-x}Cu_x)_{80}P_{20}$ metallic glasses. (From Tangonan, G., Ph.D. thesis, California Institute of Technology, Pasadena, 1976, unpublished.)

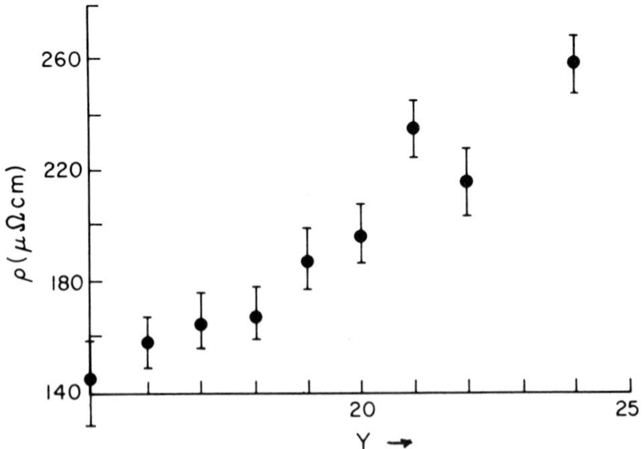

FIGURE 17. The effect of P content on the electrical resistivity of $(Pd_{0.65}Cu_{0.35})_{100-y}P_y$ metallic glasses. (From Tangonan, G., Ph.D. thesis, California Institute of Technology, Pasadena, 1976, unpublished.)

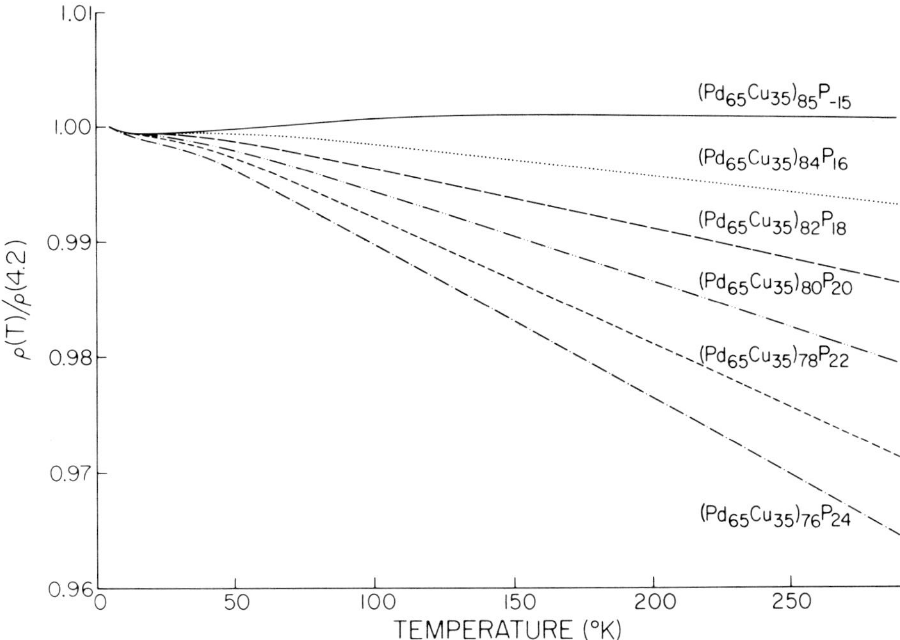

FIGURE 18. Reduced resistivity [ρ(T)/ρ(4.2)] for $(Pd_{0.65}Cu_{0.35})_{100-y}P_y$ alloys. (From Tangonan, G., Ph.D. thesis, California Institute of Technology, 1976, unpublished.)

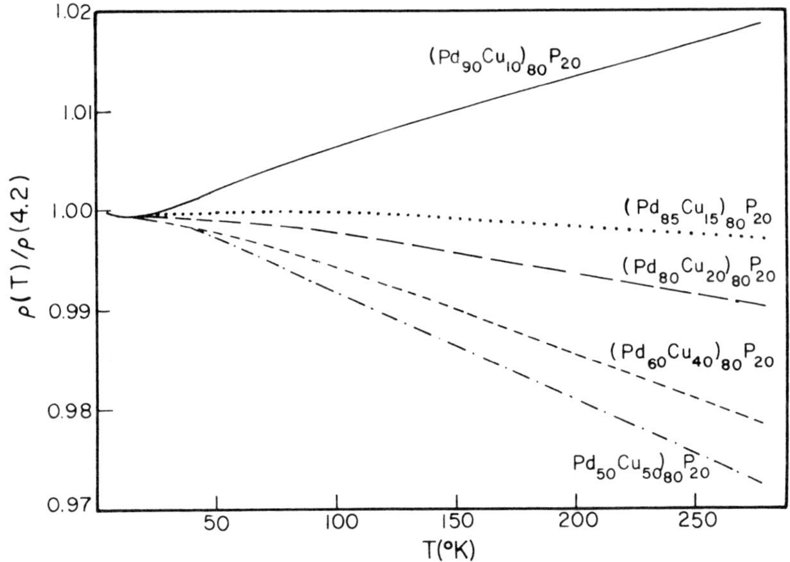

FIGURE 19. Reduced resistivity [ρ(T)/ρ(4.2)] for $(Pd_{1-x}Cu_x)_{80}P_{20}$ alloys. (From Tangonan, G., Ph.D. thesis, California Institute of Technology, Pasadena, 1976, unpublished.)

concerning these systems: (1) Why are ρ and α sensitive to the P content in both systems? and (2) Why are ρ and α sensitive to the Cu content, but not to the Ni content?

The EGL formula given before can be written as

$$\rho = f(K_f,E_f) \sin^2\eta_2 [E_f] S(2K_f,T) \tag{53}$$

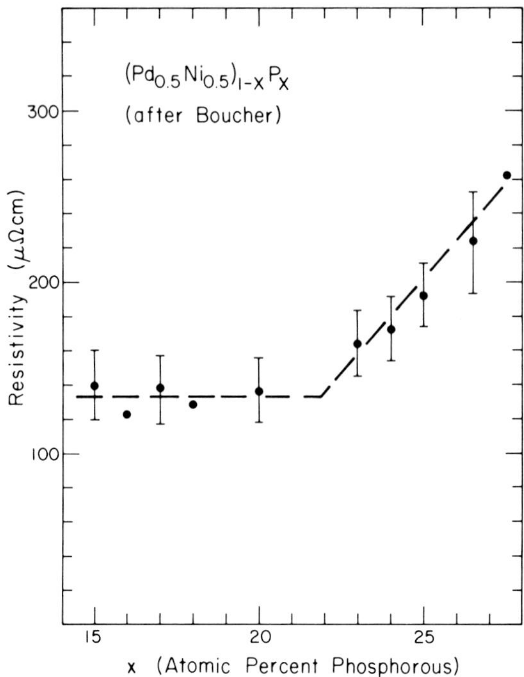

FIGURE 20. The room temperature electrical resistivity of $(Pd_{1-x}Ni_x)_{100-x}P_x$ metallic glasses. (From Boucher, B. Y., *J. Non-Cryst. Solids*, 7, 277, 1972. With permission.)

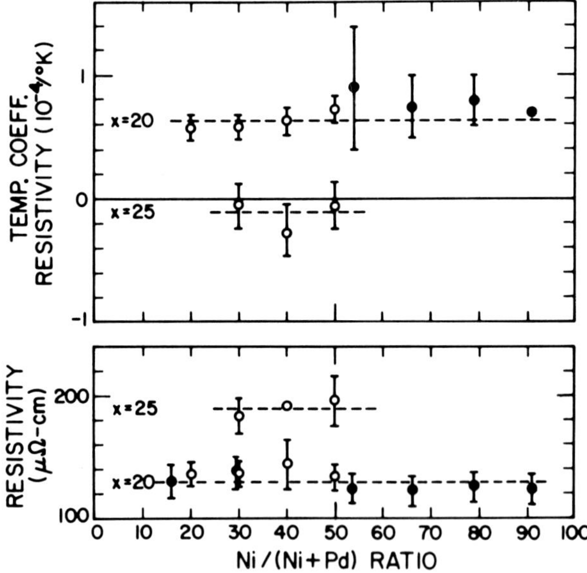

FIGURE 21. The values of ρ_N and α for $(Pd_{1-y}Ni_y)_{100-x}P_x$ metallic glasses. Values are shown as a function of [Ni/(Ni + Pd)] ratio for several values of x. (From Boucher, B. Y., *J. Non-Cryst. Solids*, 7, 277, 1972. With permission.)

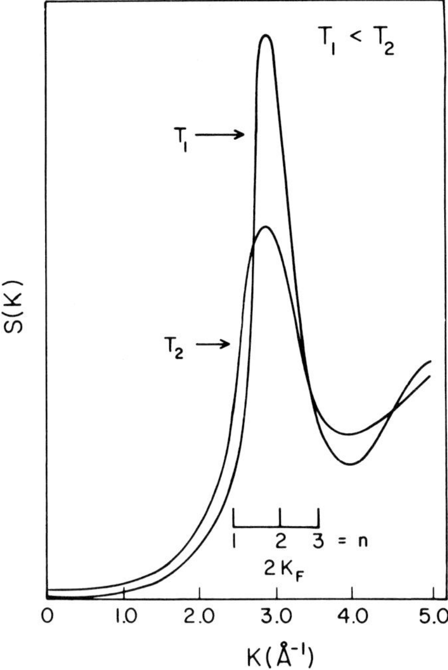

FIGURE 22. An illustration of the role of electron valence n in determining the position of $2K_f$ relative to the temperature-dependent structure factor (S(K)) in Ziman's theory of the resistivity of liquid metals.

Calculations of ρ and α using this formula are difficult due to uncertainties in K_F and E_F. Both of these quantities are to be determined by the "conduction electron" density (n) interpreted as the s and p parts of the band structure. We illustrate the problems in determining n for the related glassy system $Pd_{1-x}Si_x$. In Section III we saw that the Si 2s levels in glassy PdSi were far below E_F and should therefore be considered as bound states. This leaves the two p electrons per Si atom as conduction electrons. It is less clear what conduction electron density to assign to Pd, various authors have given it a value anywhere between 0 and 1. It is not correct to assign conduction electron contributions from the various constituents since as we saw in Section III the electronic states near E_F are the results of a strong hybridization of Si 2p and Pd 4d states. The procedure adopted by Esposito et al.[108] is certainly an improvement over the simple-minded approach above, however, as we have seen they found their method unreliable for strong scattering systems.

Despite these problems, we can give a qualitative description of the behavior of ρ and α for the glasses PdCuP and PdNiP. It is reasonable to assume that increasing the Cu or P content of the glass increases n while changing the Ni content has no effect on n since Ni is isoelectric to Pd. In Figure 22 we schematically illustrate the dependence of $S(2K_F)$ on n for two values of temperature. In the figure we see that for $n \leqslant 1.3$ or $n \geqslant 2.6$ ρ should increase with T while for $1.3 \leqslant n \leqslant 2.6$ we expect that ρ will decrease with T. With this in mind we see that adding Cu or P to these glasses causes a sharp increase in $S(2K_F)$ and takes us from a region in which $S(2K_F)$ decreases with temperature to one in which $S(2K_F)$ decreases with temperature. This explains the increase in ρ and the decrease in α we observe as the Cu and P content is increased. This has been some very recent work by Clark et al.[112,113] on experimentally determining the temperature dependence of $S(K_F)$ using X-ray and neutron diffraction. They found that for the glasses $Nb_{60}Ni_{40}$ and $Ti_{50}Zr_{10}Be_{40}$, the temperature dependence of S(K) could be well described by Equation 47. Furthermore, they

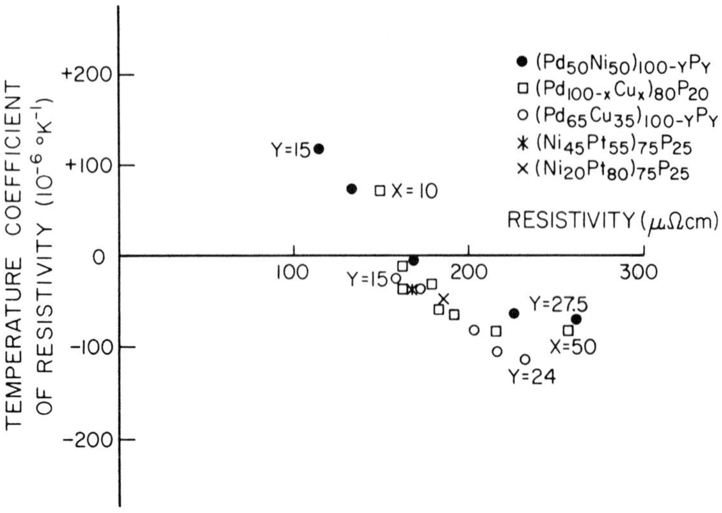

FIGURE 23. An illustration of the Mooij correlation for a number of Pd-, Pt-, Ni-, and Cu-base metallic glasses. (From Tangonan, G., Ph.D. thesis, California Institute of Technology, Pasadena, 1976, unpublished.)

found that the changes in S(K) with temperature were sufficient to explain the changes observed in ρ(T).

We can also describe the behavior of ρ and α for these glasses using the s-d scattering model of Mott.[99] When the electron mean free path (ℓ_e) becomes comparable to the interatomic spacing (a) Mott proposed that the s-d transitions become the dominant scattering mechanism. For the PdCuP and PdNiP glasses this condition would appear to be satisfied when the absolute ρ becomes larger than about 150 μΩ − cm. This is a region in which phonon scattering cannot increase ℓ_e leaving ρ(T) to be determined by s-d scattering. The models of Girvin and Jonson,[104] and Imry[105] similarly identify the region in which $\ell_e \sim a$ as being related to a change in the dominant scattering mechanism. Both of these descriptions would anticipate a negative α when ρ exceeds a certain value. We should also point out that the behavior of ρ and α in these glasses is in agreement with the Mooij correlation we discussed earlier. We illustrate this for a number of metallic glasses in Figure 23.

The approach taken by Girvin and Jonson and Imry (emphasizing the strong scattering nature of the interaction) would seem to be more reasonable for these materials than transport theories which assume $\ell_e \gg a$. The Mooij correlation seems to be upheld by most metallic glasses which suggest that one should treat the problem of ρ in more general terms. We saw above that using the EGL formula we can qualitatively understand the behavior of ρ in PdNiP and PdCuP glasses. We did so by making an assumption about the position of $2K_F$ in relation to the first peak in S(K) and the dependence of K_F on Cu, Ni, and P content. It is not clear whether we have really explained anything or if we have just used the uncertainty inherent in the various parameters to rationalize the experimental data. We feel quite uncomfortable applying this theory to glasses which are expected to be in the strong scattering regime (ρ ~ 150 μΩ − cm). Mott and Davis have pointed out that in this case $\Delta K/K \sim 1$ and the concept of a Fermi surface is not valid. How one calculates or even defines K_F in this situation is unclear.

A few measurements of the thermoelectric power (Q) in metallic glasses are available. To date, the following materials have been investigated: PtNiP,[97] BeZrTi,[114] and CuZr.[115] All these glasses were found to have positive thermopowers and the experiments on BeZrTi and CuZr in which extended temperature data are available have found that Q increases linearly with T. Nagel et al.[114] and Carini et al.[115] have argued that this observation rules out the possibility that Mott s-d scattering and the pseudo-Kondo model are the dominant scattering mechanisms. Of those theories giving predictions for Q(T) only the Ziman theory correctly explains the experimental data.

It should be clear from the preceding discussion that considerable work remains to be done before a satisfactory account of electronic transport properties of metallic glasses can be achieved. The difficulties encountered in this area are perhaps less surprising when one recalls that the theory of electronic transport in crystalline materials has been less than entirely successful. Of the various properties of crystalline solids, electrical resistivity is among the most difficult to accurately calculate from first principles. By comparison, the lack of fundamental knowledge of electronic structure in metallic glasses renders the problem one step more complex. This is clearly one of the most difficult problems facing solid state theorists in the field of amorphous solids.

REFERENCES

1. **Duwez, P.,** Metastable phases obtained by rapid quenching from the liquid state, in *Progress in Solid State Chemistry,* Vol. 3, Pergamon Press, Oxford, 1966, 377.
2. **Cargill, G. S., III,** Structure of metallic alloy glasses, *Solid State Phys.,* 30, 227, 1975.
3. **Wagner, C. N. J.,** Structure of liquid metals and alloys, *Adv. X-ray Anal.,* 12, 50, 1969.
4. **Friedel, J.,** Transition metals. Electronic structure of the d-band, in *Physics of Metals,* Vol. 1, Ziman, J. M., Ed., Cambridge University Press, London, 1969, 340.
5. **Harrison, W. A.,** *Electronic Structure and the Properties of Solids,* W. H. Freeman, San Francisco, 1980, chap. 20.
6. **Cyrot Lackmann, F.,** On the electronic structure of liquid transition metals, *Adv. Phys.,* 16, 393, 1967.
7. **Cyrot Lackmann, F.,** Expansion in moments and disordered systems, *J. Phys.,* 35, C4-109, 1974.
8. **Hodges, L., Watson, R. E., and Ehrenreich, H.,** Renormalized atoms and band theory of transition metals, *Phys. Rev. B,* 10, 3953, 1972.
9. **Soven, P.,** Coherent potential model of substitutional disordered alloys, *Phys. Rev.,* 156, 809, 1967.
10. **Velicky, B., Kirkpatrick, S., and Ehrenreich, H.,** Single site approximations in the electronic theory of simple binary alloys, *Phys. Rev.,* 175, 747, 1968.
11. **Giessen, B. C., Hong, J., Kabacoff, L., Polk, D. E., Raman, R., and St. Armand, R.,** Compositional dependence of the thermal stability and related properties of metallic glasses, in *Proc. 3rd Int. Conf. on Rapidly Quenched Metals,* Vol. 1, Cantor, B., Ed., The Metals Society, London, 1978, 249.
12. **St. Armand, R. and Giessen, B. C.,** Easy glass formation in simple metal alloys: amorphous metals containing calcium and strontium, *Scr. Metall.,* 12, 1021, 1978.
13. **Harrison, W. A.,** *Pseudopotentials in the Theory of Metals,* W. A. Benjamin, New York, 1966.
14. **Animalu, A. O. E.,** *Intermediate Quantum Theory of Solids,* Prentice-Hall, Englewood Cliffs, N. J., 1977, 185.
15. **Mott, N. F. and Jones, H.,** *The Theory of the Properties of Metals and Alloys,* Dover, New York, 1958, chap. VII.
16. **Herman, F. and Skillman, S.,** *Atomic Structure Calculations,* Prentice-Hall, Englewood Cliffs, N. J., 1963.
17. **Anderson, P. W.,** Localized magnetic states in metals, *Phys. Rev.,* 124, 41, 1961.
18. **Harrison, W. A.,** *Solid State Theory,* McGraw-Hill, New York, 1970, 480.
19. **Moriya, T.,** *Proc. Varenna School,* Marshall, W., Ed., Academic Press, New York, 1967, 206.
20. For a review of the spin fluctuation problem see: **Mills, D. L., Beal-Monod, M. T., and Lederer, P.,** *Magnetism,* Vol. V, Suhl, H., Ed., Academic Press, New York, 1973, 89.
21. **Mott, N. F. and Davis, E. A.,** *Electronic Processes in Non-Crystalline Materials,* Clarendon Press, Oxford, 1979, chap. 5.
22. **Oelhafen, P., Hauser, E., Güntherodt, H. J., and Benneman, K. H.,** New type of d-band metal alloys: the valence band structure of the metallic glasses PdZr and CuZr, *Phys. Rev. Lett.,* 43, 1134, 1979.
23. **Amamou, A.,** d-Band structure and alloying effects in crystalline and amorphous Zr-Co and Zr-Ni alloys, *Solid State Commun.,* 33, 1029, 1980.

24. **Riley, J. D., Ley, L., Azoulay, J., and Terakura, K.,** Partial densities of states in amorphous Pd-Si alloys, *Phys. Rev. B,* 20, 776, 1979.
25. **Park, R. L.,** Inner shell spectroscopy, *Phys. Today,* 28, 52, 1975.
26. **Onn, D. G., Johnson, W. D., Gleeson, P. F., Donnelly, T. A., Egami, T., and Liebermann, H. H.,** Low temperature heat capacity measurements of the metallic glasses $Fe_{80}P_{14}B_6$, $Fe_{40}Ni_{40}P_{14}B_6$, and $Ni_{80}P_{14}B_6$, *J. Phys. C,* 10, L639, 1977.
27. **Butler, W. H.,** Electron-phonon coupling in the transition metals: electronic aspects, *Phys. Rev. B,* 15, 5267, 1977.
28. **Garoche, P. and Johnson, W. L.,** An accurate determination of the excess low temperature heat capacity of a superconducting metallic glass, *Solid State Commun.,* 39, 403 1981.
29. **Mizutani, U., Hartwig, K. T., Massalski, T. B., and Hopper, R. W.,** Low temperature specific heats of glassy $Pd_{1-x-y}Si_xCu_y$ alloys, *Phys. Rev. Lett.,* 41, 661, 1978.
30. **Anderson, P. W., Halperin, B. I., and Varma, C. M.,** Anomalous low-temperature thermal properties of glasses and spin glasses, *Philos. Mag.,* 25, 1, 1972; see also **Stephens, R. B.,** Intrinsic low-temperature thermal properties of glasses, *Phys. Rev. B,* 13, 852, 1976.
31. **Matey, J. R. and Anderson, A. C.,** Phonon transport in glassy metals below 100 K, *Phys. Rev. B,* 16, 3406, 1977.
32. **Saint-James, D., Sarma, G., and Thomas, E. J.,** *Type II Superconductivity,* Pergamon Press, Oxford, 1969.
33. **Hake, R. R.,** Paramagnetic superconductivity in extreme type-II superconductors, *Phys. Rev.,* 158, 356, 1966.
34. **Anderson, P. W.,** Theory of dirty superconductors, *J. Phys. Chem. Solids,* 11, 26, 1959.
35. **Johnson, W. L., Poon, S. J., Durand, J., and Duwez, P.,** Superconductivity, magnetic susceptibility and electronic properties of amorphous $(Mo_{1-x}Ru_x)_{80}P_{20}$ alloys obtained by liquid quenching, *Phys. Rev. B,* 18, 206, 1978.
36. **Waclawski, B. J. and Boudreaux, D. S.,** Ultraviolet photoelectron spectroscopy of Pd-Si glasses, *Solid State Commun.,* 33, 589, 1980.
37. **Müller, M., Beck, H., and Güntherodt, H. J.,** Magnetic Properties of liquid Pd, Si, and Pd-Si alloys, *Phys. Rev. Lett.,* 41, 983, 1978.
38. **Güntherodt, H. J., Oelhafen, P., Lapka, R., Künzi, H. U., Indlekofer, G., Krieg, J., Laubscher, T., Rudin, H., Gubler, U., Rösel, F., Ackermann, K. P., Delley, B., Fischer, M., Greuter, F., Hauser, E., Liard, M., Müller, M., Bennemann, K. H., and Hague, C. F.,** Electronic structure of liquid and amorphous metals, *J. Phys. (Paris),* 41, C8-381, 1980.
39. **Oelhafen, P., Liard, M., Güntherodt, H. J., Berresheim, K., and Polaschegg, H. D.,** Photoemission studies of Pd-Si metallic glasses, *Solid State Commun.,* 30, 641, 1979.
40. **Meyer, F., Buster, J. H., and Bagley, B. G.,** Optical properties of the metallic glass $Pd_{0.775}$ $Cu_{0.06}$ $Si_{0.165}$, *J. Non-Cryst. Solids,* 34, 441, 1979.
41. **Mizutani, V. and Massalski, T. B.,** Hall effect measurements and electronic structure of amorphous Pd-Si-Cu alloys, *Phys. Rev. B,* 21, 3180, 1980.
42. **Matsuura, M., Nomoto, T., Itoh, F., and Suzuki, K.,** XPS study of the valence band spectrum of amorphous Fe-B alloys, *Solid State Commun.,* 33, 895, 1980.
43. **Ray, R., Hasegawa, R., Chou, C. P., and Davis, L. A.,** Iron-boron glasses: density, mechanical and thermal behavior, *Scr. Metall.,* 11, 973, 1977.
44. **Fukunaga, T., Misawa, M., Fukamichi, K., Masumoto, T., and Suzuki, K.,** Compositional study of short range structure of Pd-Si and Fe-B amorphous alloys, in *Proc. 3rd Int. Conf. Rapidly Quenched Metals,* Vol. 2, Cantor, B., Ed., The Metals Society, London, 1978, 325.
45. **Hiroyoshi, H., Fukamichi, K., Kikuchi, M., Hoshino, A., and Masumoto, T.,** Magnetic moment and high field susceptibility of amorphous FeB Invar-type alloys, *Phys. Lett. A,* 65, 163, 1978.
46. **Matsuura, M.,** Crystallization kinetics of amorphous Fe-B alloys by DTA, *Solid State Commun.,* 30, 231, 1979.
47. **Luborsky, F. E. and Liebermann, H. H.,** Crystallization kinetics of Fe-B amorphous alloys, *Appl. Phys. Lett.,* 33, 233, 1978.
48. **Polk, D. E.,** The structure of glassy metallic alloys, *Acta Metall.,* 20, 485, 1972.
49. **Chen, H. S., Krause, J. T., Shirakawa, K., and Masumoto, T.,** On the anomalies in density, Young's modulus, and glass temperature of Pd-Si glasses, *J. Non-Cryst. Solids,* 41, 79, 1980.
50. **Amamou, A. and Krill, G.,** Photoemission on a Cu-Zr glassy metal, *Solid State Commun.,* 28, 957, 1978.
51. **Oelhafen, P., Hauser, E., and Güntherodt, H. J.,** Varying d-band splitting in glassy transition metal alloys, *Solid State Commun.,* 35, 1017, 1980.
52. **Hague, C. F., Temmermann, W. M., Oelhafen, P., and Güntherodt, H. J.,** to be published.
53. **Nagel, S. R., Tauc, J., and Giessen, B. C.,** Photoemission study of Nb-Ni glasses, *Solid State Commun.,* 22, 471, 1977.

54. **Amamou, A. and Krill, G.,** Photoemission and band structure of crystalline and amorphous iron phosphoborides, *Solid State Commun.*, 33, 1087, 1980.
55. **Steiner, P., Schmidt, M., and Hüffner, S.,** Photoemission study of PdZr and CuZr alloys, *Solid State Commun.*, 35, 493, 1980.
56. **Amamou, A. and Johnson, W. L.,** Photoemission and electronic structure of Mo, Ru, and amorphous MoRuB and MoRuP alloys, *Solid State Commun.*, 35, 765, 1980.
57. **Johnson, W. L. and Williams, A. R.,** Structure and properties of transition-metal-metalloid glasses based on refractory metals, *Phys. Rev. B*, 20, 1640, 1979.
58. **Moruzzi, V. L., Janak, J. F., and Williams, A. R.,** *Calculated Electronic Properties of Metals*, Pergamon Press, Oxford, 1979.
59. **Bletry, J. and Sadoc, J. F.,** Determination of the three partial interference functions of an amorphous Co-P ferromagnet by polarized neutron scattering, *J. Phys. F*, 5, L110, 1975.
60. **Sadoc, J. F. and Dixmier, J.,** *The Structure of Non-Crystalline Materials*, Taylor and Francis, London, 1977.
61. **Sakata, M., Cowlam, N., and Davies, H. A.,** Neutron diffraction measurement of the structure factor of a CuTi metallic glass, *J. Phys. F*, 9, L235, 1979.
62. **Hayes, T. M., Allen, J. W., Tauc, J., Giessen, B. C., and Hauser, J. J.,** Short-range order in metallic glasses, *Phys. Rev. Lett.*, 40, 1282, 1978.
63. **Panissod, P., Aliaga Guerra, D., Amamou, A., Durand, J., Johnson, W. L., Carter, W. L., and Poon, S. J.,** Local symmetry around glass former sites in amorphous metallic alloys through electric quadropole effects, *Phys. Rev. Lett.*, 44, 1465, 1980.
64. **Tenhover, M.,** Chemical ordering in $Y_{66}(Fe_{1-x}Mn_x)_{34}$ metallic glasses, *J. Phys. F*, 10, L293, 1980.
65. **Tenhover, M.,** Fe^{57} Mossbauer studies of $Y_{66}(Fe_xTM_{1-x})_{34}$ metallic glasses, *J. Non-Cryst. Solids*, 1981, in press.
66. **Vincze, I., Kemèny, T., and Arajs, S.,** Short range order in transition-metal-metalloid glasses, *Phys. Rev. B*, 21, 937, 1980.
67. **Gaspard, J. P.,** Density of D states in amorphous transition metal alloys *J. Phys. (Paris)*, 41, C8-414, 1980.
68. **Pollak, R. A., Tsuei, C. C., and Johnson, R. W.,** Hybridization and bonding in transition metal compound superconductors: X-Ray photoemission from crystalline and amorphous Nb_3Ge, *Solid State Commun.*, 23, 879, 1977.
69. **Bergmann, G.,** Amorphous metals and their superconductivity, *Phys. Rep.*, 27C, 161, 1976.
70. **Johnson, W. L.,** Superconductivity in metallic glasses, in *Glassy Metals I*, Güntherodt, H. and Beck, H., Eds., Springer-Verlag, New York, 1981, chap. 9.
71. **Johnson, W. L.,** Superconducting metallic glasses, *J. Phys. (Paris)*, (41, C-8, Suppl. to No. 8), 731, 1980.
72. **McMillan, W. L.,** Transition temperature of strong-coupled superconductors, *Phys. Rev.*, 167, 331, 1968.
73. **Eliashberg, G. M.,** Interactions between electrons and phonons in a superconductor, *Sov. Phys. JETP*, 11, 696, 1960.
74. **Garland, J. W.,** unpublished results quoted in **Chan, T. T., Chen, J. T., Leslie, J. D., and Smith, H. J.,** *Phys. Rev. Lett.*, 22, 526, 1969.
75. **Barasic, S., Labbé, J., and Friedel, J.,** Strong-coupled superconductivity in V_3X Type compounds, *Phys. Rev. Lett.*, 19, 1039, 1967.
76. **Barasic, S., Labbé, J., and Friedel, J.,** Tight binding and transition-metal superconductivity, *Phys. Rev. Lett.*, 25, 919, 1970.
77. **Pettifor, D. G.,** Theory of energy bands and related properties of 4d transition metals. I. Band parameters and their volume dependence, *J. Phys. F*, F7, 613, 1977; see also *J. Phys. F*, F7, 1009, 1977.
78. **Varma, C. M. and Dynes, R. C.,** Empirical relations in transition metal superconductivity, in *Superconductivity in d- and f-Band Metals*, Douglass, D. H., Ed., Plenum Press, New York, 1976.
79. **Gomersall, I. R and Gyorffy, B. L.,** Variation of T_c with electron per atom ratio in transition metal alloys, *Phys. Rev. Lett.*, 33, 1, 1974.
80a. **Graebner, J. E., Golding, B., Schultz, R. J., Hsu, F. S. L., and Chen, H. S.,** Low temperature properties of a superconducting disordered metal, *Phys. Rev. Lett.*, 39, 1480, 1977.
80b. **Lohneysen, H. V., Platte, M., Sander, W., Schink, H. J., Minegerode, G. V., and Samwer, K.,** Low temperature thermal properties of amorphous superconducting ZrCu, *J. Phys. (Paris) Colloq.*, C8 (Suppl. to No. 8), 41, C8-745, 1980.
80c. **Tsuei, C. C., Von Molnar, S., and Coey, J. M.,** Comparison between the amorphous and crystalline (A-15) phases of Nb_3Ge, *Phys. Rev. Lett.*, 41, 664, 1978.
80d. **Hopkins, S. T. and Johnson, W. L.,** Low temperature specific heat of $(Mo_{0.6}Ru_{0.4})_{1-x}B_x$ metallic glasses: evidence for compositional variation of atomic structure, *Solid State Commun.*, 43, 537, 1982.
81. **Ziman, J. M.,** *Electrons and Phonons*, Oxford University Press, London, 1960, chap. 5.

82. **Johnson, W. L.**, Metastable superconducting alloys, in *Proc. 3rd Int. Conf. Rapidly Quenched Metals*, Vol. 2, Cantor, B., Ed., The Metals Society, London, 1978, 1.
83. **Helfand, E. and Werthamer, N. R.**, Temperature and purity dependence of the superconducting critical field H_{c2}, *Phys. Rev. Lett.*, 13, 686, 1964.
84. **Maki, K.**, The magnetic properties of superconducting alloys, *Physics*, 1, 21, 1964.
85. **Werthamer, N. R., Helfand, E., and Hohenberg, P. C.**, Temperature and purity dependence of the superconducting critical field III: electron spin and spin-orbit effects, *Phys. Rev.*, 147, 295, 1966.
86. **Maki, K.**, Effect of Pauli paramagnetism on magnetic properties of high field superconductors, *Phys. Rev.*, 148, 362, 1966.
87. **Tenhover, M., Johnson, W. L., and Tsuei, C. C.**, Upper critical fields of transition metal based amorphous superconductors, *Solid State Commun.*, 38, 53, 1981.
88. **Carter, W. L., Poon, S. J., Hull, G. W., Jr., and Geballe, T. H.**, Enhanced critical field curves of metastable superconductors, *Solid State Commun.*, 39, 41, 1981.
89. **Abrikosov, A. A. and Gorkov, L. P.**, Contribution to the theory of superconducting alloys with paramagnetic impurities, *Sov. Phys. Solid State*, 12, 1243, 1961.
90. For a general discussion, see **Maple, M. B.**, Spin-excitation effects in superconductors, *AIP Conf. Proc.*, No. 4, 175, 1972.
91. **Phillips, J. C. and Kleinman, L.**, New method of calculating wave functions in crystals and molecules, *Phys. Rev.*, 116, 287, 1959.
92. **Ziman, J. M.**, A theory of the electrical properties of liquid metals. I. The monovalent metals, *Philos. Mag.*, 6, 1013, 1961.
93. **Güntherodt, H.**, Electronic transport properties, in *Metallic Glasses*, American Society for Metals, Metals Park, O., 1978.
94. **Matey, J. R. and Anderson, A. C.**, Low temperature thermal conductivity of an amorphous palladium silicon alloy, *J. Non-Cryst. Solids*, 23, 129, 1977.
95. **Faber, T. E. and Ziman, J. M.**, A theory of the electrical properties of liquid metals. III. The resistivity of binary alloys, *Philos. Mag.*, 11, 153, 1965.
96. **Evans, R., Greenwood, D. A., and Lloyd, P.**, Calculations of the transport properties of liquid transition metals, *Phys. Lett. A*, 35, 57, 1971.
97. **Sinha, A. K.**, Electrical resistivity, thermopower, and X-ray interference function of amorphous Pt-Ni-P alloys, *Phys. Rev. B*, 1, 4541, 1970.
98. **Nagel, S. R.**, Temperature dependence of resistivity in metallic glasses, *Phys. Rev. B*, 16, 1694, 1977.
99. **Mott, N. F.**, The electrical resistivity of liquid transition metals, *Philos. Mag.*, 26, 1249, 1972.
100. **Cochrane, R. W., Harris, R., Strom-Olson, J. O., and Zuckerman, M. J.**, Structural manifestations in amorphous alloys: resistivity minima, *Phys. Rev. Lett.*, 35, 676, 1975.
101. **Kondo, J.**, Resistance minima in dilute magnetic alloys, *Prog. Theor. Phys.*, 32, 37, 1964.
102. **Tsuei, C. C.**, Anomalous electrical conduction in disordered and non-crystalline metallic conductors, *Solid State Commun.*, 27, 691, 1978.
103. **Mooij, J. H.**, Electrical conduction in concentrated disordered transition metal alloys, *Phys. Status Solidi A*, 17, 521, 1973.
104. **Girvin, S. M. and Jonson, M.**, Dynamical electron-phonon interaction and conductivity in strongly disordered metal alloys, *Phys. Rev. B*, 22, 3583, 1980.
105. **Imry, Y.**, Possible role of incipient anderson localization in the resistivities of highly disordered metals, *Phys. Rev. Lett.*, 44, 469, 1980.
106. **Anderson, P. W.**, On the absence of diffusion in certain random lattices, *Phys. Rev.*, 109, 1492, 1958.
107. **Busch, G. and Güntherodt, H. J.**, Electronic properties of liquid metals and alloys, *Solid State Phys.*, 29, 1974.
108. **Esposito, E., Ehrenreich, H., and Gelatt, C. D., Jr.**, Electrical transport in transition metal liquids and metallic glasses, *Phys. Rev. B*, 18, 1978.
109. **Fujiwara, T.**, Electronic states and transport in amorphous and liquid transition metals; Fe, Co, and Ni, *J. Phys. F*, 9, 2011, 1979.
110. **Tangonan, G. L.**, Electrical resistivity of amorphous Pd-Cu-P alloys, *Phys. Lett. A*, 54, 307, 1975.
111. **Boucher, B. Y.**, Influence of phosphorous on the electrical properties of PdNiP amorphous alloys, *J. Non-Cryst. Solids*, 7, 277, 1972.
112. **Clark, R. and Nagel, S. R.**, Observation of the temperature dependence of the structure factor in metallic glasses, *Solid State Commun.*, 27, 215, 1978.
113. **Clark, R., Nagel, S. R., Hitterman, R. L., and Mueller, M. H.**, Temperature dependence of the X-ray and neutron diffraction from a metallic glass, *Solid State Commun.*, 36, 751, 1980.
114. **Nagel, S. R.**, Thermopower and resistivity in a metallic glass, *Phys. Rev. Lett.*, 41, 990, 1978.
115. **Carini, J. P., Basak, S., and Nagel, S. R.**, Thermoelectric power of metallic glasses, *J. Phys. (Paris)*, 41, C8-463, 1980.

Chapter 4

MAGNETISM IN TRANSITION METAL BASE AMORPHOUS ALLOYS

J. Durand

TABLE OF CONTENTS

I. INTRODUCTION

This chapter will review the magnetic properties of transition-metal base amorphous alloys. These properties include both the dilute (occurrence of localized magnetic moments at zero temperature and properties of the moments at finite temperature) and the concentrated cases (long-range magnetic order) together with the intermediate regime (onset of magnetic ordering). We shall consider only the systems where the magnetic moments are related to 3d elements (Ni, Co, Fe, Mn, Cr), since the amorphous alloys containing rare-earths are reviewed in a separate chapter by Soohoo. We shall not dwell on macroscopic properties of amorphous ferromagnetic alloys, which make these materials so attractive from technological point of view, namely magnetic anisotropy, domains, magnetostriction, Invar properties, etc. Since these properties represent a rich potential of applications, they are described in the following chapter of this book. Our purpose is to concentrate on those fundamental magnetic properties most directly related to the electronic band structure of transition-metal base amorphous alloys. We shall attempt to discuss the various experimental data in the light of existing models for magnetism of 3d elements in amorphous matrices. Our main interest would be to define the role of structure in the basic magnetic properties of these materials.

With hindsight it appears that research on magnetism of amorphous alloys has been restrained for a long time by the apparent incompatibility between magnetic ordering and structural disorder. For example, König[1] reported in 1946 on Fe films obtained by evaporation on a cold substrate. In 1947, Brenner and Riddell[2] were able to produce amorphous NiP and CoP alloys by chemical reduction. In 1950, the same family of alloys was prepared by electrodeposition.[3] That these materials were amorphous and ferromagnetic at the same time was not readily ascertained and accepted. The possible existence of amorphous ferromagnetism was first discussed on theoretical grounds by Gubanov[4] in 1960, who calculated the effect of structural disorder on the Curie temperature. It took few years for the experimentalists to verify these model predictions. The first observation of magnetic domains on vapor-deposited amorphous Fe films is due to Grigson et al.[5] in 1964. In 1965, Co–Au alloys obtained by evaporation on to a cold substrate were exemplified as amorphous ferromagnets by Mader and Nowick,[6] while Judge et al.[7] and Bagley and Turnbull[8] reported on the magnetic behavior of CoP deposits produced by chemical reduction and electrolysis, respectively. In 1966, evidence of magnetic ordering was shown by Tsuei and Duwez[9] in liquid-quenched amorphous PdSi alloys containing Co and Fe. Of particular significance was the first report in 1967 by Duwez and Lin[10] of a liquid-quenched amorphous FePC alloy exhibiting high saturation magnetization, low coercive field, rather high Curie temperature and good stability at room temperature against crystallization. For several years these materials were regarded as laboratory curiosities. From 1969 to 1973, the major emphasis, particularly in the Caltech group, was on some fundamental properties of dilute magnetic amorphous alloys, such as Kondo effect, onset of magnetic ordering, while the first local magnetic studies (Mössbauer spectroscopy) on amorphous ferromagnets were performed. Since 1973, as the field of technical applications of amorphous magnetic alloys was foreseen, the number of experimental and theoretical studies has grown at a continuous rate, so that it has become a real challenge to present an exhaustive survey of the whole literature. For the period 1950 to 1977, the bibliography about amorphous magnetism and magnetic materials has been compiled by Kobe and Ferchmin.[11] Experimental and theoretical developments in this field were analyzed in a recent book by Handrich and Kobe.[12] One can usefully refer also to many recent reviews on magnetism in amorphous metallic alloys.[12-21] Particular aspects of magnetism in amorphous thin films were analyzed separately.[21] Experimental information derived from nuclear techniques applied to metallic glasses was collected in recent reviews.[22,23]

In view of the existence of such a large number of excellent bibliographical works, it appears unnecessary to undertake a comprehensive survey of the experimental literature concerned with magnetism of transition metal in amorphous alloys. We focus therefore on some aspects which seem to have been as yet somehow overlooked. First, most of the aforementioned reviews were restricted to magnetically ordered materials, thus nearly ignoring such interesting problems as the formation and properties of localized moments, or as the progressive occurrence of magnetic ordering in an amorphous matrix. Second, it seems to us that the influence of the amorphous structure on magnetic properties has not been analyzed in detail. Too often, among theoreticians and experimentalists involved in magnetism, the amorphous character of the materials under study seems to be taken for granted, so that the basic incentives of this field of research are not clearly stressed.

The amorphous structure can be regarded from two complementary points of view. At a first step, the amorphous character is defined in a negative way as the lack of long-range periodic order. But the next step for structural studies and computer simulations is to determine the average environment around a given species of atoms. It is by now commonly admitted that in amorphous metallic alloys there exists a short-range order of some sort depending upon composition and chemical nature of the constituents. Our review on magnetism of transition metals in amorphous metallic alloys will be then organized according to these two aspects of the amorphous structure. It will be divided into two sections. First, we attempt to discuss the effect of structural disorder on magnetism in dilute and concentrated amorphous alloys. We then go on to address to ourselves the question of what can be learned about the short-range order from bulk and local magnetic measurements.[24] The first section implies some knowledge of the theory of magnetism in noncrystalline solids along with some familiarity with the experimental literature on electronic band structure in amorphous metallic alloys; it is then strongly connected with Chapters 2 and 3 this book, by Kaneyoshi and Johnson, respectively. The second section is a natural appendix to Chapter 1 on structure of metallic glasses by Boudreaux.

II. INFLUENCE OF STRUCTURAL DISORDER ON MAGNETISM

Among the various transition-metal base amorphous alloys that one can obtain from different techniques, three classes can be considered: (a) the "pure" transition metal films, (b) the alloys made of two metallic elements, such as alloys of "early" and "late" transition metals (ZrNi, YNi, LaNi, NbNi) or alloys of transition and noble metals (CuZr) (We do not deal here with alloys of transition metals with rare-earths, such as RENi, RECo), and (c) the alloys between transition metals and s-p elements (NiP, CoB, FeSi, FeGe, etc.). Alloys of this last class can be produced by vacuum deposition over a very broad concentration range. The alloys of interest for magnetism will be obviously on the transition-metal-rich side.

Since we are dealing with magnetism in metallic systems, the knowledge of the electronic band structure is of particular importance, especially the symmetry and the density of electronic states at the Fermi energy E_F. For crystalline transition metals, it is well known that the Fermi level lies within the energy bands formed from the d levels of the free atoms. In spite of the overlap of the broad s-p bands with the narrower d bands, the dominant contribution of the density of states at E_F comes from regions of the Fermi surface in which the wave functions have strong d character. This is particularly true for the late transition metals.[24] For "pure" amorphous transition metals, little is known from experiment about electronic structure, due to their instability against crystallization. But it has been shown by calculation that structural disorder does not affect drastically the electronic structure of amorphous Ni and Co with respect to crystalline metals.[25] But, for metallurgical reasons, most of the amorphous materials which were studied in detail from magnetic point of view are alloys of transition metals with other metals or with s-p elements. The question arises then as to what extent the d character of the density of states at E_F is still preserved considering additional effects of alloying and of disorder. Relatively few calculations of electronic density of states in these alloys have been reported as yet.[26] From an experimental point of view, the electronic structure of these materials has become rather well documented, by means of photoemission, soft-X-ray (SXS), and γ-ray Compton scattering experiments.[27] The most interesting conclusion for our purpose is that the main contribution to the density of states at E_F still arises from d electrons in transition-metal base amorphous alloys, in spite of the important reduction due to charge transfer and/or hybridization effects. The nature of these latter alloying effects has to be discussed in detail for each family of alloys. At this point, it is of significance to note that the problems of magnetic ordering and of magnetic impurity states in transition-metal base amorphous alloys have to be treated basically in d-band hosts. The dilute impurity problem and the long-range magnetism in transition metals historically have been treated on different theoretical grounds. Thus, we analyze first the data concerning the formation and the properties of 3d moments in amorphous matrices. Then, we give a general survey of magnetically ordered transition metal base amorphous alloys. In these two subsections, we attempt to discuss the results of bulk and local properties measurements together with the magnetic contributions to the transport and calorimetric properties.

A. From Dilute Impurities to the Onset of Long-Range Magnetic Order

Atoms with partially filled d-shells, when diluted in nonmetallic hosts or when forming nonmetallic compounds, generally assume a definite ionic valence state and show the magnetic moment predicted by the corresponding Hund's rule, if crystal-field effects are neglected. In the case of metallic hosts, however, the presence of broad conduction bands makes the situation more complex. The problem of localized moments in metals has been historically regarded from two different points of view: the formation of localized moments at zero temperature and the properties of the magnetic moments at finite temperature. Our purpose is to discuss the experimental situation in transition-metal base amorphous alloys by trying to define the parameters through which the effect of structural disorder can be felt, in the light of theoretical and experimental studies in crystalline systems.[28] We analyze first the problem of moment formation and that of magnetic properties of noninteracting impurities.

1. Moment Formation; Dilute Magnetic Impurities

The dilute impurity problem in metallic crystalline alloys has been the subject of an overwhelming number of studies in the last twenty years.[29] The specific problems of transition metal hosts were reviewed by Fischer.[30] In summary, the onset of a magnetic moment on a 3d impurity can be described essentially as the result of a balance between two processes: intra-atomic interactions concerning mainly d electrons, and interaction of the impurity states with the conduction electrons. The intra-atomic interactions favor the polarization of the impurity states. Two energies play a dominant role in defining the strength of the polarization. The intra-atomic exchange energy J arises from the difference between the energies of two d-electrons with parallel and antiparallel spin. The second term is an electrostatic Coulomb repulsion U between two electrons with antiparallel spins on the same orbital, or between electrons in different orbitals for both spin directions. Of course, due to screening by conduction electrons, the parameters J and U have effective values within a metal which are not trivial to predict from atomic considerations. From Friedel's approach, the nonmagnetic state was shown[28] to be unstable for

$$2\,\epsilon_{ex}\,n_d(E_F) > 1 \qquad (1)$$

where ϵ_{ex} is an average value of the exchange parameter, and $n_d(E_F)$ is the d-density of states at the Fermi level. The quantity $n_d(E_F)$ depends on the average position E_d of the d levels with respect to E_F, and upon the width Δ of the localized states. Self-consistency condition is achieved by requiring E_d to have a value which satisfies the Friedel sum rule. The width Δ will depend on the details of the impurity potential. Qualitatively, the results obtained from numerical estimations of the condition (1) do not depend on the exact shape of $n_d(E)$. From such a simple formalism, Friedel was able to predict the magnetic and nonmagnetic states of 3d impurities in normal metals. Similarly, in the Anderson model, the criterion for the occurrence of spin magnetism reads, in the orbital degenerate case,

$$(U + 2\ell J)n_d(E_F) > 1 \tag{2}$$

where $n_d(E)$ is the nonmagnetic d density of states.

For orbital magnetism, the Anderson criterion is given by

$$(U - J)n_d(E_F) > 1 \tag{3}$$

For a given value of the Coulomb integral U, the intra-atomic exchange J helps the formation of a spin moment, but it has the opposite effect for the orbital moment. Transition-metal impurities exhibit spin-only magnetism with quenched orbital moment. These criteria for instability of the nonmagnetic state strongly resemble the Stoner condition for band ferromagnetism, as will be discussed later. They roughly determine two extreme regimes, but the transition between them is smooth. Due to the local character of the problem, one cannot speak in terms of a true transition as in macroscopic systems. In addition, the effect of local environment around a given impurity was not taken into account in the early Hartree-Fock treatments of the moment formation. The importance of local environment effects was first noted by Jaccarino and Walker[31] who interpreted the discontinuous moment formation on Co in RhPd and on Fe in NbMo alloys in terms of a phenomenological model in which the impurities were assumed to be nonmagnetic unless a minimum of 2 Pd or 7 Mo atoms, respectively, were nearest neighbors to given impurity site. More recently, similar approaches were used to account for the inhomogeneous character of the nonmagnetic to magnetic transition in many crystalline alloys.[32] These empirical treatments were subsequently justified by theoretical calculations.[33]

Finally, one must emphasize that the classification, as deduced from Hartree-Fock approximation, between magnetic and nonmagnetic impurities is ambiguous, in the sense that the Kondo phenomena are not taken into account. The fact that below the Kondo temperature T_K the magnetic moment for vanishingly small impurity concentration decreases with decreasing T and disappears at $T = 0$ was traditionally expressed in terms of an s-d exchange model, in which the impurity spin S with a well-defined value interacts with the spin s of the conduction electrons through a Heisenberg-type interaction

$$H = -J_{sd}\vec{s} \cdot \vec{S} \tag{4}$$

For magnetic alloys in the Hartree-Fock sense, the Kondo temperature defined for a negative exchange J as

$$T_K = D \exp\left[-1/n(E_F)|J_{eff}|\right] \tag{5}$$

(D being a constant of the order E_F) is very low, so that the standard experiments performed on finite impurity concentrations and over a temperature range between the helium temperature region and the room temperature will sense only the $T > T_K$ regime. The impurity is then said to carry a ''good moment''. Conversely, a nonmoment impurity in the Hartree-Fock sense could signal a T_K exceeding the temperature range investigated.

If one considers now the case of dilute impurities in an amorphous matrix, it seems to be very hard to make quantitative predictions concerning which impurity has what magnetic moment in which host above what temperature. One is reduced to describing trends. Since the amorphous alloys we are dealing with are metallic, the screening mechanisms on intra-atomic exchange and Coulomb repulsion parameters arising from conduction electrons are likely to be basically the same as in crystalline counterparts. Also, the effective local antiferromagnetic effect of the covalent admixture between localized states and conduction electrons will be essentially preserved. Presumably, the structural disorder could affect the moment formation problem mainly through the width Δ of the localized states and through local environment effects in systems where this discontinuous approach might hold. The occurrence of the local magnetic moments in amorphous metallic alloys was discussed by Cyrot[34] from a general band structure point of view. Because of a band narrowing, the appearance of magnetic moments was concluded to be easier and their values to be possibly larger in amorphous matrices than in corresponding crystalline hosts. The problems encountered when one tries to evaluate the effect of the structural disorder on the dilute impurity problem in amorphous alloys include first the fact that most of the nonmagnetic hosts available so far for such a study (PdSi, PdP, CuZr, NiPB, NiPdP) are rather complex alloys from electronic structure point of view. In addition, there is no direct experimental information as yet on such basic parameters as the width Δ of the virtual bound states in amorphous hosts. We thus will proceed in a very empirical way by reviewing first the amorphous alloys in which 3d impurities were found to have "good moments" in the Hartree-Fock sense. Then, we analyze the manifestations of a Kondo behavior. Finally, we discuss the systems where the discontinuous onset of magnetic moment can be interpreted in terms of local environment effects.

a. Dilute Impurities with Good Moments in Amorphous Hosts

The magnetic properties of dilute 3d impurities were studied in various amorphous alloys such as $Pd_{80}Si_{20}$, $Pd_{80}P_{20}$, $Ni_{79}P_{13}B_8$, $(Ni_{0.5}Pd_{0.5})_{80}P_{20}$, $(NiPt)_{80}P_{20}$, $(NiPd)_{80}B_{20}$, CuZr, and NbNi. Let us recall first what is known about the electronic structure of these amorphous hosts.

The PdSi metallic glasses have been studied by several groups through photoemission techniques.[27] The conclusions which can be drawn from these measurements are as follows. As compared with the case of pure crystalline Pd, the Fermi level E_F in PdSi glasses is raised with respect to the bottom of the Pd 4d conduction band, so that E_F intersects a region of low density of states. The ratio of the spectrum intensity at E_F to the maximum intensity is estimated to be 0.25. Nevertheless, it seems that d contributions to the density of states at E_F cannot be precluded. The shift of the Pd 4d peak with respect to E_F increases slightly with increasing Si concentration, but substitution of Cu for Si does not modify the shift at a given Pd concentration. However, from core level studies, this cannot be attributed to a simple charge transfer to or from Pd in the alloy. Measurements of partial density of states would seem to imply a strong hybridization at E_F between d states from Pd and p states from Si. As has been calculated for crystalline Pd_3Si and amorphous $Pd_{80}Si_{20}$, and as was observed also in crystalline Pd_2Si and in liquid PdSi alloys, both Pd- and Si-derived states are affected by alloying: for Pd, the 4d states are shifted below E_F; for Si, the 3p states are split into two peaks about 5 eV apart, while the 3s states remain about 10 eV below E_F and are not much affected. Si 3p antibonding states occur near E_F, while Pd 4d — Si 3p bonding states are located at about 5.5 eV below E_F. Finally, it is of interest to note that the Pd 4d bandwidth is slightly smaller in *a*-PdSi alloys than in pure Pd.

The analysis of photoemission measurements is in good agreement with the results of calorimetric studies on *a*-PdSi alloys[35] yielding a rather high value for the linear coefficient in the temperature dependence of the specific heat (γ = 1.29 mJ/mole K^2 for $Pd_{80}Si_{20}$). On

Table 1
SPECIFIC HEAT LINEAR COEFFICIENTS (γ) AND TEMPERATURE INDEPENDFENT MAGNETIC SUSCEPTIBILITY (χ) FOR AMORPHOUS ALLOYS AND RELATED CRYSTALS

Amorphous	γ(mJ/mole K^2)	χ (10^{-6} emu/g)	Ref.
$Pd_{80}Si_{20}$	1.29	−0.045	35, 36
$Ni_{78}P_{14}B_8$	1.90	1.60	81, 40
$Cu_{60}Zr_{40}$	2.80	0.52	37, 38
$Nb_{50}Ni_{50}$		2	39
Crystals			
Ni	7.04		
Pd	9.42	5.15	
Cu	0.696	−0.23	
Zr	3.03	13.20	
Ni_3B	2.80	0.95	
Ni_3P	1.44	0.40	

the other hand, the band contribution to the magnetic susceptibility is rather low, either slightly diamagnetic (χ(295 K) = $-0.045 \cdot 10^{-6}$ emu/g for $Pd_{77.5}Cu_6Si_{16.5}$) or very weakly positive (as observed in liquid $Pd_{80}Si_{20}$ alloy).[36] For comparison, let us recall that for pure Cu, $\chi = -0.23 \cdot 10^{-6}$ emu/g, and for pure Pd, χ(273 K) = $5.15 \cdot 10^{-6}$ emu/g.

Amorphous CuZr alloys constitute another family in which the impurity states have been extensively studied. Photoemission measurements showed that the d-band is split into two components: one crossing the Fermi level which arises from Zr, the other one due to Cu at 3.8 eV below E_F (for the $Cu_{60}Zr_{40}$ alloy). The relative intensities of these two components vary strongly with concentration. Such a band splitting is a well-known effect in concentrated alloys where nuclear charges or exchange fields of the components differ sufficiently. Then each alloy component has its own d-band, having minimum overlap with the 3d bands of other components. The important point is that Zr provides the main contribution to the density of states at E_F. Specific heat measurements yielded a large value for the electronic term which increases rapidly with increasing Zr content (γ = 2.80 and 3.90 mJ/mole K^2 in a-$Cu_{60}Zr_{40}$ and $Cu_{35}Zr_{65}$, respectively[37]). The temperature-independent Pauli-paramagnetism is rather high ($\chi = 0.52 \cdot 10^{-6}$ emu/g for $Cu_{60}Zr_{40}$).[38]

A split valence band was observed also by photoemission study of a-NbNi alloys. The density of states at E_F is reduced as compared with that of pure Nb, but still it is rather high. The Pauli-paramagnetic susceptibility for a-$Nb_{50}Ni_{50}$ was found to be $2 \cdot 10^{-6}$ emu/g.[39] Little experimental information is available on the electronic structure of other nonmagnetic matrices, such as $Pd_{80}P_{20}$, $(Ni_{0.5}Pd_{0.5})_{80}P_{20}$, $(Ni_{0.5}Pt_{0.5})_{80}P_{20}$, or $(Ni_{0.5}Pd_{0.5})_{80}B_{20}$, in which solute magnetic impurities were studied. The values of the band magnetic susceptibilities are listed in Table 1, together with the other experimental values for the specific heat linear coefficients.

More interesting is the case of the NiPB alloys in the critical concentration range for the onset of magnetism (i.e., 78, 79, and 80 at % Ni), because the magnetic states of Fe, Co, Mn, Cr diluted in this system were studied in some detail. The data concerning the approach to magnetic ordering with increasing Ni concentration in a-NiPB alloys will be discussed later. We briefly summarize the experimental information about the electronic structure of a-$Ni_{78}P_{14}B_8$. Photoemission measurements were performed on this alloy as well as on crystalline pure Ni, Ni_3P, and Ni_3B. The valence band spectra in crystalline and amorphous compounds are very similar to those observed for pure Ni. Electronic d states arising from

Ni are the dominant contribution at E_F. The temperature-independent susceptibility in *a*-$Ni_{78}P_{14}B_8$ ($\chi = 1.60 \cdot 10^{-6}$ emu/g)[40] is rather high. But from the concentration dependence of this susceptibility in a series of *a*-NiPB alloys, one can obtain for *a*-$Ni_{75}P_{17}B_8$ an extrapolated value of 0.65 ($\pm$ 0.10)$\cdot 10^{-6}$ emu/g, which lies between the experimental values for *c*-Ni_3P and Ni_3B (0.40 and $0.95 \cdot 10^{-6}$ emu/g, respectively).[41] This is also in agreement with the estimate of the density of states yielded by the linear term in the specific heat: for amorphous $Ni_{78}P_{14}B_8$, $\gamma = 1.9$ mJ/mole K^2 while $\gamma = 1.44$, 2.80, and 7.04 for *c*-Ni_3P, Ni_3B, and pure Ni, respectively.[41]

In summary, the various transition-metal base amorphous alloys in which 3d magnetic impurities were diluted so far have a density of states at the Fermi level which is more or less depressed as compared with pure transition metals. Nevertheless, the d-character of these electronic states is basically preserved. Now, we review briefly the experimental work on dilute magnetic moments in these matrices.

From this point of view, the amorphous $Pd_{80}Si_{20}$ alloy has been the most extensively studied since 1969, mainly by Tsuei and Hasegawa,[42,43] Zentko et al.,[44,45] and more recently by Dublon et al.[46] The lowest impurity concentration which was studied was 0.1 at. % Fe.[43] Even in that case, it does not seem that the dilute limit of noninteracting impurities was really reached. In spite of the resulting difficulties in analyzing properly the Kondo problem independent of RKKY-type interactions in these alloys, it can be said that Co, Fe, Mn, Cr have "good moments" in the Hartree-Fock sense when diluted in *a*-PdSi, in contrast to no-moment impurities such as Cu and Ni.[45] We will analyze further these above-mentioned data when we discuss the combined effects of Kondo and spin-glass behavior observed in these alloys.

Amorphous $Pd_{80}P_{20}$ alloys containing rather large amount of Fe, Co, Mn were studied by means of bulk and local magnetic measurements.[47,48] From the magnetic phase diagram thus obtained, one can guess that dilute Fe has a "good moment", but metallurgical problems precluded measurements in alloys with less than 13 at.% Fe. These difficulties were overcome by Hasegawa[49] who used the amorphous ternary alloy $(Ni_{0.5}Pd_{0.5})_{80}P_{20}$ as a matrix in which small amounts of Fe or Mn could be dissolved. In that matrix, which exhibits a very small Pauli-paramagnetism ($\chi \simeq 10^{-7}$ emu/g), Fe and Mn can be classified as good moment bearers. The situation seems to be the same for Fe and Cr in $Pd_{41}Ni_{41}B_{18}$[50] and for Cr and Mn in $(Ni_{0.3}Pt_{0.7})_{75}P_{25}$.[51] Dilute 3d impurities were also studied in NiPB alloys. Dilute Co was found to be nonmagnetic[52] as well as dilute Cr and V,[53] while 0.2 at.% Fe and 0.5 at.% Mn exhibit good moments.[54,55]

The experimental situation from CuZr alloys appears to be more complicated. According to Mizoguchi and Kubo,[56] iron has no moment when diluted (2 at.%) in $Cu_{57}Zr_{43}$, while Gruzalski and Sellmyer[57] reported an effective moment of 0.8 μ_B/Fe atom for 5 at.% Fe in $Cu_{60}Zr_{40}$. Discrepancies between these authors have not been resolved. It seems that for Mn the agreement is better. Mn has an effective moment of 1.6 μ_B/Mn atom in CuZr. In a more detailed study,[58] Amamou analyzed the concentration dependence of magnetic properties for Mn varying from 0.5 to 7 at.% Mn in $Cu_{60}Zr_{40}$. Up to 2 at.%, Mn was found to be in a dilute noninteracting regime. A fit of magnetization curves to a Brillouin function yielded an impurity spin value of $S = 2.2$ for one out of 6.5 Mn atoms, the other Mn atoms being nonmagnetic. By comparing with crystalline $(Cu_{0.59}\,Zr_{0.41})_{99}Mn_1$, it was deduced that one out of 4.5 Mn is magnetic in the crystalline case. These findings were suggested to arise from various local environment effects in glassy compounds. An interesting point is that the concentration range for the dilute and the spin-glass regimes could be determined without any ambiguity. The Kondo regime (with a T_K of the order of 1 K) holds up to an effective concentration of magnetic impurities of about 0.3 at.%. A spin-glass regime characterized by scaling laws[59] for magnetization is observed over an effective concentration range of about 0.3 to 1 at.% Mn. These concentration regimes are very much the same as those

Table 2
VALUES OF THE "EFFECTIVE SPIN" S FOR 3d IMPURITIES IN VARIOUS AMORPHOUS MATRICES (THE "EFFECTIVE MOMENT" $\mu_{eff} = g\sqrt{S(S+1)}$ IS OBTAINED FROM THE CURIE-WEISS CONSTANT)

Matrix	Cr	Mn	Fe	Co	Ref.
$Pd_{80}Si_{20}$	1.30	2.45	2.40	0.85	42
$Ni_{78}P_{14}B_8$	—	2.50	2.25	—	52, 55
$Ni_{40}Pd_{40}P_{20}$			1.70		49
$Cu_{60}Zr_{40}$		0.45	1.14		56, 57
$Ni_{50}Nb_{50}$			0.50		57
$(Mo_{0.6}Ru_{0.4})_{80}P_{20}$			0.25		60

determined for canonical crystalline spin glasses, such as AuFe. This casts some doubt about the conclusions drawn from the early studies of *a*-PdSi with 3d elements, namely that structural disorder would favor the existence of a single impurity regime over a concentration range much broader than in crystalline counterparts owing to such factors as weakening of d-d exchange interaction and shortening of electronic mean-free-path. These effects do exist in amorphous metallic alloys, but their importance has been often overestimated.

Finally, let us mention that iron was found to carry a small moment (1.7 μ_B/Fe atom) in *a*-$Nb_{50}Ni_{45}Fe_5$[57] and in *a*-$(Mo,Ru)_{80}P_{20}$[60] (0.5 μ_B/Fe atom). These low values of the iron moment in amorphous alloys containing transition metals of the second series have to be compared with the experimental variation of dilute iron moment in crystalline metals and alloys of the second-row transition elements.[61]

In summary, experimental results on the moment formation for 3d elements in amorphous matrices remain scarce to date. Systematic studies on dilute impurities in amorphous alloys of the same composition as a single phase crystal compound are clearly needed for a proper evaluation of the effect of structural disorder. Experimental results obtained so far in various systems are listed in Table 2. One can notice that Fe has a rather large moment in alloys containing Ni or Pd and a rather low moment in alloys containing elements from the second-transition series, which is not unexpected from the point of view of electronic band structure.

b. Kondo Behavior

It was the striking logarithmic rise in resistivity with decreasing temperature in crystalline noble metals containing 3d impurities which led Kondo to show evidence for a divergence in the third-order perturbation term in the simple s-d exchange model. This in turn has generated a subsequent interest in the local moment problem. The $-\log(T/T_K)$ behavior of the low temperature resistivity, coupled with the increasing scattering of conduction electrons by phonons at higher temperatures, produces a minimum in the total resistivity of the alloy. This resistivity minimum has become synonymous with the Kondo effect, although the gradual transition from paramagnetic to nonmagnetic regime which occurs for the impurity around T_K is probed not only by resistivity as a function of temperature, but also by a negative magnetoresistivity, a peak in the thermoelectric power, a broad Schottky-like peak in the specific heat, a modification in the temperature dependence of the initial susceptibility ($\chi \propto (T + \theta)^{-1}$), and departures of the impurity magnetization from free-spin behavior.[62] Depending on which of these latter techniques is used, the value of T_K can differ by a factor of 2 or 3 (at best) for a given system with a given impurity concentration. Thus a great significance should never be attached to the numerical values of T_K. Very small impurity concentrations are of particular importance to obtain results which are genuinely single-impurity effects, that is, characteristic of interactions between the matrix and the impurity rather than of interactions between impurities. The borderline between the two concentration regimes is given by the empirical rule "50 ppm times T_K". In most cases, this criterion is not reached, so that interactions between impurities are present whose importance depends upon the value of T_K and upon the sensitivity of the experimental technique to single-impurity phenomena or to long-range interaction effects.

The manifestations of the Kondo effect in amorphous structures were first evidenced by means of resistivity and magnetoresistivity measurements in a series of $Pd_{80}Si_{20}$ alloys containing Cr, Fe, Mn, and Co.[42,43] Actually, the matrix $Pd_{80}Si_{20}$ was probably the only alloy to deserve properly the name of Kondo alloy, owing to the fact that the palladium contained iron in the ppm range. The resistivity for $Pd_{80}Si_{20}$ exhibited a resistivity minimum at about 10 K. Later on, this anomaly was not found for alloys, of same composition, made of ultra-high purity constituents under the same quenching conditions.[63] This Kondo effect in PdSi was further substantiated by recent magnetoresistivity measurements[64,65] which showed evidence for negative contribution obviously magnetic in origin. Kästner et al.[65] noted that the Kondo anomaly was more depressed by an applied field in crystalline than in amorphous PdSi, while Dierker et al.[64] could suppress the resistivity upturn by a mild heat-treatment of the amorphous sample. Although the resistivity minimum in PdSi alloys is admitted to be definitely magnetic in origin, it seems that the exact interplay between disorder and Kondo effect is not fully understood as yet.

PdSi alloys containing 3d elements were studied over a concentration range of 0.5 to about 10 at.%, except for Fe, where the study was extended to alloys with 0.1 at.% impurities. For all these alloys, it seems that we are dealing with the interacting impurity regime. In fact, as demonstrated by more recent observations[44] these ternary alloys are rather inhomogeneous from both magnetic and atomic-scale structure points of view, which may result in the coexistence of magnetic clusters interacting like in canonical spin glasses and of isolated impurities acting as Kondo impurities. Magnetic properties are sensitive primarily to the interactions, so that they will be analyzed later on as spin-glasses, while the resistivity exhibits mainly the Kondo behavior. It is well known, even in crystalline systems, that transport properties, particularly resistivity, provide a very sensitive, often detailed, source of information. This is amplified in amorphous metallic alloys where the electron mean-free-path is of the same order of magnitude as the interatomic distance. Thus, the electrical properties can preferentially probe the details of the atomic environment over the long-range properties. For rather concentrated alloys, the magnetic contribution to resistivity was found to exhibit the Kondo logarithmic dependence around T_K while, for $T \ll T_K$, it becomes proportional to $1-(T/T_K)^2$ (see Figure 1 for a-$Cr_5Pd_{75}Si_{20}$.[63]

For seven $Ni_{41}Pd_{41}B_{18}$ alloys containing between 0.5 and 4 at.% Cr, the incremental magnetic resistivity was shown to scale with impurity concentration,[50] so that the reduced resistivity could be fitted to Hamman's formula[66]

$$\Delta\rho/\rho_0 = \frac{1}{2}\left[1 - \frac{\log(T/T_K)}{[\log^2(T/T_K) + S(S+1)\pi^2]^{1/2}}\right] \qquad (6)$$

From the value of T_K determined from the logarithmic slope, an estimate of an effective s-d interaction was obtained by using Equation 5. These values for Co, Fe, Mn, and Cr in PdSi were found to range typically between -0.20 and -0.60 eV. On the other hand, Hasegawa was able to show experimental evidence for a $T^{-1/2}$ singularity in resistivity vs. temperature of amorphous $Ni_{40}Pd_{40}P_{20}$ alloys, where small amount of Fe or Mn substituted for Ni.[49] This term was interpreted as arising from interference between multiple scattering

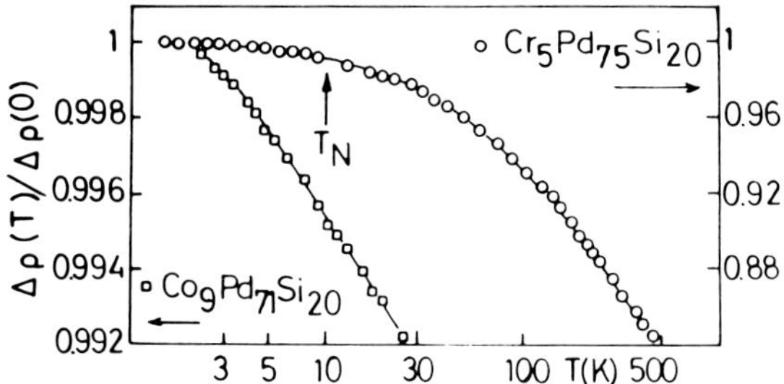

FIGURE 1. Transition metal contribution to the total resistivity of the amorphous alloys $Cr_5Pd_{75}Si_{20}$ and $Co_9Pd_{71}Si_{20}$. The values of $\Delta\rho(o)$ for $Cr_5Pd_{75}Si_{20}$ and $Co_9Pd_{71}Si_{20}$ are 149.74 and 125.87 μΩ·cm, respectively. The arrow indicates the temperature (10 K) of the susceptibility maximum for the Cr-substituted alloy. (From Tsuei, C. C. and Hasegawa, R., *Solid State Commun.*, 7, 1581, 1969. By permission of Pergamon Press Ltd, London.)

of conduction electrons by atoms in magnetic and nonmagnetic states. An incoming conduction electron first interacts with the nonmagnetic impurities to form a spin-density fluctuation system which interacts with a magnetic impurity. Observation of this additional term in low-temperature resistivity is made possible in an amorphous matrix as a consequence of a short electron mean-free-path and breakdown of conservation of electron momenta.

Several questions have to be raised about applying a simple Kondo treatment to resistivity anomalies in these concentrated magnetic alloys. When high enough to be determined unambiguously from ρ vs. T, the Kondo temperature T_K was found to be concentration dependent (roughly linearly). Such a dependence is hardly compatible with a mere Kondo situation. On the other hand, the singularities observed in the concentration dependence of the Kondo-like anomalies preclude that they could be attributed only to macroscopic inhomogeneities, and, thus, to metallurgical accidents. Several questions are involved, namely whether the Kondo anomalies are true properties of the spin-glass regime or only due to more or less isolated impurities which would remain free out of the spin-glass freezing. Further investigations have to be performed about the correlated effects of the macroscopic inhomogeneities and impurity concentration dependence in these nondiluted magnetic alloys. This will be discussed further along with magnetic properties of spin-glass and cluster-glass systems.

c. Local Environment Effects on Moment Formation

It is generally recognized that the occurrence of local moments in normal metal or transition-metal-base alloys can be strongly influenced by the local environment. This has been verified also in amorphous alloys, especially for the onset of Co moment. For example, in *a*-$(NiCo)_{78}P_{14}B_8$, Amamou[52] showed that Co atoms have no moment when they have 0 or 1 Co as nearest neighbors. When Co has 2 or 3 Co atoms as first neighbors, it carries a small moment of 0.50 μ_B/at. When the number of Co first neighbors exceeds 3, the Co carries its full moment, i.e., 1.15 μ_B/atom, as in *a*-$Co_{78}P_{14}B_8$. This result compares rather well to what was deduced for crystalline Au–Co alloys:[67] isolated Co and pairs of Co atoms are nonmagnetic; only those Co atoms included in groups of at least 3 Co atoms are magnetic. The Jaccarino-Walker model[31] was successfully applied also to the amorphous YCo alloys,[68] where the Co atoms were found to be magnetic when surrounded by at least 8 Co atoms as nearest neighbors. The requirement is weaker for Co in *a*-GdCo (7 Co first neighbors) and for Fe in *a*-FeSi alloys (6 to 7 Fe n.n.).[69] Unfortunately, comparison with crystalline counterparts is not easy, due to the difficulties of varying continuously the concentration in the crystalline case. An evaluation of the role of fluctuations in number of first neighbors for the onset of local moment in an amorphous matrix would require such a comparison, which can be achieved only in ternary compounds, such as $La(Cu_{1-x}Co_x)_5$ and $Y(Ni_{1-x}Co)_5$, for example.[68]

2. *Spin Glasses and Cluster Glasses*

The interactions between localized moments dissolved in a crystalline matrix with infinite electron mean-free-path can be readily derived by starting from the s-d exchange Hamiltonian given by Equation 4. A single impurity creates in its neighborhood an oscillatory conduction-electron spin density proportional to $(\cos 2k_Fr)/r^3$ in its asymptotic form (k_F being the Fermi momentum and r the distance away from the impurity). A second impurity (of spin S_2) will interact with the first impurity (S_1) via the polarization of the conduction electrons. This interaction, assumed to be isotropic and of infinite range, is called the Ruderman-Kittel-Kasuya-Yosida (RKKY) interaction and is expressed as:

$$H_{RKKY} \sim -\frac{J^2 S_1 S_2}{r^3} \cos(2k_F r) \tag{7}$$

Due to its oscillatory nature, the RKKY interaction between two impurities will be positive or negative depending on their distance r. For moderately concentrated magnetic alloys, this admixture of interactions of opposite sign can give rise to an original type of short-range magnetic order which is called spin glass. The ordering or freezing temperature T_M is usually defined by a sharp cusp in zero-field (ac) susceptibility. Below T_M, field-cooling effects and time relaxation phenomena are observed. Experimentalists have accumulated evidences for the complex nature of the spin-glass ordering process, namely, both the shape and the temperature of the susceptibility maximum were found to be extremely field sensitive and, in most cases, frequency dependent. Moreover, specific heat measurements exhibit no singularity at T_M, but a broad maximum, instead, at a temperature T which varies as a function of applied field. Considerable effort has developed among theorists and experimentalists in order to decide whether these phenomena are related to a true phase transition. This is still a domain of very active research.[70,71]

Before analyzing the spin-glass properties in amorphous metallic alloys, it might be of some use to characterize more accurately the different concentration regimes encountered with increasing the impurity content from the Kondo limit up to the onset of long-range magnetic order. In canonical dilute magnetic systems, the regime of interacting single spins holds for concentrations of about 1 at.%. In this regime, to which the term "spin glasses" is most appropriate, the bulk magnetic properties (e.g., magnetization and remanence) and the specific heat scaled to concentration x follow universal functions of T/x, H_{ext}/x and, in addition, T_M scales with x.[59] These scaling laws are consequences of the invariance of xr^3. For impurity concentrations larger than 1 at.%, T_M is still roughly proportional to x, and the remanence is still a function of T/x, but the magnetization M/x is no longer a function of T/x and H/x. This is due to the occurrence of small magnetic clusters, since there is a finite probability of finding two magnetic impurities as first neighbors. Magnetization measurements can still be analyzed by assuming that all the impurities are involved in clusters of average size S* and of effective concentration $x^*(x^*S^* = xS)$.[72] As long as the clusters do not overlap, the reduced magnetization is a function of T/x and H/x*, and the strength of the RKKY interaction ($V_0 \propto J_{sd}^2$) can be determined from the approach to saturation according to the model of Larkin-Khmel'nistkii-Matho.[73] The strength V_0^* of the effective interaction between nonoverlapping clusters decreases with increasing the average size of the clusters in such a way that $V_0S = V_0^*S^*$. This regime could be termed cluster glass . When the percolation concentration is approached, the clusters start to overlap and the

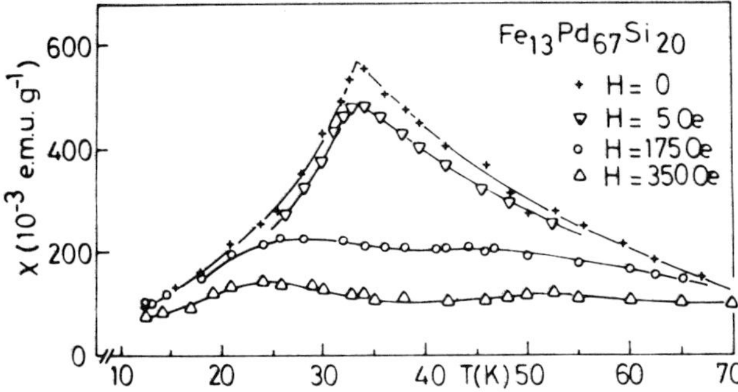

FIGURE 2. Susceptibility vs. temperature for amorphous $Fe_{13}Pd_{67}Si_{20}$. (From Zentkova, A., Duhaj, P., Zentko, A., and Tima, T., *Physica (Utrecht)*, 86—88B, 787, 1977. With permission.)

aforementioned regularities are not observed anymore. This is typically the regime of mictomagnetism.[74] Finally, for concentrations larger than the percolation, one reaches the regime of inhomogeneous long-range order, where cluster-glass or mictomagnetic effects can still be observed at low temperature. Such a progression toward a long-range homogeneous magnetic order as a function of concentration of magnetic impurities is typically found in a system such as Au–Fe, where the dilute impurity has a low T_K and the concentrated regime is ferromagnetic. Spin glass regimes occur also in systems where the dilute impurity does not have a well-defined moment (high T_K or spin-fluctuation temperature) or in systems where the magnetic moment occurs on a group of atoms (giant moments) instead of individual atoms (NiCu, VFe) or in systems evolving toward antiferromagnetism with increasing concentration of magnetic impurities. In these latter cases, distinction between spin-glass, cluster-glass, and mictomagnetic regimes would be meaningless. The term "cluster glass" would often be the most appropriate one.[75]

We first summarize the spin-glass or cluster-glass phenomena reported on various transition-metal-base amorphous alloys. We then discuss the theoretical problems raised by these experimental results.

a. Experimental Data

The first experimental work[42,43] on spin glasses in amorphous PdSi alloyed with 3d elements was performed before attention was drawn back to the spin glass problem by the discovery of a sharp zero-field susceptibility cusp in canonical Au–Fe alloys. Nevertheless, broad susceptibility maxima were observed in dc susceptibility at finite field (especially for alloys containing Cr and Mn), whose concentration dependence was poorly defined. On the other hand, the similarity of the magnetic phase diagram for amorphous FePdP alloys with that for crystalline AuFe alloys over the concentration range for emergence of ferromagnetism was underlined by Sharon and Tsuei.[47] Later on, ac susceptibility measurements on PdSi containing Fe and Co yielded sharp maxima[44] characteristic of spin-glass or cluster-glass regime (Figure 2). Subsequent Mössbauer measurements determined the concentration dependence of the short-range ordering for $1 \leq x_{Fe} \leq 7$ at.% dissolved in $Pd_{80}Si_{20}$.[76] More recent studies were devoted to the determination of the magnetic phase diagram of the amorphous $(Fe)Pd_{82}Si_{18}$ systems.[46] In all the alloys of PdSi with Fe studied so far, bulk magnetization measurements showed evidence for rather strong magnetic clustering effect at low temperature, even over the concentration range 0.1 to 1 at.% Fe.[43] It can be deduced that the true single-spin interacting regime lies below 0.1 at.% Fe and then that all these alloys belong to the family of cluster glasses. Consequently, the value V_0, the strength of the RKKY interaction, as calculated for 2 at.% Fe in $Pd_{82}Si_{18}$[46] from the approach to saturation

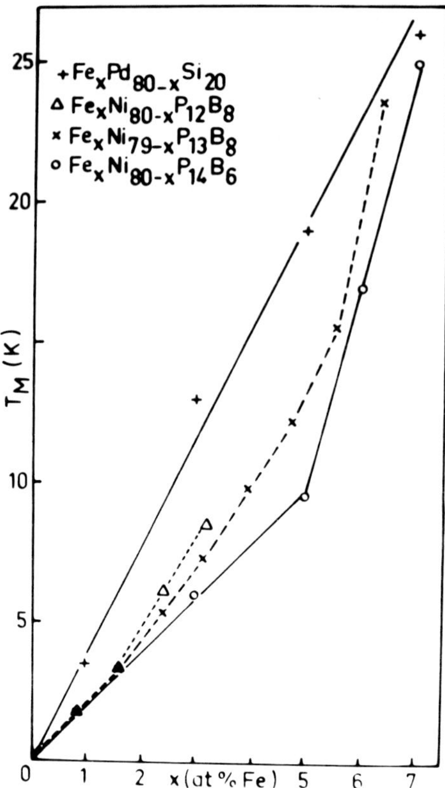

FIGURE 3. Spin-glass (cluster glass) ordering temperature vs. Fe content for iron-substituted amorphous alloys, $Pd_{80}Si_{20}$ (data from Reference 76), $Ni_{80}P_{12}B_8$, $Ni_{79}P_{13}B_8$ (data from Reference 77), and $Ni_{80}P_{14}B_6$ (data from Reference 80).

according to the Larkin's model ($V_0 \simeq 7.10^{-39}$ erg cm^3) is related in fact to an effective RKKY intercluster interaction V_0*. This can be seen clearly by comparing this value with that obtained for 0.2 at.% Fe in NiPB alloys ($V_0 = 2.45 \cdot 10^{-37}$ erg cm^3).[54] A larger value would be expected in PdSi as compared with NiPB, since the slope, dT_M/dx, of the characteristic temperatures vs. Fe concentration is twice as large in PdSi as in NiPB[77] and it has been shown that V_0 scales with dT_M/dx in spin glasses.[72] The apparent discrepancy arises from the fact that $Fe_2Pd_{80}Si_{18}$ is a cluster glass, so that the effective interaction strength V_0* is much smaller than the true single interacting spin value V_0 (V_0*S* = V_0S). The concentration dependence of susceptibility cusp temperature for Fe in PdSi and in NiPB alloys is illustrated in Figure 3.

At this point, it is interesting to refer to the studies performed by the Bratislava group through electron microscopy and Mössbauer spectroscopy measurements on (Fe)PdSi and (Co)PdSi.[44] Two sites for Fe were evidenced by Mössbauer studies, suggesting the presence of two phases in the amorphous alloys as a result of a sort of spinodal decomposition. Possible phase separation in amorphous metallic alloys will be discussed more extensively in the Section III of this chapter along with the role of eutectic composition in atomic-scale structure of transition-metal metalloid base amorphous alloys. At this point we make the following two remarks. First, the relationship between atomic-scale structure and spin-glass properties in canonical crystalline systems is an area of intense investigation through Mössbauer, EXAFS, or diffuse X-ray scattering techniques. There is no reason why amorphous

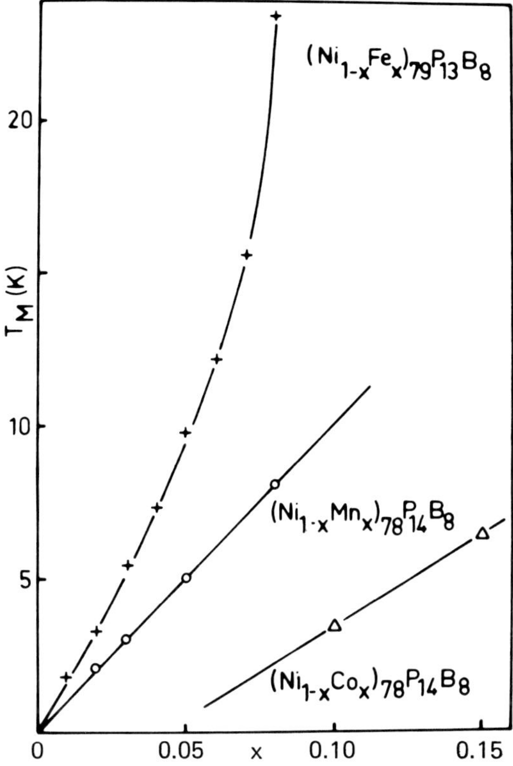

FIGURE 4. Spin (cluster)-glass temperature vs. transition metal content for Fe (Reference 77), Mn (Reference 55), and Co (Reference 52) dissolved in amorphous NiPB alloys.

systems would be free of these precipitations, chemical ordering phenomena, phase separation commonly observed in crystalline spin glasses. Second, the amorphous PdSi alloys are especially prone to alloy mixing effects.[78] The observations made in PdSi base alloys have to be extended to a broader range of amorphous alloys, for these macroscopic concentration fluctuations in an amorphous matrix might yield a clue to understand their formation and stability as well as some of their physical properties.

Spin-glass and cluster-glass regimes were studied in some detail also for NiPB alloys containing Fe or Mn. Sharp susceptibility maxima in ac field were observed whose concentration dependence is linear up to 4 at.% Fe in $Ni_{79}P_{13}B_8$[77] and up to 7 at.% Mn in $Ni_{78}P_{14}B_8$[55] (see Figure 4). Again, we are dealing with cluster glasses for Fe concentrations of the order of 1 at.%. This is evidenced by the strong concentration dependence of the strength of the RKKY interaction[54] and by the superparamagnetic behavior of the magnetization.[79] The average size of the magnetic clouds can be determined consistently either from a fit of magnetization to a modified Brillouin function or from the Souletie-Tournier scaling laws modified for cluster glasses, $M/x = f(T/x, H/x^*)$. For the magnetic concentrations between 4 and 7 at.% in $Ni_{79}P_{13}B_8$, the magnetic clusters overlap more and more as the critical concentration for ferromagnetism is approached. The same behavior was observed for dilute concentrations of Fe in slightly different matrices, such as $Ni_{80}P_{12}B_8$ or $Ni_{80}P_{14}B_6$.[80] Let us note that Chien (unpublished, quoted in Reference 76) observed the onset of magnetic hyperfine interaction in $Fe_3Ni_{77}P_{14}B_6$ and in $Fe_6Ni_{74}P_{14}B_6$ at temperatures in agreement with those determined by ac susceptibility. Specific heat measurements[81] also exhibited rounded maxima in alloys with 1 at.% Fe at temperatures close to those of the susceptibility cusps. Interestingly enough, for Fe concentrations larger than 1 at.%, a low-temperature linear coefficient in the incremental specific heat is determined and found to scale with the fraction of isolated spins up to 7 at.% iron. At higher concentrations, collective effects are clearly seen.

Finally, let us mention the experimental studies performed on spin- and cluster-glasses in $La_{80}Au_{20}$ containing Gd. From our point of view, Gd differs from 3d impurities only by the small value of J_{df} (owing to the highly localized character of the electronic f shells) and also by the fact that J_{df} is positive (as distinguished from the negative J_{sd} giving rise to the Kondo effect). For $x \leq 1$ at.%, the amorphous $La_{80-x}Gd_xAu_{20}$ alloys were found to satisfy the criteria for RKKY spin glasses (scaling laws for susceptibility and magnetization) with a value for the strength of the RKKY interaction very close to that obtained for Gd in various crystalline compounds.[72] Between 1 and about 30 at.% of gadolinium, the La–Gd–Au alloys can be classified as cluster glasses. The susceptibility cusp temperatures are roughly proportional to Gd content and the remanence scales with T/x. The average size of the magnetic clusters as a function of Gd concentration corresponds to the statistical number of Gd–Gd nearest neighbors, and the RKKY-type intercluster interactions prevail over the intracluster exchange. When the Gd concentration exceeds 30 at.%, a mictomagnetic regime occurs, where the effects of large clusters are predominant.

Summarizing, the phenomenology of spin-glass or cluster-glass phenomena is basically the same for 3d (and Gd) impurities in amorphous and in crystalline matrices. For impurity concentrations smaller than 1 at.%, single spins interact via conduction electrons according to the RKKY model. For moderately dilute alloys, spin-clustering effects are felt, but these clusters act as "superspins" up to a given concentration depending upon the matrix and upon the nature of the impurity (4 at.% for Fe in NiPB, 30 at.% for Gd in $La_{80}Au_{20}$). Beyond this concentration, percolation is progressively reached through a mictomagnetic regime. The great similarities between crystalline and amorphous systems with respect to spin-glass mechanisms raise the question as to the range of the RKKY interaction in an amorphous alloy.

b. Range of the RKKY Interaction in Amorphous Alloys

The RKKY interaction cannot be expressed in its asymptotic form (Equation 7) when the electron mean-free-path is finite. De Gennes' calculations[82] showed that the effect of the finite mean-free-path ℓ on the charge (or spin) polarization in an alloy is complex in nature and could not be reduced to an intuitive damping factor of the form $\exp(-r/\ell)$ (abusively named the "De Gennes damping factor"). Nevertheless, this exponential factor has been widely used in the literature to account for the experimental effects of finite mean-free-path in magnetically dilute disordered systems.

Such damping effects should be drastic in amorphous alloys, if the range of indirect interaction were directly related to the electron mean-free-path. It can be deduced from resistivity measurements that the mean-free-path in amorphous alloys is of the order of the nearest-neighbor atomic distance. In fact, from the existence and the properties of spin glasses in amorphous alloys, it was concluded[72] that the range and the spatial dependence (r^{-3}) of the indirect exchange interaction are not exponentially damped, and that the RKKY interaction is longer ranged than the electron mean-free-path.[83] Recent calculation of De Chatel[84] using LCAO wavefunctions showed that an exponentially damped RKKY interaction is inadequate for amorphous alloys and that in such alloys the range of the indirect interaction should far exceed the electron mean-free-path.

3. Emergence of Long-Range Inhomogeneous Magnetic Order

In most of crystalline alloys, the onset of long-range magnetic ordering as a function of magnetic impurity concentration is approached in an inhomogeneous way. Some examples include the following. In exchange-enhanced transition metals (e.g., Pd), small amounts of magnetic impurities create polarization clouds in the matrix so that magnetic ordering is obtained for very low impurity concentrations. When the impurity has a "good moment", the alloy undergoes, before long-range ordering, the sequence of intermediate magnetic regimes already described (spin glass, cluster glass, and mictomagnetism). Even when the isolated impurities do not carry localized moments (high T_K or spin fluctuation temperature), a moment results on a group of atoms once critical conditions for local environment around impurities are satisfied. In this latter case (e.g., CuNi and VFe), the emergence of long-range magnetic ordering is preceeded by a cluster-glass regime. In the aforementioned cases, the magnetic phase diagrams are phenomenologically very much the same in the vicinity of the critical concentration, whatever the real value of this concentration is. Before reviewing the various magnetic phase diagrams obtained in amorphous metallic systems, we give a broad outline on various concentrated spin-glass or mictomagnetic amorphous alloys reported in the literature. Then we will return to the problem of resistivity in magnetically inhomogeneous amorphous alloys.

a. Various Concentrated Spin-Glass or Mictomagnetic Amorphous Alloys

Spin-glass-like properties have been investigated in detail in several concentrated magnetic amorphous alloys based on transition metals. Among these amorphous compounds, MnSi provides the opportunity of an interesting comparison with the properties of crystalline MnSi.[85] In both crystalline and amorphous systems, the extrapolated paramagnetic Curie temperature is about 30 K, showing that the average interaction in both is ferromagnetic and of the same amplitude. But magnetic ordering in the two cases is different. In its cubic crystalline form, MnSi orders helimagnetically at 30 K, while amorphous MnSi undergoes a spin-glass transition at 22 K. The concentration dependence of the spin-glass temperature was determined in a-$Mn_{100-x}Si_x$ alloys for x varying between 25 and 75 at.% Si.

Amorphous YFe_2 has been the subject of many investigations including magnetic susceptibility, Mössbauer, spin resonance, neutron scattering studies.[86] Forester et al.[86] focused on the complex nature of the spin-glass behavior in amorphous YFe_2. A truly thermodynamic spin-glass transition is defined at 58 K. At a lower temperature ($T \simeq 20$ K), the susceptibility vs. T curve shows a break which is believed to arise from magnetic blocking of individual spins or regions of correlated spins. This latter anomaly would be a "pseudo-transition" of cluster — freezing type. Spin-resonance measurements on bulk sputtered and thin films of coevaporated YFe_2 suggest that the surface ferromagnetic layer is another amorphous phase of the alloy.[87]

b. Inhomogeneous Character of the Approach to Long-Range Magnetic Ordering in Amorphous Alloys Where Dilute Ni, Co, and Fe Carry No Localized Moments

The critical concentration range for the onset of long-range magnetic ordering was determined for amorphous Ni base-(NiP,[88] NiAg,[89] NiY,[90] NiPB[40]), Co-base(CoP,[91] CoAg,[89] CoSn,[92] CoY,[68] CoLa, CoZr,[93] and CoTi,[94]) and Fe-base-(FeLa, FeLu, FeZr,[93] FeGe,[95] FeSi,[69] FeSn,[96] FeTh,[97] FeY,[98] and FeB[99]) amorphous alloys. The approximate critical concentrations for these various amorphous systems were interpreted according to different authors either in terms of charge transfer model or within a local environment picture. In most cases, especially in the NiY, NiPB, and FeSn amorphous alloys, a cluster glass regime was clearly identified before going through the percolation region. In the NiY alloys, for example, cluster glass anomalies were characterized by a susceptibility cusp and a specific heat maximum at low temperature.[90] Magnetic phase diagrams around the critical concentration for emergence of long range magnetic ordering are thus expected to be formally similar to those obtained in amorphous alloys where dilute Ni, Co, or Fe carry localized moments. Finally, let us mention another family of cluster glasses, namely amorphous alloys

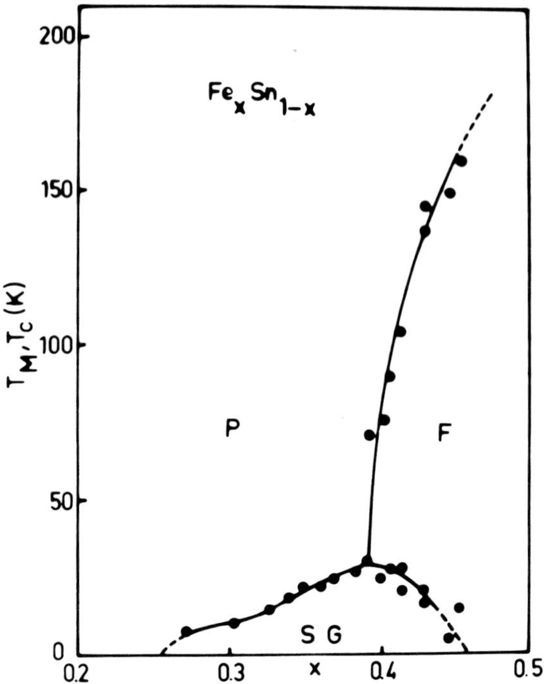

FIGURE 5A. Magnetic phase diagram for amorphous FeSn alloys. (From Teirlinck, D., Piecuch, M., Geny, J. F., Marchal, G., Mangin, Ph., and Janot, Ch., *IEEE Trans. Magn.*, MAG-17, 3079, 1981. With permission.)

with a random admixture of ferromagnetic and antiferromagnetic bonds, such as (Fe,Mn)PC or (Fe,Mn)PB alloys over the critical range for disappearance of long-range magnetic order.

c. Magnetic Phase Diagrams around the Percolation Region

Ferromagnetic spin-glass transitions around the critical concentration for onset of ferromagnetism were observed through low-field susceptibility measurements in three families of amorphous magnetic alloys, namely alloys with good moments (Ni–Fe base alloys), alloys with discontinuous appearance of moments (Ni–Co base alloys and FeSn[96]) and alloys with competing interactions (Fe–Cr, Fe–Mn and Co–Mn base alloys).[100] The magnetic phase diagrams obtained in these three typical cases are plotted in Figures 5A, B, and C. The ordering temperatures (T_C and T_M) were determined from ac or dc susceptibility measurements. As to spin-glass ferromagnetic transitions observed for concentrations larger than the critical concentration, additional microscopic measurements are clearly needed to understand the specific character of this new low-temperature phase which has all the attributes of a cluster-glass (dc and ac susceptibility and hysteresis behavior), but little resemblance, probably, to the canonical dilute spin-glass. However, it is noteworthy that low-temperature specific heat data[101] on Fe–Ni base amorphous alloys exhibited a considerable enhancement of the linear term for concentrations above the percolation (Fe content = 11, 13, 16, and 20). Such an enhancement is reminiscent of that observed for spin or cluster glasses. It was interpreted as arising from remaining nonmagnetic clusters within an otherwise magnetic matrix. Note also that no spin-wave contribution was detectable in the alloy with 20 at.% Fe. In view of the relative insensitivity of calorimetric measurements to the spin freezing for dilute canonical spin glasses, more detailed information about the spin-glass ferromagnetic transition is not expected to be obtained from specific heat data. Additional evidence for

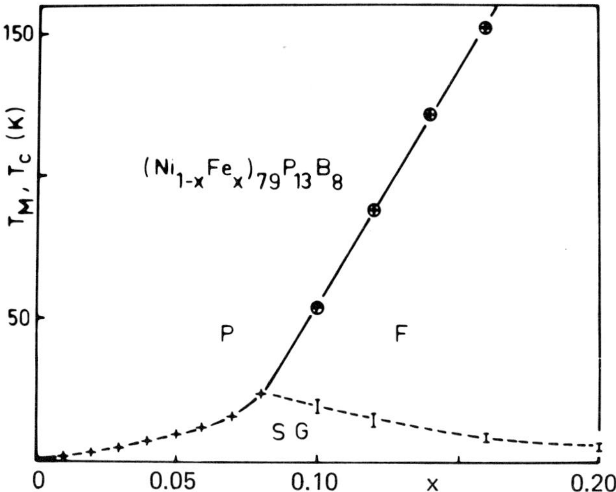

FIGURE 5B. Magnetic phase diagram for $(Ni_{1-x}Fe_x)_{79}P_{13}B_8$ alloys over the critical concentration region for onset of ferromagnetism. (Data from Reference 77 and Loegel and Durand quoted in Reference 24.)

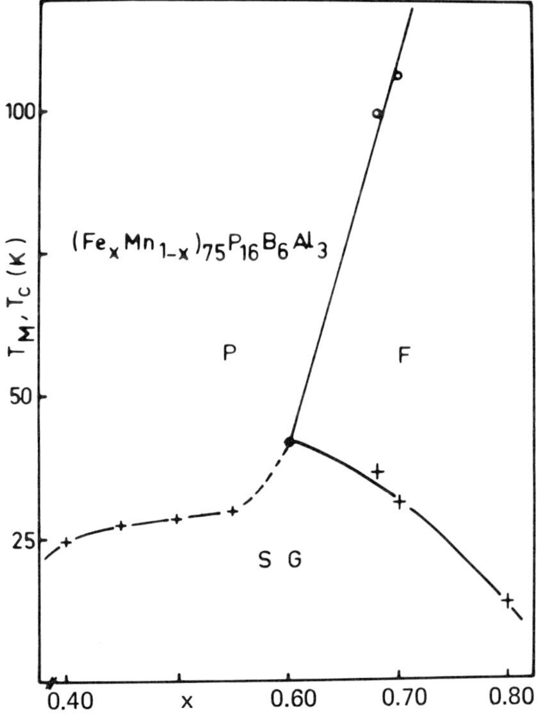

FIGURE 5C. Magnetic phase diagram for $(Fe_xMn_{1-x})_{75}P_{16}B_6Al_3$ alloys around the critical concentration region for onset of ferromagnetism. (From Yeshurun, Y., Salamon, M. B., Rao, V. K., and Chen, H. S., *Phys. Rev.*, 24, 1536, 1981. With permission.)

reentrant ferromagnetism is these amorphous alloys was yielded by recent measurements including high pressure, temperature dependence of the spin-wave stiffness constant through ferromagnetic resonance, and inelastic neutron scattering experiments.[100]

The magnetic phase diagrams obtained for various amorphous magnetic systems are schematically the same as those determined recently for different kinds of crystalline alloys.[102] One is then led towards an uniform magnetic phase diagram for magnetic alloys with mixed types or order, be they crystalline or amorphous. These double magnetic transitions have some theoretical support. They were predicted for the exactly soluble infinite range Ising model of spin-glasses and for a Heisenberg model with interactions restricted to a certain number of nearest neighbors. The conditions for the existence of the paramagnetic, ferromagnetic, and spin-glass phases were analyzed for amorphous systems, where the exchange is approximated by an extended molecular field and where the structural disorder is described by a random uniaxial anistropy.[103]

Amorphous transition-metal base alloys were found to constitute ideal three-dimensional random-alloy systems for a detailed experimental study of a multicritical behavior. Indeed, from scaling analysis of magnetization measurements in Fe(Mn)PBAl alloys,[100] multicritical, spin-glass, and critical exponents were determined along with the critical properties of the spin-glass-ferromagnetic line. It was found experimentally and justified from calculation that the slope of the spin-glass-paramagnetic phase boundary around the tricritical point varies depending on the transition elements involved in the appearance (or disappearance) of ferromagnetism.[104] Nevertheless, several points are still open to discussion. Is the paramagnetic spin-glass line really a line of phase transitions? What is the exact nature of the low-temperature phase of the reentrant ferromagnets as compared with a true spin-glass regime? This is an area of current investigations.

d. Implications for Resistivity of Magnetic Amorphous Alloys

As can be seen above from magnetic phase diagrams, alloys whose magnetic content is slightly larger than the critical concentration for ferromagnetism are metastable at low temperature against reversal of individual isolated spins or misaligned spin clusters. For example, in the $(Ni_{1-x}Fe_x)_{79}P_{13}B_8$, such a situation holds for iron concentrations up to $x = 0.20$ at least. A natural question arises as to the ground state of alloys with $x = 0.3$ to 1.0. At variance with the case of non-S rare-earth base amorphous alloys where electric field gradient effects are of particular importance,[103] the demagnetized state in most of the concentrated magnetic alloys based on transition metals is likely to be ferromagnetic. But in view of the preceeding discussions, it also seems rather natural that these alloys would remain magnetically inhomogeneous over the whole concentration range. Resistivity vs. temperature could be the most sensitive property to probe these inhomogeneities inside the otherwise ferromagnetic matrix.

It was noted very early that the Kondo-like minimum in resistivity survives when impurity concentration increases beyond the spin-glass (cluster-glass) regime in "good moment" amorphous alloys.[42] A resistivity minimum was observed also in the ferromagnetic $Fe_{80}P_{13}C_7$ amorphous alloy[105] at a temperature (10 to 20 K) far below T_c. Kondo-type anomalies were reported also for amorphous systems where magnetic moment appears in form of magnetic clusters in the vicinity of the concentration for the onset of ferromagnetism[88] (NiP). Resistivity minima in amorphous alloys[63,106] have proliferated in such a way that they were suggested to be a signature of the structural disorder resulting in low-lying energy tunnelling states.[107] However, it is an almost universal rule that alloys containing an element capable of carrying a moment show these low temperature anomalies, while those with nonmagnetic constituents show no such anomalies. This argues strongly for a magnetic origin.

Before discussing the specific character of these anomalies in amorphous metallic alloys, let us recall that resistivity minima were reported (besides the trivial Kondo-case) for a large variety of crystalline magnetic systems, namely in transition metals with moderate exchange-enhancement (PtMn, PtCo), in "less-dilute" magnetic alloys when the dilute impurity exhibits broad spin-fluctuation behavior (RhFe), in alloys with competing interactions (NiCr), in concentrated alloys of the NiCu type (NiCu, NiV, VFe), in disordered compounds in the vicinity of appearance of ferromagnetism (NiAl, CoAl, FeAl), and in off-stoichiometry ferromagnetic compounds with high T_c (Fe_3Si,Fe_5Si_3).[108]

In view of the experimental situation in crystalline systems, the amorphous magnetic alloys can be characterized, as far as the transport properties are concerned, by several typical aspects. First, from structural point of view, the topological disorder is expected to favor the existence of magnetic inhomogeneities over a larger concentration scale than in the crystalline magnetic phase diagrams, together with local concentration fluctuations. Second, from a chemical point of view, most of the ferromagnetic amorphous alloys are off-stoichiometric with respect to the concentration of stable crystalline compounds. Third, the electron mean-free-path in amorphous materials is extremely short, so that the electrons will be practically ignorant of magnetic phenomena involving distances larger than about 5 Å, and, consequently they will be very sensitive to scattering off isolated impurities and small magnetic clusters.

This last point is clearly illustrated by the concentration dependence of the logarithmic resistivity coefficient in amorphous $(Ni_{1-x}Fe_x)_{79}P_{13}B_8$, in connection with the magnetic phase diagram.[77,108] The logarithmic coefficient increases up to about $x = 0.04$, which corresponds to the upper limit of small nonoverlapping Fe clusters. Beyond $x = 0.04$, the logarithmic term decreases steeply as the magnetic clusters overlap progressively to form an infinite cluster around the percolation concentration. The same concentration dependence was found for the resistivity minimum T_m in $(La_{1-x}Gd_x)_{80}Au_{20}$ alloys.[72,109] The value of T_m is maximum for a Gd concentration corresponding to the limit of the simple cluster-glass regime. It is reduced to a constant value over the ferromagnetic regime where isolated impurities and clusters seem to remain constant in quantity and in size. Such a maximum correlated to mean-free-path vs. magnetic cluster size is reminiscent of the maximum observed in resistivity during clustering in a phase separation process, where the maximum is known to occur when the electron mean-free-path equals twice the cluster radius.[110] In both of the above-mentioned diagrams, the resistivity minimum occurs at temperatures (T_m) higher than those of the susceptibility maxima T_M when the impurity concentration is below the percolation. Above the percolation, $T_m < T_c$, T_m and T_M coincide roughly for $x = 0.08$ and $x = 0.60$, in $(Ni_{1-x}Fe_x)_{79}P_{13}B_8$ and in $(La_{1-x}Gd_x)_{80}Au_{20}$ alloys, respectively; those are the concentrations which correspond to emerging long-range ferromagnetic order (Figures 6 and 7).

Concentration dependence of low-temperature resistivity anomalies was studied also in several other metallic glass systems containing Fe and Ni.[108] The authors are not unanimous in interpreting the data. Rao et al.[111] proposed an alternative explanation to the Kondo-like behavior in terms of a magnetic Ziman scattering model suggested before by Fert et al.[112] For others,[113] the magnetic contribution might be superimposed with two-level-system phenomena. At this stage, it is important to refer to the strong compositional effects observed also in ferromagnetic Fe-base metallic glasses containing Mn, Cr, and V.[114] The concentration effect observed in low-temperature resistivity of amorphous Fe_xAu_{1-x} evaporated alloys is of particular significance.[115] Indeed, resistivity minima were observed only over the concentration range where Fe is progressively losing its moment.

All these results seem to favor an explanation of these low-temperature anomalies in terms of magnetic inhomogeneities as initially suggested by Tsuei and Hasegawa.[12] Several attempts at a theoretical formalism to describe the electron scattering from these isolated spins or clusters inside a ferromagnetic matrix, were reviewed by Tsuei.[63] A more recent model[116] assumes superexchange interactions between two next-nearest neighbor magnetic atoms which are separated by a metalloid atom. Such an "ad hoc" explanation does not seem to hold for resistivity minima in ferromagnetic alloys containing no metalloids. Resistivity minima in amorphous ferromagnets were also produced within the multiple scattering approximation.[117] Fert et al.[112] showed that a positive contribution to the resistivity from

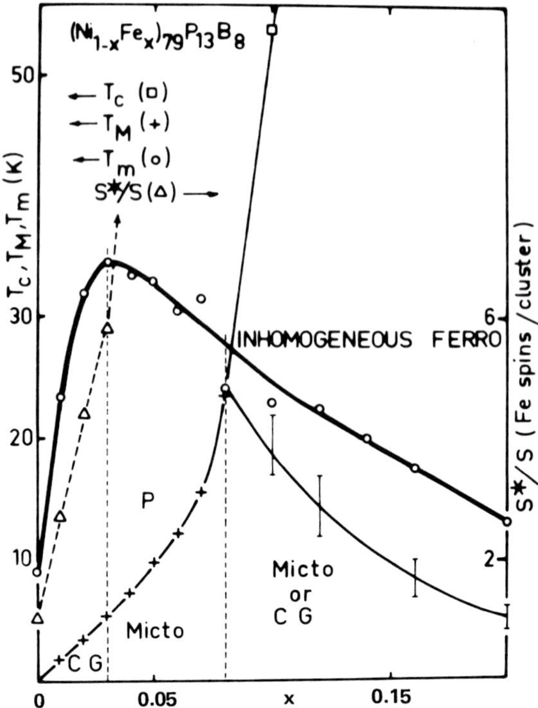

FIGURE 6. Resistivity minimum temperature (T_m), susceptibility maximum temperature (T_M), Curie temperature (T_c), vs. iron content in amorphous $(Ni_{1-x}Fe_x)_{79}P_{13}B_8$ alloys.[77,24,108] Dashed line indicates the concentration dependence of the iron cluster size S*/S over the cluster glass regime.[54] (Above $x_{Fe} = 0.04$, the assumption of nonoverlapping Fe clusters does not hold anymore.[54])

magnetic ordering can be ascribed to coherent exchange scattering by neighboring magnetic ions. One can usefully refer also to the theoretical studies of Kondo-type anomalies in less-diluted or concentrated crystalline systems.[118]

In summary, the proper effect of structural disorder on magnetic properties of dilute or concentrated amorphous alloys cannot be easily distinguished from chemical or alloying effects. In particular, some early conclusions about the attenuation of exchange interaction, or about the range of the RKKY indirect coupling, or about a drastic reduction of the paramagnetic Curie temperature[119] in amorphous dilute or less-dilute magnetic alloys need to be taken with care. However, it seems to be clear that, in the amorphous alloys, the inhomogeneous character of the magnetism is amplified and plays a bigger role than in the equivalent crystalline systems. This is particularly evidenced in concentrated alloys. The resistivity constitutes a very sensitive probe for these magnetic inhomogeneities, owing to the very short electron mean-free-path.

B. Magnetically Ordered Amorphous Alloys

Ferromagnetism in amorphous transition-metal base alloys is related to d-electrons which are distributed in energy bands defined by their energy density of states N(E). As in any metallic systems, the Stoner criterion for ferromagnetism,

$$U_{eff} N(E_F) < 1 \tag{8}$$

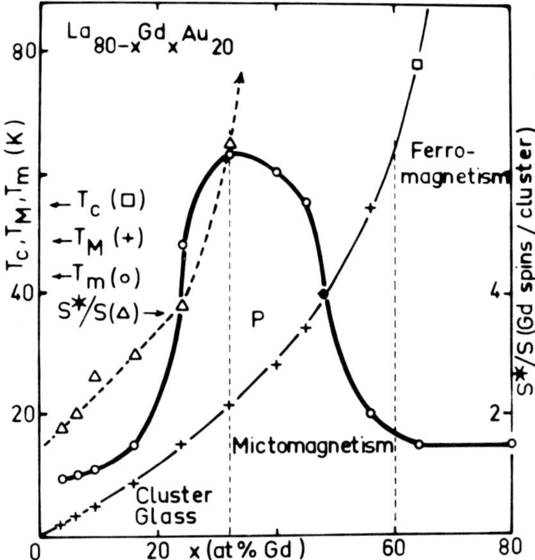

FIGURE 7. Temperatures for resistivity minimum (T_m), susceptibility maximum (T_M) and Curie transition (T_c) vs. Gd content in amorphous $La_{80-x}Gd_xAu_{20}$ alloys.[72,109] Dashed line indicates the concentration dependence of the average size of Gd clusters (S*/S) over the cluster-glass region.[72] (For alloys with more than 32 at.% Gd, the Gd clusters extend beyond the first atomic shell around a Gd atom, so that the Gd clusters overlap.)

is relevant as a first order approximation in amorphous metals as well. Application of the band (itinerant electron) model of ferromagnetism to amorphous alloys require a knowledge of the influence of structural disorder on the density of states curve and on the effective electron interaction U_{eff}.

1. Most of the stable, amorphous ferromagnets of interest are made of Ni, Co, and Fe alloyed with 10 to 30% of s-p elements. Recent photoemission studies performed on this family of alloys were very conclusive about the strong d-character of the electronic states at E_F. On the other hand, the lack of periodicity results in a general smoothing out of the singularities occurring in the density of d-states curves for crystalline transition metals, thus leading to slight increase or decrease of N(E). Finally, as noted before, all the authors agree that alloying effects result in a hybridization of the 3d electrons with the outer p electrons of the solute atom, rather than in a charge transfer mechanism.
2. Most likely, the effective interaction U_{eff} is not every sensitive to amorphicity in these alloys. The intra-atomic Coulomb interaction U is independent of the atomic environment. The many body correlations which reduce U to U_{eff} depend on the band structure which is slightly affected by the topological disorder. As emphasized by Wohlfarth,[11] there is little doubt that magnetic properties of transition-metal base amorphous alloys have to be discussed within the framework of the itinerant magnetism model. Since many aspects of ferromagnetism in these alloys have been reviewed previously, we only give a broad survey of fundamental bulk and local properties at zero temperature and at finite temperature.

1. Magnetic Properties at Zero Temperature

a. Saturation Moments

The study of "pure" amorphous Ni, Co, and Fe obtained in form of thin films was undertaken very early.[11]

Accurate magnetic measurements were made difficult by the unstable character of these materials against crystallization and by the presence (especially for Ni) of undesirable, but unavoidable, gaseous impurities. Complete references can be found in the review of Grundy.[21] In spite of recent efforts, the experimental state of affairs can be summarized as follows. Structural disorder has little, if any, effect on magnetic moment of amorphous Co. For amorphous Ni, the attenuation of saturation moment as compared with crystalline Ni seems to be about of 20 to 30%. As expected, the reduction is more drastic for the amorphous Fe moment (at least 50%).[120,121] It is well-known indeed that the band structure of Fe is very sensitive to the crystal structure (interatomic distances, number of Fe first neighbors). Bcc α-Fe is ferromagnetic with a moment of 2.2 μ_B, while hexagonal ϵ-fe, obtained by applying pressures of the order of 130 kbar, is not magnetic and, possibly, superconducting.[122] Pure fcc γ-Fe is usually regarded as an antiferromagnet, although small variations of lattice parameters can make γ-Fe a weak ferromagnet ($\mu = 0.50\ \mu_B$, $T_c = 900$ K) analogous to crystalline Ni.[122] On the other hand, Fe in solution in a fcc environment is found to carry a very large moment (2.8 to 2.9 μ_B/at). This extreme sensitivity of the Fe moment to crystal modifications is the key for understanding the Invar anomalies. It can be conjectured then that amorphous Fe (with fluctuations around an average number of 12 n.n.) will lie somewhere in between α and γ-Fe (values reported for μ in *a*-Fe range between 0 and 1 μ_B/at). As yet, the magnetic ground state of amorphous Fe has not been ascertained.

Let us consider now the concentration dependence of the moment in binary amorphous Ni, Co, or Fe base alloys. We briefly recall the information about the electronic structure of an alloy that one can expect to obtain from magnetization measurements. Then, we discuss the data reported on amorphous alloys of Ni, Co, and Fe with s-p elements. Finally, we will analyze the influence of substitution of one transition metal for another.

For moderately concentrated alloys, the mean magnetic moment $\bar{\mu}$ (in Bohr magnetons per atom of alloy) varies linearly with the concentration x of nonmagnetic impurity:

$$\bar{\mu} = \mu_0 + ax \qquad (9)$$

μ_0 being the matrix moment and "a" a constant.

If alloying results in a mere dilution effect, one has $a = -\mu_0$. If the impurity causes some change in the electronic structure of the matrix, one has $a = -\mu_0 - p$.

The physical meaning of "p" is not trivial. Very often, for alloys of transition metals with s-p elements, p is taken as an indication of the number of s-p electrons filling up the d bands of the transition metals. In most cases, the best recommendation for such a charge transfer evaluation is simplicity. But, such an interpretation is not always compatible with the electronegativity difference between the host and the solute. In addition, this picture relies on a rigid band approximation, whose basic assumptions and shortcomings were critically discussed a long time ago.[123] Furthermore, photoemission measurements failed to provide any evidence of a charge transfer of the magnitude apparently required by the concentration dependence of the alloy magnetic moment. In fact, the value of "p" obtained from magnetization studies can yield useful qualitative information about the charges $Z\uparrow$ and $Z\downarrow$ displaced by the solute in the host subbands, by writing $p = Z\uparrow - Z\downarrow$ and by using the Friedel's sum rule ($Z = Z\uparrow + Z\downarrow$) which states that the excess charge Z introduced by the solute is completely screened to save local charge neutrality in a metal.[123] Screening of s-p elements in transition metals still is a subject of theoretical investigations.

In most of the binary amorphous alloys of transition metals with sp elements studied so far, the dilute impurity region was not reached. One is then forced to describe trends. For Ni base systems, it seems that the depression of the Ni moment over the concentrate impurity concentration range is less severe in amorphous alloys than in crystalline counterparts. In crystalline NiP, the magnetization vanishes for 15 at.% P,[124] while for the same P concentration, the amorphous equivalent is still weakly ferromagnetic.[88] The same phenomena were observed in the NiY system, where YNi_5 is a Pauli paramagnet in its crystalline phase, while it has a T_c of 27 K in its amorphous modification.[90] No satisfactory explanations for these phenomena were proposed as yet. Let us note also that for amorphous NiB alloys[125] and for the metal-metal NiZr amorphous alloys[126] the magnetization disappears for x_B = 27 at.% and for x_{Zr} = 10 at.%, respectively. The concentration dependence of the magnetic moment was found to be very much the same in amorphous and crystalline phases of CoSi and CoP alloys.[91] In amorphous CoB alloys, the effect of structural disorder on magnetic moment seems to be negligibly small.[127] The moment variation rate in the amorphous CoZr system[126] is practically the same as in CoB alloys. As was first noted by Felsch,[120] binary Fe base amorphous alloys have a particular behavior with respect to the impurity-concentration dependence of magnetic properties. At low concentrations (less than about 10 at.% of s-p elements), the Fe moment starts to collapse with increasing Fe content. This was first observed in amorphous FeO, FeSi, FeAu alloys,[120] and more recently in FeZr alloys,[126] where the Fe moment obtained for x_{Zr} between 9 and 12 at.% was found to extrapolate linearly to zero for pure amorphous Fe. The variation of Fe moment with B content was studied in detail by many authors in both splat-cooled[18,128] and sputtered[99] amorphous FeB alloys. When analyzed in detail, the concentration dependence for Fe moment in the FeB system is rather complex. Up to 12 at.% B, metastable solid solutions were obtained, where B was found to be substitutional in the bcc Fe lattice.[129] For B concentrations larger than 12 at.%, the alloys were amorphous, without any drastic change in the moment versus boron content curve. Some anomalies were observed within the concentration range of eutectic composition. Analysis of the moment depression within the crystalline range shows that the impurity screening occurs mainly with the d$\downarrow$ Fe subbands, similar to the case of P in crystalline Fe.[130] The effect of P[24], C[131] and Sb[132] on Fe moment in binary amorphous alloys was also investigated over concentration ranges which do not allow firm conclusions in terms of electronic structure. In amorphous FeSi, FeGe, and FeSn alloys,[69,96] the Fe moment was shown to increase with Fe content above the critical concentration (about 40 at.% for all three alloys) at a rate similar to that observed for dilute Si, Ge, and Sn in solid solution with bcc Fe.

Most of the transition-metal base amorphous alloys obtained by liquid quenching contain at least two glass-formers. The different influence of various s-p elements on the *TM* moment can be readily evidenced by substituting one glass-former for the other one with a constant *TM* content, or by changing only the respective concentrations of *TM* and one glass former (the other one remaining unchanged). This was first investigated in amorphous FeP alloys containing Ga, Ge and As.[133] Later on, the FePB system was studied in some detail.[134] More recently, interest on "metalloid" influence has spread over a very large variety of ternary amorphous alloys with at least two glass-formers,[17,18,135,136] including alloys such as FeBC, FeBAl, FeBSi, FeBGa, FeBSiC, FePC, FePSi, FePGe, FePAl, CoBC, CoBSi, CoBP, CoBGe, and CoBAs. Several attempts were made toward understanding these "metalloid" effects in terms of charge transfer or atomic size influence (Figure 8). None of them are entirely successful. This might be due to several reasons. First, the substitution of one glass-former for another one does not result in a simple linear variation of the moment as was demonstrated for amorphous FePB.[134] This is due to changes in the short-range order as a function of the respective concentrations of glass-formers. Second, this nonlinear variation of the moment might explain the discrepancies between various published data. Thus, for s-p elements having the same outer electronic configuration s^2p^2, the atomic size effect was

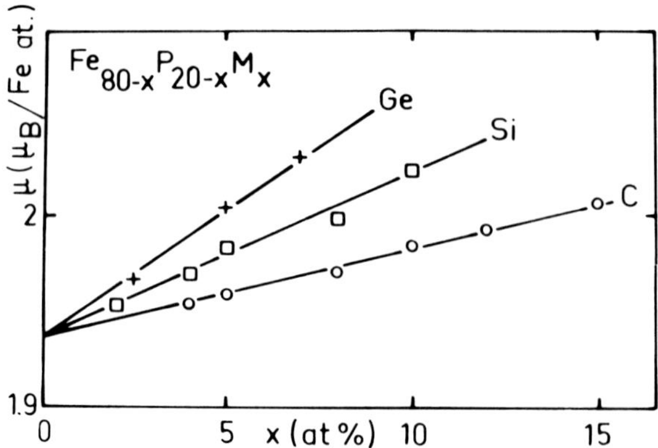

FIGURE 8. Concentration dependence of the magnetic moment per Fe atom in amorphous $Fe_{80}P_{20-x}M_x$ alloys, with M = Ge, Si, C substitute for P.[135]

reported[135] to be predominant on the moment depression in FeBM (M = Si, Ge, C) alloys according to the sequence C < Si < Ge, while Luborsky et al.[137] obtained a stronger depression for Si, than for C, giving then the sequence Si < Ge < C. Nevertheless, some preliminary conclusions can be drawn. First, the moment attenuation is somehow proportional to the number of outer electrons, for example, the ordered set P < C < B follows the sequence of the electronic configuration $s^2p^3 < s^2p^2 < s^2p^1$. But a quantitative correlation cannot be expressed by simple rules. Second, size effects have to play some role. But, the determination of atomic radii for s-p elements in these amorphous alloys is not straightforward, as has been pointed out by Johnson and Williams.[138] Third, in the absence of any accurate knowledge of the atomic-scale structure in these complex materials, any simple explanation of electronic processes will remain conjectural.

It is interesting to note at this point that the moment per Fe (Co) atom was found practically identical in amorphous and crystalline modifications of Fe (Co) compounds such as Fe_3B, Fe_3P,Co_3B, Co_2B.[135] The equilibrium phases Fe_3Si and Fe_5Si_3 exhibit the same Fe moment as in the amorphous counterparts. For less concentrated Fe compounds, the local environment seems to play a predominant role: crystalline FeSi is nonmagnetic, while in amorphous FeSi, Fe carries a moment.[69] Differences are more striking for FeSn and $FeSn_2$ owing to the antiferromagnetism which dominates in the crystalline compounds.[96]

The concentration dependence of the magnetic moment in amorphous alloys where a transition metal substitutes for another transition metal was studied by many authors. Mizoguchi et al.[139] performed the first systematic investigation of the effect of TM substitution on a series of amorphous $(Fe_{1-x}TM_x)_{80}P_{10}B_{10}$ (TM = Ni, Co, Mn, Cr, V) alloys. Values for the magnetic moment per TM atom were plotted as a function of the number N of TM outer electrons. The Slater-Pauling curve thus obtained was very similar to that obtained for transition metal borides.[140] Main differences between these Slater-Pauling curves and those of pure transition metals included (1) a shift of the maximum toward lower value of the number N of outer electrons ($\Delta N \simeq 0.3$) and (2) the absence of the Invar depression occurring over the concentration range between bcc and fcc phases. Otherwise, one observes in the case of (crystalline or amorphous) borides the same linear decrease of the moment between Fe and Ni as in the pure TM case, along with the same moment depression on the low N side of the maximum (Mn, Cr, V). More recently, impurity states of dilute transition metals (x = 0.01 to 0.04) substituted for Fe in $Fe_{79}P_{13}B_8$ were studied by means of magnetization measurements.[114] Analysis of the data within the framework of Friedel's model[123] suggested that the strong depression of Fe moment caused by Mn could be explained by a half-empty

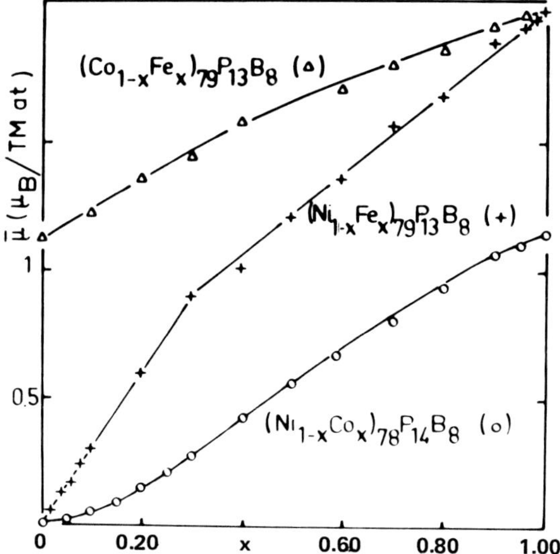

FIGURE 9. Mean magnetic moment per transition-metal atom as a function of x in amorphous $(Co_{1-x}Fe_x)_{79}P_{13}B_8$ (data from Reference 174), $(Ni_{1-x}Fe_x)_{79}P_{13}B_8$ (data from Reference 77), and $(Ni_{1-x}Co_x)_{78}P_{14}B_8$ (data from Reference 52) alloys.

half-empty virtual-bound-state at the Fermi level. For Cr and V, the virtual-bound-state should be practically empty and lie above E_F.

The (FeNi) base amorphous alloys have been the subject of much interest. In the $(Ni_{1-x}Fe_x)_{79}P_{13}B_8$, the concentration dependence of the mean amount per metal atom was accounted for[77] by assuming that Fe carries a constant moment of about 2 μ_B, while Ni is progressively polarized to reach, at around x = 0.30, a constant value of about 0.4 μ_B/atom. The variation of magnetic moment was also studied in (Fe, Ni) alloys with $P_{16}B_6Al_3$, with $(P,B)_{20}$, with $(SiB)_{\sim 20}$, with B_{20}, with P_{20} and with B_{14}.[17,18] These last alloys are of particular interest, since the concentration dependence of the magnetic moment is reminiscent of the Invar anomaly observed in crystalline FeNi alloys.[141]

In (Fe,Co) base amorphous alloys, the moment per metal atom deviates slightly from a dilution law assuming about 2 μ_B/Fe atom and about 1 μ_B/Co atom. Such a substitution was studied with various glass-former combinations, including, in addition to those mentioned above: $P_{13}C_7$, $(SiB)_{\sim 20}$ and $P_{13}B_8$.[17,18] Interest has been focused on Co-substituted amorphous alloys mainly because magnetostriction vanishes on the Co rich side. The concentration dependence of the moment was studied in detail for $(Ni,Co)_{78}P_{14}B_8$ alloys[52] (Figure 9). A complete substitution study of Mn for Fe in $(Fe,Mn)_{75}P_{15}C_{10}$ alloys raised the question as to the possibility of having an antiferromagnetic ground state in an amorphous matrix.[142] The effect of Mn on Co moment was studied in the CoSiB system.[143] Aso et al.[144] compared the effect of Cr on Fe moment in various Fe-based amorphous alloys and concluded that Cr was likely to carry no moment in these systems. Finally, let us refer to some detailed studies of the (Fe,Mo) substitution on magnetic properties of amorphous alloys with B_{20} and $P_{16}B_6Al_3$.[145] The effect of Mo in depressing Fe moment was found to be comparable to that observed for Cr.

b. Local Magnetic Properties

Low-temperature local measurements such as neutron scattering, Mössbauer and NMR spectroscopy are of particular importance to probe the ground state of a magnetic system.

Neutron experiments on amorphous alloys have remained rather scarce to date.[146] A complete review of NMR and Mössbauer investigations[22,23,147] on transition-metal base amorphous alloys would be far beyond the scope of this chapter. We rather concentrate on three points, namely the magnetic correlations, the proportionality between the average hyperfine field (hff) and the average moment of the transition metal and the distribution of hff.

Neutron measurements of magnetic correlations were recently[148] carried out on amorphous $Fe_{83}B_{17}$ and PdSi alloys containing Fe and Co. From the magnetic reduced radial distribution function, three parameters were obtained:

1. The position of the first peak at r = 2.58(± 0.05) Å, in good agreement with the value from X-ray measurement (r = 2.573 Å).
2. Oscillations indicating ferromagnetic alignment are shown over a range of about 12 Å, which is commensurate with the range of variations in atomic structure ($r_e \simeq 15$ Å).
3. The periodicity of the oscillations is 2.06 Å, which can be regarded as the magnetic structural correlation length. Rather surprisingly, the magnetic correlations appear to have a structure based on fcc γ-Fe (n.n. distance 2.57 Å, planar spacing 2.02 Å).

Magnetic disorder was concluded to be small. PdSi base magnetic alloys exhibited an antiparallel moment correlation at low temperature. This feature might help reconcile the high value of the moment (5.7 μ_B/Fe atom) deduced from neutron scattering in the paramagnetic region with the low temperature saturation moment ($< 1.8\ \mu_B$). Actually these latter alloys were found to be on the borderline between mictomagnetism and ferromagnetism in the magnetic phase diagrams.

In inelastic neutron scattering experiments on amorphous CoP alloys, it was observed that the hyperfine splitting at ^{59}Co was proportional to the average moment per Co atom. This was confirmed later by ^{59}Co NMR measurements on the same family of alloys. It was noted that such a proportionality (with a value for hyperfine coupling constant very close to that for pure transition metal) was a common feature occurring in a very broad variety of transition-metal base crystal compounds with s-p elements, in marked contrast to dilute disordered transition-metal base alloys. It was then suggested that electronic band structure and hyperfine field mechanisms should be very similar in concentrated amorphous alloys and in crystalline compounds of equivalent composition.[149] Indeed, nuclear orientation experiments on ^{60}Co in amorphous $Co_{80}P_{20}$ showed that the hff is negative, indicating that core polarization is the dominant contribution to the hyperfine interaction, as is known to be true for crystalline cases. For the crystalline compound Fe_3P, it was verified (for references, see the review of Durand et al.[147]) from combined neutron and Mössbauer spectroscopy experiments that the proportionality between hff and Fe moment holds for individual inequivalent Fe sites within the crystalline compound. It can be reasonably assumed that such a proportionality found between the average values of hff and magnetic moment would hold for the various transition-metal environments existing in the disordered structure as well, allowing the moment distributions to be inferred from the hff distributions measured by NMR or Mössbauer spectroscopy.

As for a correct evaluation of the hff distribution P(H), difficulties were encountered in the early NMR and Mössbauer studies of amorphous alloys. For spin-echo NMR experiments, P(H) is in principle directly obtained from the echo intensity vs. frequency. But difficulties arise from the sensitivity of the spectrum to excitation conditions and from the frequency dependence of the signal to be correctly taken into account over a very broad frequency range. For Mössbauer spectroscopy, P(H) is calculated from the experimental spectra, relying then on fitting procedures. The first P(H) obtained for ^{57}Fe in amorphous alloys were biased by the analytic expressions for P(H) which were arbitrarily introduced into the fitting procedures.[47] Distributions were more realistically evaluated later by using sophisticated analytical methods. The most widely-studied system is probably the FeB alloys. There seems to be an agreement among Mössbauer spectroscopists about a fairly symmetrical shape for P(H) in $Fe_{80}B_{20}$ alloys. When B content varies from 14 to 25 at.%, the symmetry of P(H) is not much altered, but the center of gravity follows the concentration dependence of the Fe moment. The symmetrical character of P(H) observed for FeB alloys is more an exception than a rule as compared with other Fe base amorphous alloys. However, the smoothness of the curve which seems to be perfectly preserved when varying the B content is questionable in view of the NMR spectra on amorphous alloys, where such regularities do not occur except for poor signal to noise ratio or for inadequate excitation condition, as will be discussed later. One is led to conjecture that such a smoothness in P(H) obtained from Mössbauer spectra is somehow related to the fitting procedure. As to the interpretation of the distribution, there are two main schools of thought. Either it would reflect the probability for Fe to have different numbers of Fe n.n. in a dense random packing of hard sphere model, or it would arise from the broadened lines corresponding to the different Fe sites in the metastable Fe_3B crystalline compound. The second interpretation is favored by the conclusions drawn out from Mössbauer studies on Ni-substituted amorphous FeB or FePB alloys, namely that the shape of P(H) is not very sensitive to the number of Fe n.n. On the other hand, the shape of P(H) and its mean value were shown to be strongly affected by the number and the nature of s-p atoms around the Fe sites. Indeed, asymmetrical P(H) was obtained for amorphous $Fe_{80}P_{20}$. The degree of asymmetry becomes larger when another metalloid is added, as in FePC, FeBSi, and FeBGe alloys. Such an influence is so pronounced that one might think that a change in short-range order would be induced by metalloid-metalloid substitution. The hff field distribution exhibited also a strong asymmetry in Co- and Cr-substituted Fe base amorphous alloys. In Mn-substituted alloys,[114] antiferromagnetic interactions related to Mn atoms develop already for low Mn content so as to affect the otherwise ferromagnetic structure. The structures observed in the hyperfine field distributions of these latter alloys with an important P(0) contribution is a rather common feature in the P(H) for amorphous FeSi, FeGe, FeSn, and YFe alloys. Experiments carried out under external applied field showed[96] that the Mössbauer spectra have to be analyzed as a sum of magnetic sextuplets and "nonmagnetic" doublets. The uncoupled Fe atoms would experience a weak dipolar field produced by the magnetic neighborhood. That the concentration of these "nonmagnetic" Fe atoms increases in the vicinity of the critical region for disappearance of ferromagnetism was attributed to local environment effects, with a critical number of 6 to 7 Fe n.n. needed for Fe to be magnetic.[96] Alternatively, there also might be some clustering effects occurring over this critical concentration range, so that the Fe atoms having too many Fe n.n. (11 to 12) could be "nonmagnetic", as was inferred for amorphous Fe.

In summary, it appears that the effects of structural disorder on the transition metal hff (and moment) distributions can vary on a very large scale from one amorphous alloy to another one depending upon composition. A simple case seems to be that of amorphous Fe_3B where P(H) is rather narrow (half-width $\simeq$ 45 kOe for an average value H = 286 kOe) and symmetrical. P(H) encompasses the hff values for the three Fe sites in crystalline Fe_3B, and can be generated by assuming a gaussian broadening of the three hff crystalline lines with a distribution half-width of about 30 kOe.[150] Another extreme case is that of amorphous alloys where the P(H) exhibits structures suggesting a sort of magnetic phase separation. In between, all degrees of broadening or asymmetry can be found, so that the concept of structural disorder does not seem to be unequivocal.

2. *Magnetic Properties at Finite Temperature*

We first discuss the data on the temperature dependence of the spontaneous magnetization in transition-metal base amorphous alloys, with a special emphasis on the low-temperature range where magnetic excitations can be sensed. We then review the problem of critical phenomena around the Curie temperature T_c. Finally, we comment on the effects of structural disorder on the values of T_c.

a. Temperature Dependence of the Spontaneous Magnetization

Magnetic excitations constitute the dominant mechanism for the decrease of the spontaneous magnetization at low temperature in crystalline magnets. Accumulating evidence indicates that this is also the case for most amorphous magnets. This result was not *a priori* evident, since other low-energy excitations of local character are expected in disordered systems. There are two important elementary excitations inherent in the model of itinerant ferromagnetism, namely the Stoner excitations, or single-particle excitations which are governed by Fermi statistics and can be described basically as one electron going from a spin up to a spin down state, and the spin waves, which are collective electron-hole pair excitations whose superpositions correspond to a wave-like change in spin direction with position. Stoner excitations result in a quadratic term for the temperature dependence of the spontaneous magnetization, while in a simple spin-wave formalism the reduced spontaneous magnetization obeys

$$M(T)/M(0) = 1 - BT^{3/2} + \ldots \tag{10}$$

where B is related to the spin-wave stiffness constant D. The prefactor B can be also determined from the temperature dependence of the hff obtained through Mössbauer spectroscopy. Ferromagnetic resonance (FMR) technique has been widely used for that purpose as well. Spin-wave stiffness is more readily determined by low-angle inelastic neutron scattering experiments which allow to verify the ferromagnetic dispersion relation

$$\hbar\omega = D(T)q^2 + E(T)q^4 \tag{11}$$

From the temperature dependence of the spontaneous magnetization alone, it is often rather conjectural to discriminate between the Stoner and the spin-wave behavior. Thus it is of special interest to compare the values for B yielded by different techniques. Spin-waves in an amorphous alloy were first ascertained by means of magnetization and inelastic neutron scattering experiments on amorphous Co_4P and FePC.[146] A puzzling result was that the coefficient B obtained from magnetization measurements was too large, yielding then for D a calculated value considerably smaller than that measured by neutron technique. However, this discrepancy was shown to be smaller when magnetization and spin-wave dispersion through ferromagnetic resonance were measured on the same CoP sample.[151] Finally, most of the previous difficulties were overcome by a series of combined experiments performed on the same samples $(Fe_xNi_{1-x})_{75}P_{16}B_6Al_3$ with x varying from 0.3 to 1; from neutron scattering and from the Bloch law for bulk magnetization and Fe hff (Mössbauer spectroscopy), the same spin-wave stiffness constant was obtained for the most concentrated alloys ($x = 1$, 0.65, 0.5), while discrepancies were observed between magnetic and neutron measurements on the $x = 0.30$ alloy.[152] It was proposed as a possible explanation for too-high values of the prefactor B in the case of amorphous alloys containing 30 to 40 at.% Fe that a true Bloch law for magnetization was not really obeyed.[153] This can be easily understood, in view of our previous discussion about the inhomogeneous character of magnetism in these alloys with Fe concentration just above the critical concentration for ferromagnetism. Such a magnetic clustering effect was observed in Co_4P through small angle magnetic scattering.[154] In conclusion, it seems that in "good" amorphous alloys, which otherwise do not exhibit Invar anomalies, values of the spin-wave stiffness constant can be obtained consistently from neutron scattering and magnetization measurements[155] as well as from ferromagnetic resonance and static magnetic experiments.[156] But discrepancies between values of the spin-wave stiffness constant as obtained from neutron scattering and magnetization measurements are systematically observed in amorphous and crystalline Invar alloys.[157] In most of amorphous alloys, the value of D/T_c, which gives an approximate measure of the range of the exchange interaction, was found to be near 0.2 meV.Å.K^{-1} (see Table 1 in Reference 155). Finally, we will just mention the puzzling observation reported for FePC and Co_4P[154] about excitations reaching a minimum energy at the maximum in the static structure function (q = 3 $Å^{-1}$). The interpretation of this minimum by analogy with rotons in ^{4}He was critically discussed by Alben et al.[13] More theoretical and experimental work is clearly desirable on this matter.

At this point, it seems appropriate to discuss briefly the over-all temperature dependence of magnetization in transition-metal base amorphous alloys. It has been noted by many authors that the reduced magnetization M(T)/M(0) versus reduced temperature T/T_c in amorphous alloys falls below the Brillouin function behavior fairly well observed for pure crystalline Ni, Co, and Fe. That such a "flattening" was not the signature of structural disorder could have been guessed by remembering that a similar effect occurs in crystalline Invar alloys. Nevertheless, it has been inferred from the temperature dependence of the magnetization that fluctuations in the exchange interaction should be very large in most of the amorphous ferromagnets. These estimates were relying on a molecular-field approach of the Heisenberg model,[12] where fluctuating environments in an amorphous matrix were assumed to result in a distribution of exchange integrals. By studying the temperature dependence of the hff distribution in an amorphous ferromagnet, it was demonstrated that the temperature decrease of the individual magnetic moments is governed by long-range correlations.[158] While the distribution of hff (and of magnetic moments) is rather large in an amorphous ferromagnet, the distribution in exchange interactions has to be rather narrow. The temperature dependence of the magnetization in amorphous alloys is practically identical to that of crystalline counterparts, as observed for crystalline and amorphous $(Fe_{0.5}Ni_{0.5})_3B$ and for amorphous $Fe_{78}B_{12}Si_{10}$ compared with crystalline Fe_3Si.

b. Critical Phenomena

An interesting question in magnetism of amorphous alloys is whether amorphous magnets can display a well-defined magnetic transition like the second-order phase transition occurring in most crystalline magnets. Accumulating experimental evidence has shown that it is also true in many amorphous magnetic alloys. Qualitative indications for a sharp transition at T_c were provided by early Mössbauer and specific heat measurements on amorphous FePC.[78,159] The first quantitative analysis of the critical behavior around T_c in an amorphous ferromagnet is due to Mizoguchi et al.[160] Values for the critical exponents β, γ, δ were determined from magnetic measurements on $Co_{70}B_{20}P_{10}$. These exponents were shown to obey the standard scaling law and the data around T_c were fitted to an equation of state previously derived for a second-order phase transition in fluid systems. Similar bulk magnetic investigations were reported on $Fe_{80}P_{13}C_7$,[12] $Fe_{32}Ni_{36}Cr_{14}B_{12}P_6$,[161] $Fe_{29}Ni_{49}P_{14}B_6Si_2$,[12] $Fe_{20}Ni_{60}P_{14}B_6$,[162] $Fe_{27}Ni_{53}P_{14}B_6$,[163] $Fe_{40}Ni_{40}P_{14}B_6$[162] and on Gd base amorphous alloys such as $Gd_{80}Au_{20}$[12] and $Gd_{67}Co_{33}$.[164] Special care has to be taken in determining critical exponents in ternary (Fe,Ni) base alloys, owing to strong superparamagnetic effects around T_c related to compositional short-range-order phenomena. Thus, more careful experiments on $Fe_{32}Ni_{36}Cr_{14}B_{12}P_6$ corrected[161] the "anomalous" values for the critical exponents which were initially reported. As can be seen in Table 3, the values listed for β, γ, δ in various amorphous alloys are very close to those obtained for crystalline Fe, Co, Ni, and Gd, and rather near to the predictions for a 3D (three-dimensional) Heisenberg ferromagnet. However, one can observe a trend for β in amorphous alloys to be enhanced ($\beta \geq 0.40$) as compared with the experimental values for pure crystalline metals and with the value (0.37) commonly accepted for

Table 3
CURIE TEMPERATURES AND CRITICAL EXPONENTS FOR SOME AMORPHOUS AND CRYSTALLINE FERROMAGNETS. (*) VALUES DETERMINED FROM SPECIFIC HEAT MEASUREMENTS; THE OTHER LISTED VALUES FOR α ARE OBTAINED THROUGH $\alpha = 2 - 2\beta - \gamma$

Amorphous	T_c (K)	β	γ	δ	α	Ref.
$Co_{70}B_{20}P_{10}$	453.01 ± 0.07	0.402 ± 0.007	1.342 ± 0.025	4.39 ± 0.05	−0.14 ± 0.05	160
$Fe_{80}P_{13}C_7$	586.55 ± 0.10	0.38 ± 0.002	1.30 ± 0.05	4.47 ± 0.05	−0.06 ± 0.06	12
$Fe_{29}Ni_{49}P_{14}B_6Si_2$	384.5	0.40 ± 0.01	1.7 ± 0.1	5.25 ± 0.10	−0.50 ± 0.12	12
$Fe_{32}Ni_{36}Cr_{14}B_{12}P_6$	250 ± 1	0.43 ± 0.02	1.33 ± 0.05	4.43 ± 0.17	−0.19 ± 0.09	161
	249.0 ± 0.1	0.42 ± 0.04	1.35 ± 0.03	4.30 ± 0.20	−0.19 ± 0.11	161
$Fe_{20}Ni_{60}P_{14}B_6$	227.7 ± 0.5	0.39 ± 0.02	1.33 ± 0.05	4.45 ± 0.07	−0.11 ± 0.09	162
$Fe_{40}Ni_{40}P_{14}B_6$	519.9 ± 0.2	0.38 ± 0.01	1.31 ± 0.02	4.46 ± 0.04	−0.07 ± 0.04	162
$Fe_{75}P_{15}C_{10}$					−0.18 ± 0.04(*)	166
$Fe_{34}Pd_{46}P_{20}$					−0.065 ± 0.05(*)	167
$Gd_{80}Au_{20}$	149.45 ± 0.20	0.44 ± 0.02	1.29 ± 0.05	3.96 ± 0.03	−0.17 ± 0.09	12
$Gd_{67}Co_{33}$	169.9 ± 0.2	0.41 ± 0.02	1.16 ± 0.05	360 ± 0.10	+0.02 ± 0.07	164
Crystals						
Fe	1042.0 ± 0.3	0.37 ± 0.03	1.30 ± 0.06		−0.06 ± 0.12	12
Ni	627.4	0.37 ± 0.04	1.34 ± 0.01	4.58 ± 0.05	−0.08 ± 0.02	12
Co	1043	0.36 ± 0.08	1.23 ± 0.05		0.05 ± 0.21	12
Gd	289	0.38	1.19	3.61	−0.20 ± 0.02(*)	12
3D Heisenberg		0.33 − 0.37	1.36 (S = 1/2) 1.33 (S = ∞)	4.2		12

3D Heisenberg model. This might not be related to structural disorder but rather to a dilution effect, as predicted by the missing-cluster model of Müller-Krumbhaar. That these amorphous alloys display a sharp T_c transition with standard β, γ, δ exponents implies that the critical exponent α is negative. Indeed, it was shown by a simple physical argument and by renormalization group calculations[165] that $\alpha < 0$ is a necessary condition for a homogeneous phase transition to occur in a disordered material. It can be verified from the relation $\alpha = 2\text{-}2\beta\text{-}\gamma$ that the $\alpha < 0$ condition is satisfied.

Of particular importance, then, is the direct evaluation of α from specific heat or resistivity measurements. The first quantitative study of the critical specific heat is due to Schowalter et al.[166] Again, typical Heisenberg value was obtained for α (-0.18 ± 0.04) in $Fe_{75}P_{15}C_{10}$. Specific heat and resistivity experiments also gave negative α values for $Fe_{34}Pd_{46}P_{20}$, $Fe_{20}Ni_{60}B_{20}$ and $Fe_{80}P_{13}C_7$ alloyed with Ni, Mn, Cr.[167]

The spin-wave-stiffness constant was studied over the critical temperature range by inelastic neutron scattering experiments.[168] The constant D was found to follow the power law $D \sim (1 - T/T_c)^{0.5\pm0.1}$ for $0.02 \leq (1 - T/t_c) \leq 0.2$, while at $T = T_c$ the energy width varies as $q^{2.7\pm0.2}$ for $0.5 \leq q \leq 0.18$ Å^{-1}. These results agree fairly well with dynamical scaling theory for the Heisenberg ferromagnet and furthermore they strongly resemble those obtained for crystalline Fe, Co, and Ni.

The order parameter of interest for the phase transition is the spontaneous magnetization. If one looks now at the temperature and field dependence of the magnetization in the critical temperature region, strong anomalies can occur in the Arrott plots for ferromagnetic amorphous alloys. An attempt was proposed to account for these deviations by calculating the influence of spatially random fluctuations of the Landau parameters as well as the influence of inhomogeneous stress fields.[169]

c. Curie Temperatures

The Curie temperature was found in most cases to be affected by the structural disorder more severely than the saturation moment. We give a brief survey of the Curie temperatures evaluated for "pure" amorphous Ni and Fe, and measured for amorphous alloys of the same composition as the single-phase crystalline compounds. Then, we discuss the concentration dependence of T_c in binary alloys and in ternary alloys where a glass-former substitutes for another one, or where a transition metal substitutes for another transition metal. Various models proposed to account for these effects will be referred to.

The Curie temperature for "pure" amorphous Ni was estimated rather consistently from the temperature dependence of the magnetization in Ni thin films[21] and by extrapolating the T_c of amorphous NiAg alloys.[89] T_c for amorphous Ni was thus determined to be 540 K (about 14% lower than T_c of crystalline Ni, 627 K). Extrapolations made for binary or ternary Fe base amorphous alloys indicate a very low T_c for pure amorphous Fe. First extrapolations made in binary and ternary[128] Fe base alloys yielded a T_c of about 300 K. Most recent binary amorphous ZrFe alloys[126] suggest that T_c of pure amorphous Fe would be practically zero. These results are consistent with the values inferred for the magnetic moment as has been discussed before. Comparison between T_c's in crystalline and amorphous modifications is seldom possible. Co base amorphous alloys undergo crystallization before reaching T_c. In Fe base amorphous alloys, the effect of structural disorder on T_c can be evaluated in a few cases. In amorphous $Fe_{75}P_6B_{19}$, the T_c value was found to be lower by 12% as compared with the single-phase ternary compound.[134] For Fe_3B, the reduction of T_c can be estimated to be of about 10% as compared with the metastable compound.[128] For Fe_3P, the effect of structural disorder seems to be larger (about 20%). In the case of bimetallic alloys, the change between the values of T_c in the amorphous and the crystalline phase can be drastic. For example, amorphous YFe_2 is a concentrated spin-glass with a susceptibility maximum at 58 K, while the Laves phase is ferromagnetic with a T_c of 530 K.

Since the first approach by Gubanov,[4] many model calculations have tried to estimate the effect of structural disorder on the Curie temperature. Very few calculations were made in the framework of the itinerant magnetism model, except for the particular case of weak ferromagnetism.[137] As far as one deals with an effect of the order of 10%, a theoretical justification is easily found, such as the effect of smaller density (2 or 3%) in the amorphous state affecting the exchange integral in a molecular field model. The reader will refer to Handrich and Kobe's book[12] or to Chapter 2 of this book. But, when a drastic effect occurs (as in YFe_2), the differences between amorphous and crystalline phases cannot be accounted for without going through a detailed band structure calculation.

When some s-p elements are added to Fe in binary Fe base amorphous alloys, T_c increases with the "metalloid" content. Such a variation was first tentatively attributed[134] to two different magnetic states of Fe in these alloys, by analogy with the Weiss approach to the Invar problem. Later, such a similarity to the behavior observed in Invar alloys was treated in the framework of the Wohlfarth model.[137] The general trends of the T_c variation in binary Fe base amorphous alloys were fairly well reproduced by using reasonable values of the pressure dependence of T_c together with values of the compressibility; extrapolating to zero concentration of s-p elements, T_c of amorphous Fe was then estimated to be 140 K.

The change in T_c due to substitution of one glass-former for another one in a ternary amorphous alloy was first reported by Chen[170] for FePC alloys. Such a variation was shown to be rather similar in amorphous FePB alloys and in crystalline $Fe_3P_{1-x}B_x$ compounds.[134] Systematic studies on the effects of the different s-p elements on the T_c or ternary amorphous alloys were reported[17,18,135-137] (Figures 10 and 11). The relative importance of the effects of various elements on T_c is not clear. A tentative rationale, however, was presented in terms of size effects[135] (Figure 12). Earlier measurements by Chen et al.[170] were interpreted by invoking the various expansion of the Fe–Fe distance caused by admixture of Fe with different s-p elements in the framework of the Slater-Bethe curve (Figure 13).

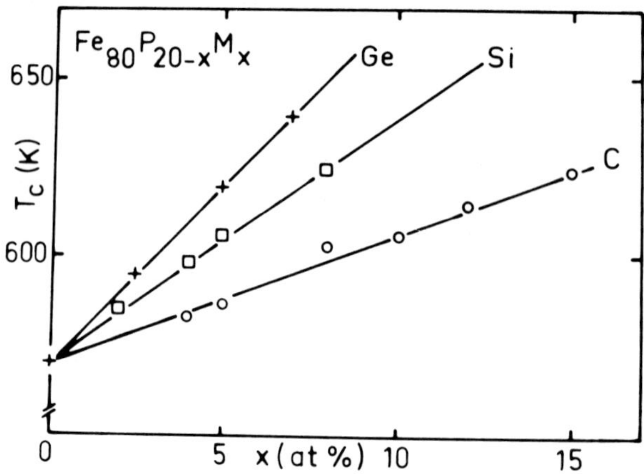

FIGURE 10. Curie temperature vs. x in amorphous $Fe_{80}P_{20-x}M_x$ alloys (M = Ge, Si, C).[135]

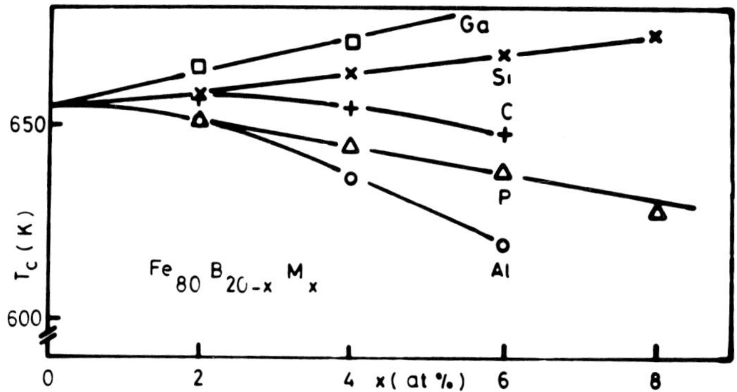

FIGURE 11. Curie temperature vs.x in amorphous $Fe_{80}B_{20-x}M_x$ alloys (M = Ga, Si, C, P, Al).[24]

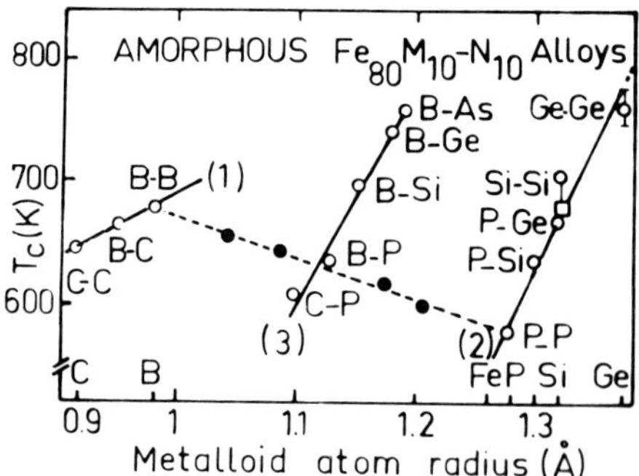

FIGURE 12. Curie temperature of amorphous $Fe_{80}M_{10}N_{10}$ alloys vs. the average atomic radius of the metalloids r = |r (M) + r(N) |/2. (From Kazama, N., Masumoto, T., and Mitera, M., *J. Magn. Magn. Mater.*, 15—18, 1331, 1980. With permission.)

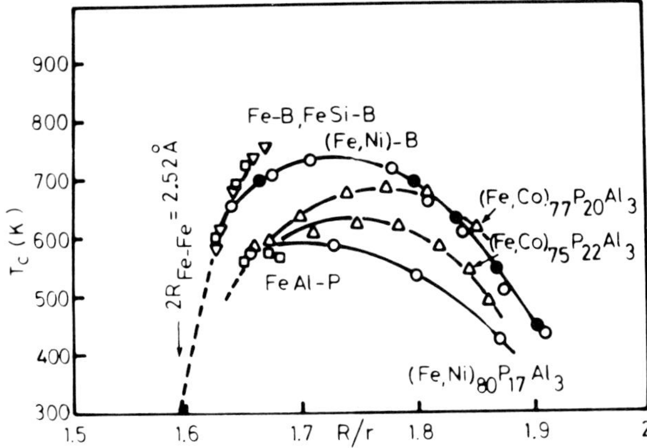

FIGURE 13. Slater-Bethe curves for metallic glasses, with 2R being the average interatomic distance among transition metals, and 2r being the diameter of incomplete inner shells of transition metals. (Chen, H. S., Sherwood, R. C., and Gyorgy, E. M., *IEEE Trans. Magn.*, MAG-13, 1538, 1977. With permission.)

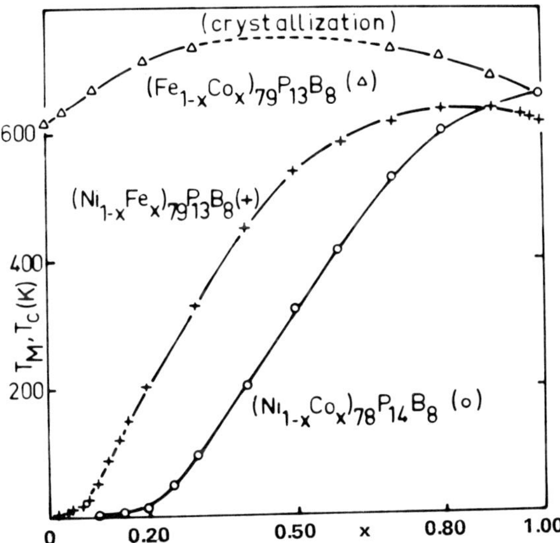

FIGURE 14. Curie temperature as a function of x in amorphous $(Fe_{1-x}Co_x)_{79}P_{13}B_8$ (data from Reference 174), $(Ni_{1-x}Fe_x)_{79}P_{13}B_8$ (data from Reference 77), and $(Ni_{1-x}Co_x)_{78}P_{14}B_8$ (data from Reference 52) alloys.

The effect of substitution of various transition metals for Fe in Fe base amorphous alloys was also studied in great detail[17,18,126,143-145,170] (Figure 14). For Ni- and Co-substituted alloys, T_c goes through a maximum for $x_{Ni} \simeq 15$ at.% and $x_{Co} \simeq 50$ at.%, while the magnetic moment decreases rather linearly with increasing Ni or Co content. Such a behavior is reminiscent of crystalline Invar alloys. Several attempts were proposed to account for these variations in a local-magnetism approach.[137] From these systematics, the influence of structural disorder T_c is not readily distinguishable from the composition or chemical effect.

III. MAGNETIC PROPERTIES AND SHORT-RANGE ORDER

While surveying the effects of structural disorder on magnetic properties of transition-metal base amorphous alloys, the words "structural disorder" were understood in a rather loose way to mean the lack of periodicity in atomic arrangement over distances exceeding the first two or three atomic shells in a system which was assumed to be otherwise metallurgically homogeneous. At the same time, the expression "short-range order" (SRO) was invoked without specifically stating as to this regular arrangement for the first atomic shells around a given atom whether it is due to the combined effects of geometry and randomness (e.g., canonical Bernal holes) or it results from chemical affinities. During the course of our discussion, we were led several times to depart from this commonly admitted, implicit picture. It appeared that corrections should be brought to the standard definitions of SRO and long-range disorder as usually adopted from geometrical random packing models for the structure of amorphous alloys. On one hand, the SRO in some amorphous systems was inferred to be rigorously defined by chemical bonding considerations, and then strongly correlated to the SRO prevailing in the crystalline counterpart (Fe_3B, Ni_3B-type alloys). On the other hand, the structural disorder was concluded in some other amorphous alloys to be so severe that the system should be better described in terms of phase separation on a finite scale than in homogeneous disordered model. Local and bulk properties were found to be so sensitive to the local environment, that in many cases they failed to yield much information about the topological disorder. Thus, it seems to be of interest in analyzing the relation between magnetism and structure of amorphous alloys to deliberately adopt the following point of view, that magnetic measurements can be used as investigative tools for the atomic-scale structure of amorphous alloys. This will be done in the present section by first considering magnetic measurements on amorphous alloys of the same composition as single-phase crystalline compounds. Then, we summarize the magnetic data obtained for "off-stoichiometry" amorphous compounds. Finally, we discuss the magnetic investigations indicating the special role played by the eutectic composition as well as those suggesting a "phase separation" in transition metal base amorphous alloys.

A. Local Symmetry in Stoichiometric Amorphous Alloys

Major controversies concerning amorphous solid structure and its relation to crystal structure center around the SRO. A complete description of the SRO in a binary amorphous alloy made of about 80 at.% of transition metal (TM) with about 20 at.% of glass formers (GF) atoms would include the knowledge of (1) the average coordination numbers $\overline{Z}_{TM}(TM)$, $\overline{Z}_{GF}(TM)$, $\overline{Z}_{GF}(GF)$, $\overline{Z}_{TM}(GF)$ around both the TM and GF atoms together with the fluctuations about these average values, (2) the mean interatomic distances r_{TM}, r_{GF} and their fluctuations for the first atomic shell surrounding a TM(GF) atom and, (3) the average symmetry of the environment about a given atomic species. The knowledge of the first two points is achieved (at least for the TM component, and also for the minority component when the atoms are relatively heavy) to some extent through X-ray, electron, and neutron scattering measurements. The problem is not trivial for glass formers constituted of light atoms (B). In any case, the knowledge of the average local symmetry about transition-metal and glass-former atoms can be reached through experimental investigations of the electric field gradients (EFG). The EFG can be determined most directly by local magnetic measurements, such as Mössbauer or NMR spectroscopy.

It might be of some use to summarize the formalism commonly used in these experiments. The nuclear charge distribution for nuclei having a spin $I > 1/2$ is of nonspherical symmetry. These nuclei have an electric quadrupole moment Q which interacts with the EFG around the nuclear sites. They can be used then as a probe of the local symmetry. Assuming that higher order terms are negligible, the EFG can be expressed as a tensor, whose eigenvalues

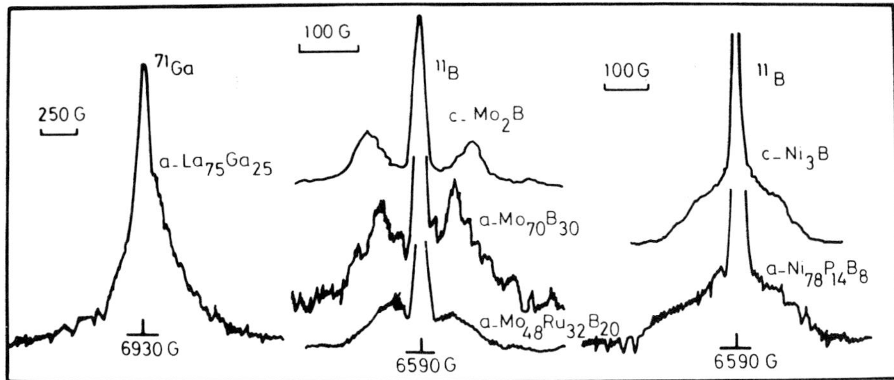

FIGURE 15. NMR spectra on ^{71}Ga and ^{11}B in *a* -$La_{75}Ga_{25}$ (T = 10 K) (left), in *c* -Mo_2B, *a* -$Mo_{70}B_{30}$ and *a* -$Mo_{48}Ru_{32}B_{20}$ (T = 10 K) (middle), in *c* -Ni_3B and *a* -$Ni_{78}P_{14}B_8$ (right).[171]

in a principal axis system are $|V_{zz}| > |V_{yy}| > |V_{xx}|$. The quadrupolar frequency ν_Q deduced from NMR spectra is proportional to V_{zz}. V_{zz} is zero for a local spherical symmetry (cubic compounds). Thus, ν_Q (or ΔE in Mössbauer spectroscopy) measures the departure from a spherical symmetry. If the local symmetry is axial, one has $V_{xx} = V_{yy}$. Then, the asymmetry parameter $\eta = |V_{xx} - V_{yy}|/|V_{zz}|$ measures the departure from an axial symmetry. We first summarize recent data obtained through NMR measurements about the local symmetry around the GF sites in transition-metal base amorphous alloys. We then discuss the information that can be yielded by magnetic measurements concerning the average symmetry about the transition metal sites.[22,24,147]

1. Local Symmetry Around the Glass-Former Sites

The EFG parameters (ν_Q, η) were determined by NMR on s-p elements (^{71}Ga and ^{11}B, both having a nuclear spin I = 3/2) in some typical amorphous alloys, such as splat-cooled $La_{75}Ga_{25}$, sputtered $Mo_{70}B_{30}$, splat-cooled $Mo_{48}Ru_{32}B_{20}$ and $Ni_{78}P_{14}B_8$.[171] Experimental spectra are shown on Figure 15 together with spectra obtained on compositionally related compounds displaying a local symmetry of spherical (La_3Ga), axial (Mo_2B), nonaxial (Ni_3B) type, respectively. Values for the quadrupole parameters obtained by computer simulations are listed in Table 4. The conclusions drawn from this study can be summarized as follows.

1. There exist different types of local symmetry around the GF atoms in amorphous alloys of the *TM*-GF family.
2. The nature of the local symmetry (η) prevailing in crystalline compounds is basically preserved in their amorphous modifications.
3. The structural disorder can be described by a distribution of quadrupole frequencies (characterized by σ in a gaussian approximation) around a mean value $\bar{\nu}_Q$ and by an average value of the asymmetry parameter η, the ratio σ/ν_Q being a measure of the degree of randomness.
4. The knowledge of $\bar{\nu}_Q$ alone is not sufficient to characterize the local symmetry; ν_Q is about the same for ^{11}B in Ni_3B and Mo_2B, but striking differences in the NMR spectra arise from the different values of η.
5. The degree of randomness is the same in sputtered $Mo_{70}B_{30}$ and in splat-cooled NiPB. Thus the fabrication technique does not affect much the nature and the distribution of the local symmetry.
6. The degree of randomness is sensitive to the supplementary chemical disorder resulting from substitution of Ru for Mo in amorphous $(Mo_{0.48}Ru_{0.32})_{80}B_{20}$ as compared with amorphous $Mo_{70}B_{30}$.

Table 4
EFG PARAMETERS IN AMORPHOUS ALLOYS AND IN RELATED CRYSTALLINE COMPOUNDS

Compounds	$\lvert\bar{\nu}_Q\rvert$ (KHz)	η	$\sigma/\lvert\bar{\nu}_Q\rvert$
c-La_3Ga (cubic)	0	0	—
a-$La_{75}Ga_{25}$	<100	—	>2
c-Mo_2B (tetragonal)	240 ± 10	0	—
a-$Mo_{70}B_{30}$	210 ± 20	0 to 0.2	0.2 to 3
a-$Mo_{48}Ru_{32}B_{20}$	180 ± 20	0 to 0.3	0.4 to 0.6
c-Ni_3B (orthorhombic)	180 ± 10	0.6	—
a-$Ni_{78}P_{14}B_8$	220 ± 20	0.5 to 0.8	0.2 to 0.3

From Panissod, P., Aliaga-Guerra, D., Amamou, A., Durand, J., Johnson, W. L., Carter, W. L., and Poon, S. J., *Phys. Rev. Lett.*, 44, 1465, 1980. With permission.

7. That the local symmetry is preserved in transition-metal base amorphous alloys is not related to the directional character of the bonding due to metalloid atoms, since Ga has a definite metallic character. This point was further ascertained by the fact that ^{27}Al (I = 5/2) quadrupole spectra are practically identical in splat-cooled amorphous $La_{75}Al_{25}$ and in hexagonal La_3Al.[147]

2. *Local Symmetry Around the Transition-Metal Sites*

Attempts were made to characterize the local symmetry around the transition metal sites through EFG studies. A complete determination of the quadrupole parameters (ν_Q or ΔE, and η) has not been possible as yet. Due to the fact that the transition metals in an amorphous alloy have a large variety of possible environments, while the s-p elements have only transition metal atoms as their first neighbors, the distribution of quadrupole parameters is expected to be larger at the transition metal sites than at the glass-former sites. Well-defined quadrupole splitting in Mössbauer spectra are observed in the paramagnetic state of amorphous alloys of the Fe_3B family and the value for ΔE compares rather well with that for crystalline compound of the same composition.[172] Unfortunately ^{57}Fe is not a suitable isotope for the evaluation of the asymmetry parameter which characterizes properly the nature of the symmetry. That the short-range order is basically the same around the transition-metal sites in an amorphous alloy and in its crystalline counterpart can be inferred by studying the hyperfine field distribution in the magnetically ordered state.[147]

The ^{59}Co NMR spectrum in amorphous Co_3B does display a structure,[173] but it is not quadrupolar in origin. Neither does it correspond to any of the structures observed in the ^{59}Co spin-echo spectrum for crystalline Co_3B, which are related to inequivalent sites for Co. From a thorough NMR study of the amorphous CoB alloys as a function of concentration, for various annealing treatments and under different excitation conditions, it appears that the structures observed in the ^{59}Co NMR spectrum for amorphous Co_3B arise from Co nuclei located in domain walls having different motion regimes and different orientations in the sample. This point will be discussed further in the following. To summarize, a local study of amorphous Co_3B shows that this alloy cannot be regarded as a simple disordered Co_3B crystalline compound.

Another approach for the local symmetry of transition metals in an amorphous alloy consists in studying the electric quadrupole effects on a small amount of suitable isotope substituted for the majority transition metal element. This was done by substituting Co for Fe in FePB alloys.[174] Well-defined quadrupolar spectra were observed by ^{59}Co NMR spectroscopy in the dilute Co limit, yielding anomalously large values for ν_Q. Similar values, however, were obtained in a Co-substituted crystalline Fe_3P compound. This might be related to the preferential substitution reported for transition metals in the Fe_3P crystalline compound.[175]

B. Short-Range Order in "Off-Stoichiometry" Amorphous Alloys

Among the large variety of transition-metal base amorphous alloys, those having the same compositions as single-phase crystalline compounds represent the exception rather than the rule. The question arises then as to what is the SRO in off-stoichiometry amorphous alloys. Some information may be obtained from local and bulk magnetic measurements about the SRO in binary and ternary amorphous alloys with compositions which depart from that of a single-phase compound.

1. Binary Amorphous Alloys

The SRO in amorphous FeB alloys was described by Vincze et al.[172] on the basis of the P(H) distribution calculated from Mössbauer spectra. The glass structure for alloys with B concentration down to 15 at.% was described as a locally distorted nonstoichiometric Fe_3B quasicrystalline structure with a distribution of lattice parameters around the mean value of the crystalline intermetallic compound.

A more complex picture was obtained from ^{59}Co NMR on amorphous electrodeposited CoP alloys.[149] Structures in the spin-echo spectra were interpreted in terms of two coexisting "amorphous phases", namely a Co-rich phase (supersaturated CoP solid solution) and a Co_3P-type phase. The relative integrated intensities of the signal arising from two "amorphous phases" were found to vary consistently with concentration (Figure 16). Structureless ^{59}Co spin-echo spectra were reported subsequently on amorphous CoP alloys. More recent experiments demonstrated that these discrepancies arise from the fact that the shape of the Co NMR spectra in these amorphous alloys is very sensitive to the experimental excitation conditions.[173] The structures of the hff distribution are due to different motion regimes for the walls where most of the resonant Co nuclei are located. By using too strong a rf field, these different regimes are not seen anymore, and the structures on the spin-echo spectrum are washed out.

Ferromagnetic resonance was used to measure the average magnetization on the surface (1 μm thick) of electrodeposited plates of amorphous CoP.[176] The P concentration was evaluated from the linear relation between composition and bulk magnetization. Thus, composition profiles could be obtained by stepwise thinning of the samples. Large fluctuations on the phosphorus concentration through the sample thickness were then evidenced, while the average concentration of the sample was found to be unaffected. These results together with the NMR results[149] might be related to the large-scale composition fluctuations (10 to 10^4 Å) observed by X-ray SAS on amorphous $Co_{78}P_{22}$ samples.[177] The composition fluctuation regions were found to be rodlike shaped with their major axis (>2000 Å) perpendicular to the foil plane and their minor axes (~ 60 Å) in the foil plane.

That such inhomogeneities are not basically due to the electrodeposition process was inferred from ^{59}Co NMR measurements on chill-cast CoB ribbons with B concentrations ranging between 14 and 28 at.%.[173] By carefully varying the experimental parameters (rf field strength, orientation of the foils vs. excitation field and thermal annealing treatments), the origin of the structures displayed by the Co NMR spectra together with their composition dependence was reasonably explained. Cobalt-rich regions along with regions of Co_3B and Co_2B type are tentatively assigned to the observed structures. Complementary measurements such as X-ray SAS and compositional profiles vs. thickness would be highly desirable for these CoB alloys.

2. Ternary Amorphous Alloys

Progressive change in SRO of amorphous alloys can be produced either by glass-former

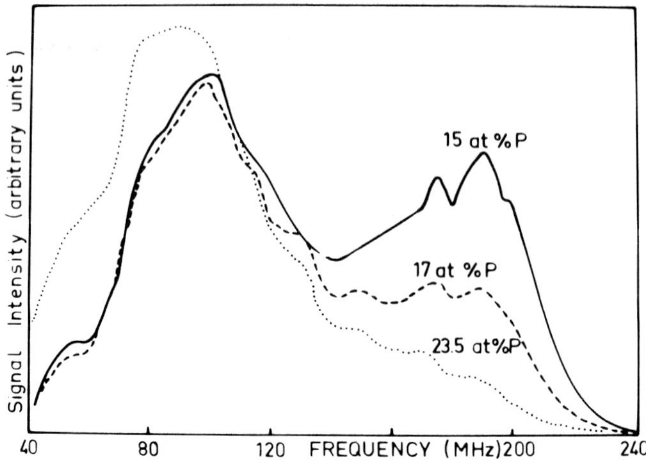

FIGURE 16. ^{59}Co spin echo NMR spectra (T = 4.2 K) in amorphous electrodeposited CoP alloys for various concentration.[149]

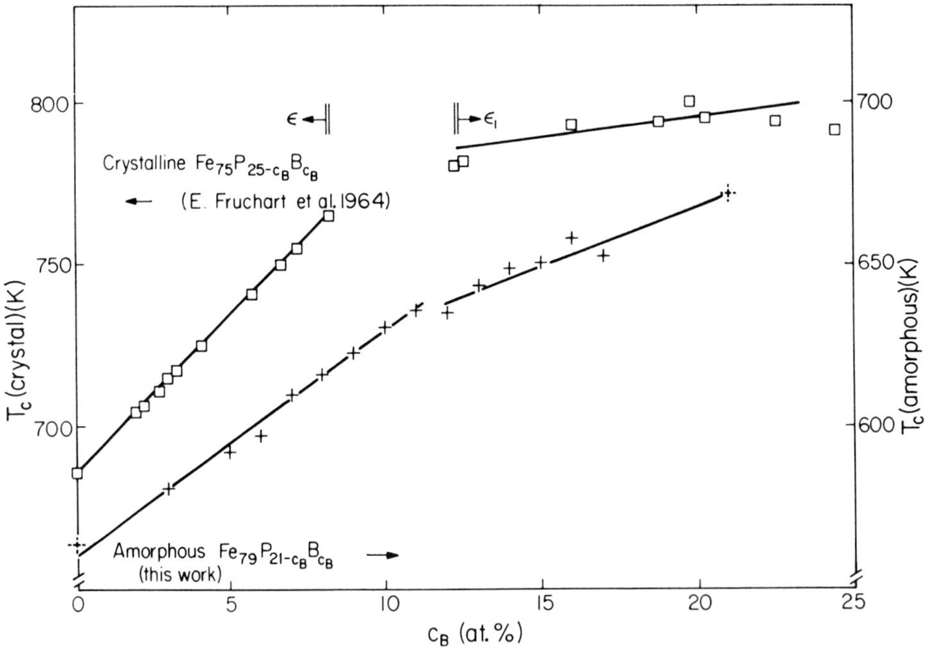

FIGURE 17. Curie temperature vs. B content in amorphous $Fe_{79}P_{21-x}B_x$ alloys and in crystalline $Fe_{75}P_{25-x}B_x$ compounds. (From Durand, J., *IEEE Trans. Magn.*, MAG-12, 945, 1976. With permission.)

substitution or by transition-metal substitution. By replacing P by B in amorphous $Fe_{79}P_{21-x}B_x$, one observes a change of slope in the concentration dependence of T_c and of the magnetic moment.[134] Similar behavior occurs in the single-phase crystalline compound $Fe_3P_{1-x}B_x$, where it is related to a change from an ϵ-tetragonal phase (Fe_3P and low B side) to an ϵ_1-tetragonal phase (rich B side), both tetragonal phases having the same coordination numbers for the inequivalent sites, but differing only by a contraction of the Fe_{III}-Fe_{III} distances in the ϵ_1 modification. Magnetic measurements suggested then that amorphous alloys exhibited a similar change of SRO when B content increased. This suggestion was qualitatively confirmed by Mössbauer spectroscopy studies[147] (Figure 17).

Similar change of SRO can be detected by measuring, by Mössbauer spectroscopy, the concentration dependence of the quadrupole splitting when Ni replaces Fe in amorphous and crystalline $(Fe_{1-x}Ni_x)_3B$.[172] The crystal structures for Fe_3B and Ni_3B are tetragonal and orthorhombic, respectively.

In other cases, the replacement of one metalloid by another one induces strong inhomogeneities, as observed by Mössbauer spectroscopy on $Fe_{84}B_{16-x}C_x$ alloys.[147] We already discussed the magnetic properties of amorphous PdSi (Co), PdSi (Fe) alloys[44] where a strong trend to phase separation was reported. The magnetic behavior of CuZr(Mn) alloys suggests the existence of two inequivalent sites for Mn in this amorphous system.[58] Composite amorphous alloys such as Metglas® 2826 A containing Fe, Ni, and Cr were reported to contain two kinds of Fe environments, namely a "FeNi phase" and a "FeCr phase".[161]

C. Eutectic Amorphous Alloys; Phase Separation

In the course of the preceeding discussion, we mentioned several amorphous alloys for which phase separation was reported or suggested. We want now to summarize what can be inferred from magnetic measurements about the special role played by the eutectic composition in the atomic-scale structure of amorphous alloys. Then, we discuss the demixing phenomena which seem to be related to this eutectic composition.

1. Role of Eutectic Composition

The eutectic composition has been known for a long time as indicating a composition range particularly favorable for glassification of alloys. Around this composition, viscosity of liquids and T∫, of metallic glasses are depressed, suggesting the occurrence of atomic clustering.[78] Indeed, anomalies in density, Young's modulus and glass temperature were observed in the region of eutectic composition for the archetypal PdSi alloys.[78] These effects were interpreted as indicative of an alloy mixing effect. This interpretation is in good agreement with the conclusions obtained on Co- and Fe-substituted PdSi alloys from electron microscopy and magnetic studies.[44] Other effects were reported to occur over the same concentration range, such as a change of slope in the concentration dependence of the linear term in specific heat[35] and a change in the Hall-effect coefficient,[178] which might suggest a change in SRO around the eutectic composition. Mössbauer measurements on ^{57}Fe-substituted PdSi alloys exhibited also a steep drop of the isomer shift in this concentration region.[179] Some drastic effects were observed around the eutectic composition on $(Mo_{0.6}Ru_{0.4})_{1-x}B_x$ alloys.[180] These effects include changes in electrical resistivity, in linear coefficient for low-temperature specific heat, suggestive of some dramatic structural change. Magnetic measurements might be useful to characterize what happens in amorphous alloys in the eutectic composition region.

Indeed, some striking change occurs in the magnetic behavior of electrodeposited $Ni_{100-x}P_x$ alloys for $x \simeq 18$ at.%.[88] Below this concentration, the magnetic properties are typically those of a weak amorphous ferromagnet, indicative of a rather homogeneous system. In alloys with P concentrations larger than 18 at.% a strong Ni clustering develops in an otherwise paramagnetic matrix. These two magnetic regimes on both sides of the eutectic composition suggest a change of SRO: solid-solution-like on the Ni-rich side, atomic clustering, demixing effects on the other side (18 to 22 at.% P).

The eutectic composition in electrodeposited FeP alloys is accompanied by a series of phenomena in the concentration dependence of various physical properties, such as a minimum in the quadrupole splitting observed in Mössbauer spectroscopy, a minimum in the width of the Curie transition[24] (Figure 18) and a minimum in the logarithmic term in resistivity.[181] Mössbauer investigations by Takahashi and Koshimura[182] on liquid-quenched amorphous FeP alloys exhibited at the eutectic composition an abrupt change of the angle of inclination of the magnetization relative to the sample plane, suggestive of some change

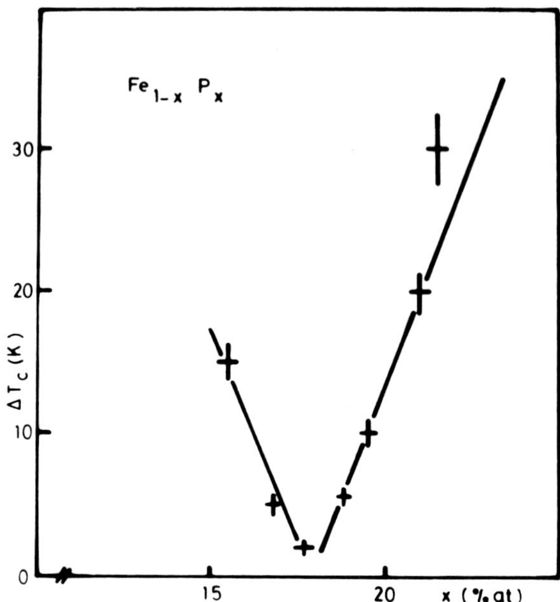

FIGURE 18. The temperature width ΔT_c of the Curie transition vs. P content in amorphous electrodeposited FeP alloys.[24] (T_c was measured with a standard ac bridge; the width was determined as the temperature interval corresponding to 0.1 and 0.9 times the signal difference between the paramagnetic and ferromagnetic states.)

in the local structure. Two remarks can be made about the electrodeposited FeP alloys. First, it is interesting to note that local and bulk magnetic measurements were sensitive to structural effects which were not detected by large angle X-ray scattering experiments performed on the same samples.[181] Second, the concentration effects occurring around the eutectic composition suggest a well-defined atomic arrangement rather than the atomic clustering commonly observed in other amorphous alloys. One is led to guess that a metastable crystalline compound could possibly exist in the vicinity of the eutectic composition in the FeP system.

A change of slope in the density and the temperature dependence of the bulk magnetization was observed around the eutectic composition in splat-cooled FeB amorphous alloys.[128] This was attributed to a change of SRO from a loose-packed (of the bcc Fe type) to a dense-packed type of structure. Subsequent investigations of the magnetic properties and of the concentration dependence of the density performed on Ni-substituted $Fe_{86}B_{14}$ alloys seem to support this interpretation.[141] That no corresponding change in the hff Fe distribution was detected from Mössbauer spectroscopy studies on the FeB series remains rather puzzling.

Detailed studies of magnetic properties of the ternary FeSiB alloys were reported by Narita et al.[136] A change of slope and a maximum in the concentration dependence of the magnetization and of the Curie temperature, respectively, was observed for metalloid content between 20 and 25 at.%. In view of the other data on the influence of metalloid substitution,[138] it seems that Narita's results could be related to a change of SRO occurring over the eutectic concentration region (Figure 19).

Finally, let us refer to recent Mössbauer studies of the splat-cooled amorphous YFe alloys around the eutectic composition, and of eutectic $Y_{66}Fe_{34}$ alloys with substitution of transition metals and metalloids for iron.[183] From the concentration dependence of the quadrupole splitting and of the isomer shifts, it is suggested that these eutectic alloys are intrinsically inhomogeneous, the glass consisting of two amorphous phases having different types of short-range order.

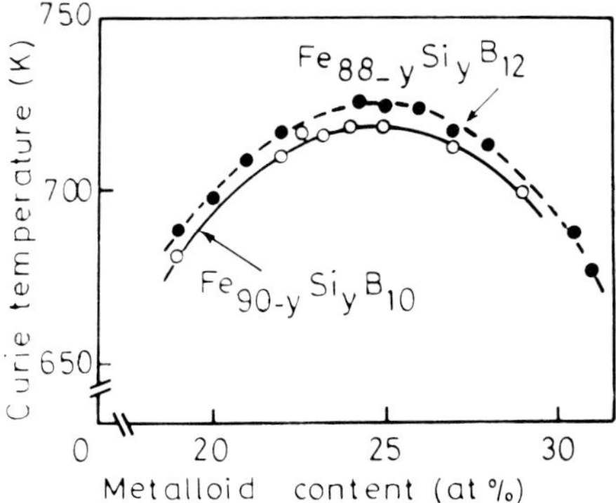

FIGURE 19. Curie temperature vs. metalloid content in amorphous $Fe_{88-y}Si_yB_{12}$ and $Fe_{90-y}Si_yB_{10}$ alloys. (From Narita, K., Fukunaga, H., and Yamasaki, J., *Jpn. J. Appl. Phys.*, 16, 2063, 1977. With permission.)

All the above observations made on various eutectic alloys suggest that demixing effects occur around the eutectic composition of transition-metal base amorphous alloys. When these inhomogeneities are extended on a large enough scale, they can be described in terms of different amorphous phases which would have comparable proportions at the eutectic composition. The scale of these inhomogeneities is better determined by SAS measurements. But bulk and local magnetic measurements can help assigning the nature of these amorphous phases.

2. Phase Separation

Demixing in the glassy state is known to occur in nonmetallic glasses, such as silicate glasses, where this phenomenon of phase separation is applied to the production of Vycor-type glasses. Interest in similar features for metallic glasses originates from the early work of Chen et al. (references can be found in Chen's review[78]). This interest was renewed by recent thermodynamic[184] and static studies of a large variety of amorphous alloys.[185-187] Depending upon the scale over which these amorphous phases develop, they are observable by large angle X-ray scattering (split first peak in the X-ray scanning) or by X-ray SAS. For example, in PbSbAu alloys, phase separation occurs on a scale of ~ 1000 Å,[186] while in MoRuB alloys[180] it is believed that phases segregate into domains of characteristic size ~ 10 to 15 Å with a typical interdomain distance of 20 to 30 Å. The transmission electron microscopy images obtained on ZrTiBe metallic glasses[187] strongly resemble those observed in crystalline spinodal structures.

We already discussed the existence of the magnetic phases in such materials as CoP and CoB. A problem has still to be solved, as to what extent these inhomogeneities depend upon the fabrication technique. The implications of these concentration fluctuations together with their geometric arrangement inside the amorphous sample is of paramount importance for understanding of macroscopic properties (e.g., anisotropy, magnetostriction, etc.) as well as for the anomalous transport properties of amorphous alloys. This might also be true for the mechanical properties. Indeed, it has been proposed by Chen[78] that local structural and compositional fluctuations somewhat analogous to phase separation in regions of ~ 20 Å accompanied by structural relaxation may be responsible for the enhanced embrittlement in alloy glasses. The inhomogeneous character of amorphous alloys may also have some implications for the Invar problem. Many metallic glasses are known to exhibit interesting Invar and Elinvar properties,[188] in particular the FeB base alloys. It would be of particular importance for the basic understanding of these phenomena to know the degree of homogeneity of these alloys. The existence of two amorphous phases in amorphous $Fe_{84}B_{16}$ was reported by Walter and Bartram.[189] But, it seems that these transmission electron microscopy measurements have remained somehow controversial. SAS experiments are clearly desirable over the eutectic composition range of amorphous FeB alloys. Indeed, a basic controversy has been long standing about the interpretation of Invar properties in crystalline materials. Interpretation in terms of homogeneous weak itinerant magnetism[14] or in terms of magnetically or metallurgically inhomogeneous ferromagnets[190] were alternatively proposed for FeNi crystalline Invar alloys. A better characterization of structural properties of those amorphous alloys which display most striking Invar behavior could help clarifying the theoretical situation.

In summary, we briefly outlined the structural information that can be yielded by bulk and local magnetic measurements. This information can be unique and hardly accessible by any other structural investigation technique, as was shown for the local symmetry around the glass-former sites. In many cases, the information gained from magnetic measurements seems to be complementary to that obtained from SAS-X-ray or neutron techniques.

IV. CONCLUSION

A logical sequence for this survey would have been to start with a description as detailed as possible of the atomic-scale structure of amorphous alloys, and then to discriminate the effects of topological and chemical disorder together with the manifestations of SRO on bulk and local magnetic properties. Instead, we decided to better follow the historical sequence according to which the problems have been raised. Although empirical, this latter method was not arbitrary. First, structural order at short or medium range in transition-metal base amorphous alloys is a field which has been in complete evolution for the last few years, so that one can conjecture that some new approaches we mentioned in the above section concerning the role of eutectic composition along with microscale or large scale phase separations will be clarified or revised in the future. Second, although trivial, it is important to emphasize that, at variance with crystalline materials, the field of structural studies in amorphous alloys cannot develop independent of the field of physical properties. Solid state physicists working on metallic glasses do not normally deal with materials whose structure can be fully characterized as a preliminary step of research. Information concerning the static and dynamic properties of these materials must then be extracted using many different experimental techniques. Indeed, investigations of physical properties, including magnetism, especially sensitive to local environment have proven very informative about SRO and medium-range order in amorphous alloys, so giving incentives for new structural investigations.

Search for spectacular consequences of structural disorder on fundamental magnetic properties of transition-metal base amorphous alloys may have looked somehow like a disappointing enterprise. One starts with a transition-metal base nonmagnetic amorphous alloy, whose electronic band structure is basically the same, within the limits of experimental accuracy of photoemission measurements, as in compositionally related crystalline compounds. This material is assumed to be metallurgically homogeneous. One introduces in this matrix a progressively increasing amount of 3d magnetic impurities, so as to approach the critical concentration region for emergence of long-range magnetic order. Meanwhile, one meets a sequence of old interesting problems within an apparently new context: Kondo effect, spin-glasses, cluster-glasses, mictomagnetism, and emerging long-range inhomogeneous ferromagnetism. A major effect is expected to occur within the long-range interacting impurity concentration regime, owing to a severe damping of the RKKY interaction in a system where the electron mean-free-path is very short. Rather deceivingly, no drastic attenuation is observed, which implies that the range of the RKKY interaction is notably larger than the electron mean-free-path. But it must be remarked that the study of spin glasses in amorphous alloys remains in its infancy. Most of the dynamic aspects of the spin glasses recently studied in crystalline alloys such as frequency dependence of susceptibility maxima, energy relaxation, time-dependent effects and magnetization jumps have not been investigated at all in amorphous systems as yet. More specific manifestations of the structural disorder can reasonably be expected along this line.

The most interesting manifestations of structural disorder encountered in this approach toward ferromagnetism are given by transport properties. This is attributable to the electron mean-free-path being shorter than the range of magnetic interactions. In these intermediate magnetic regimes, inhomogeneous in nature, the shortsighted conduction electrons remain blind in front of the over-all magnetic situation so as to sense preferentially the minority elements. Thus, when the magnetic alloy as a whole behaves as a cluster glass, the transport properties still act as in a Kondo system. And, possibly, when an amorphous alloy manifests itself as a long-range ferromagnet with respect to standard magnetic measurements, the conduction electrons probe mainly the uncoupled entities at low temperature.

Zero-temperature magnetism is primarily governed by the situation in the first atomic coordination shells in these crystalline and amorphous transition-metal-base compounds, so that the structural disorder does not much affect the basic magnetic properties. However, the opportunity is given to study the screening effect of a large variety of s-p elements dissolved in amorphous transition metal base alloys over a much broader concentration range than in crystalline systems. Structural disorder for homogeneous magnetic amorphous alloys results in a distribution of hff and local moments. But, this distribution is quite comparable to what can be found in crystalline systems where crystal field effects have little influence.

Magnetic properties at finite temperature in transition-metal base amorphous alloys do not appear to strongly reflect the effects of structural disorder. Spin-wave excitations dominate the low-temperature dependence of magnetization without giving much possibility to observe the low-energy excitations of local character which are expected in disordered systems. However, the steep minimum in spin wave energy observed around $q = 3\ \text{Å}^{-1}$ in amorphous CoP and FePC alloys is likely to spur more neutron scattering investigation in the future. On the other hand, in amorphous transition-metal base alloys studied so far, the disorder is averaged out by long-range fluctuations near the Curie transition. Thus, the phase transition is classically defined with critical exponents which do not significantly depart from 3D Heisenberg model predictions.

Of special interest among magnetic amorphous alloys based on transition metals are the Fe-base amorphous alloys. Iron properties are known to be extremely sensitive to Fe environment and interatomic distances, so as to span all the possible magnetic situations between antiferromagnetism and strong ferromagnetism. Thus, it has become possible through various fabrication techniques to create new families of amorphous Fe alloys which were unknown in the crystalline state such as new Fe-base Invar alloys. But, iron itself more than structural disorder is responsible for these new interesting properties.

We started this review with the working definition of an amorphous metallic alloy, as a perfectly disordered material where atomic positions were defined only by the effects of randomness and atomic sizes. It progressively has become evident from magnetic properties that chemical ordering effects play an important role in modeling the amorphous structure. The local symmetry around the glass-former sites is strictly defined and does not vary much between amorphous and crystalline modifications. The SRO around transition-metal atoms was found to be looser than around s-p elements, which can give rise in off-stoichiometric amorphous alloys to demixing effects, or, eventually, to phase separation. To what extent these inhomogeneous states, which constitute more than an exception, are linked to fabrication processes is a problem which remains unsolved as yet. It is then premature to evaluate the exact role of these inhomogeneities in the formation and stability of the amorphous phase. But it seems already clear that they strongly affect the magnetic, transport, and mechanical properties of amorphous alloys.

ACKNOWLEDGMENTS

This review germinated in August 1980 during a summer stay with Prof. W. L. Johnson's group at Caltech. It is my pleasure to thank Prof. W. L. Johnson, Prof. P. Duwez, and their colleagues of the ''amorphous'' group for their hospitality and for stimulating discussions. I am indebted also to my colleagues of University Louis Pasteur, Strasbourg, and University of Nancy I for enlightening conversations. Gordon Fish helped me correct my English. The skilled assistance of Christiane Brun and Gilbert Kaul in preparing the manuscript is gratefully acknowledged.

REFERENCES

1. **König, H.,** The smallest ferromagnetic regions of iron, *Naturwissenschaften,* 33, 71, 1946.
2. **Brenner, A. and Riddell, G.,** Deposition of nickel and cobalt by chemical reduction, *J. Res. Nat. Bur. Stand.,* 39, 385, 1947.
3. **Brenner, A., Couch, D. E., and Williams, E. K.,** Electrodeposition of alloys of phosphorus with nickel or cobalt, *J. Res. Nat. Bur. Stand.,* 44, 109, 1950.
4. **Gubanov, A. I.,** Quasi-classical theory of amorphous ferromagnetics, *Sov. Phys. Solid State,* 2, 468, 1960.
5. **Grigson, C. W. B., Dove, D. B., and Stilwell, G. R.,** Amorphous magnetic films, *Nature (London),* 204, 173, 1964.
6. **Mader, S. and Nowick, A. S.,** Metastable Co-Au alloys. Example of an amorphous ferromagnet, *Appl. Phys. Lett.,* 7, 57, 1965.
7. **Judge, J. S., Morrison, J. R., Speliotis, D. E., and Bate, G.,** Magnetic properties and corrosion behaviour of thin electroless CoP deposits, *J. Electrochem. Soc.,* 112, 681, 1965.
8. **Bagley, B. G. and Turnbull, D.,** Formation and magnetic behaviour of amorphous CoP and NiP alloys, *Bull. Am. Phys. Soc.,* Series II(10), 1101, 1965.
9. **Tsuei, C. C. and Duwez, P.,** Metastable amorphous ferromagnetic phases in palladium-base alloys, *J. Appl. Phys.,* 37, 435, 1966.
10. **Duwez, P. and Lin, S. C. H.,** Amorphous ferromagnetic phase in FePC alloys, *J. Appl. Phys.,* 38, 4096, 1967.
11. **Kobe, S. and Ferchmin, A. R.,** Amorphous magnetism and magnetic materials. Bibliography 1950—1976, *J. Mater. Sci.,* 12, 1713, 1977; **Ferchmin, A. R.,** Bibliography 1976—1977 (unpublished), presented at Amorphous Metallic Materials, Smolenice (Czechoslovakia), September 26 to 29, 1978.
12. **Handrich, K. and Kobe, S.,** *Amorphous Ferro- and Ferrimagnetika,* Akademie Verlag, Berlin, 1980.
13. **Alben, R., Budnick, J. I., and Cargill, G. S., III,** Magnetic structures, in *Metallic Glasses,* Gilmann, J. J. and Leamy, H. J., Eds., American Society for Metals, Metals Park, O., 1978, 304.
14. **Wohlfarth, E. P.,** Magnetic properties of crystalline and amorphous alloys, *J. Magn. Magn. Mater.,* 7, 113, 1978; **Wohlfarth, E. P.,** Some fundamental aspects of amorphous magnetism, *IEEE Trans. Magn.,* MAG-14, 933, 1978.
15. **Methfessel, S.,** Ferromagnetism in amorphous metals, in *Liquid and Amorphous Metals,* Lüscher, E. and Coufal, H., Eds., Sijthoff and Noordhoff, The Netherlands, 1980, 501.
16. **Mizoguchi, T.,** Magnetism in amorphous alloys, *AIP Conf. Proc.,* 34, 286, 1976; **Mizoguchi, T.,** Basic magnetic properties of amorphous alloys, *Sci. Rep. Res. Inst. Tohoku Univ. Ser. A,* 117(Suppl.), 1978.
17. **Graham, C. D. and Egami, T.,** Magnetic properties of amorphous alloys, *Annu. Rev. Mater. Sci.,* 8, 423, 1978; **Graham, C. D. and Egami, T.,** in *Proc. 3rd Conf. Rapidly Quenched Metals,* Vol. 2, Cantor, B., Ed., The Metals Society, London, 1978, 96; **Graham, C. D. and Egami, T.,** *IEEE Trans. Magn.,* MAG-15, 1398, 1979; **Graham, C. D. and Egami, T.,** *J. Magn. Magn. Mater.,* 15—18, 1325, 1980.
18. **Luborsky, F. E.,** Amorphous ferromagnets, in *Ferromagnetic Materials,* Vol. 1, Wohlfarth, E. P., Ed., North Holland, Amsterdam, 1980, 451.

19. **Zaveta, K. and Schneider, J.,** Magnetic properties of amorphous metallic materials, in *Proc. Conf. Amorphous Metallic Materials,* Smolenice (Czechoslovakia), Duhaj, P. and Mrafko, P., Eds., Veda, Bratislava, 1980, 367.
20. **Kronmüller, H., Fähnle, M., Domann, M., Grimm, H., Grimm, R., and Gröger, B.,** Magnetic properties of amorphous ferromagnetic alloys, *J. Magn. Magn. Mater.,* 13, 53, 1979; **Kronmüller, H.,** Magnetization processes and the microstructure in amorphous metals, *J. Phys. (Paris),* 41 (Colloq. C8-618), 1980.
21. **Wright, J. G.,** Amorphous transition metal films, *IEEE Trans. Magn.,* MAG-12, 95, 1976; **Dietz, G.,** Amorphous ferromagnetic materials deposited from vapor or liquids, *J. Magn. Magn. Mater.,* 6, 27, 1977; **Grundy, P. J.,** The structure and magnetic properties of amorphous magnetic thin films, *J. Magn. Magn. Mater.,* 21, 1, 1980.
22. **Gonser, U., Ed.,** *Application of Nuclear Techniques to the the Studies of Amorphous Metals,* Atomic Energy Review, Suppl. 1, International Atomic Energy Agency, Vienna, 1981.
23. **Kajcsos, Zs. et al., Eds.,** *Proc. Int. Conf. Amorphous Systems Investigated by Nuclear Methods,* Central Research Institute for Physics, udapest, 1981.
24. For short reviews along the same line, see **Durand, J.,** Magnetic properties of amorphous metallic alloys related to their atomic scale structure, *J. Phys. (Paris),* 41(Colloq. C8-609), 1980; **Durand, J.,** Les propriétés magnétiques moyennes des métaux et alliages amorphes, *Rev. Phys. Appl. (Paris),* 15, 1036, 1980.
25. **Khanna, S. N. and Cyrot-Lackmann, F.,** On the bulk and surface electronic structure of amorphous transition metals, *J. Phys. (Paris),* 41(Colloq. C8-423), 1980; **Gaspard, J. P.,** Density of d-states in amorphous transition metal alloys, *J. Phys. (Paris),* 41(Colloq. C8-414), 1980.
26. **Gaspard, J. P.,** Theory of electronic structure of metallic glasses, in *Glassy Metals II,* Beck, H. and Güntherodt, H. J., Eds., Springer Verlag, Berlin (to be published).
27. For a review, see **Güntherodt, H. J. et al.,** Electronic structure of liquid and amorphous metals, *J. Phys. (Paris),* 41(Colloq. C8-381), 1980 and **Oelhafen, P.,** Electron spectroscopy of metallic glasses, in *Glassy Metals II,* Beck, H. and Güntherodt, H. J., Eds., Springer Verlag, (to be published).
28. **Wohlleben, D. K. and Coles, B. R.,** Formation of local magnetic moments in metals, in *Magnetism V,* Suhl, H., Ed., Academic Press, New York, 1973, chap. 1; **Blandin, A.,** *in Magnetism V,* Suhl, H., Ed., Academic Press, New York, 1973, 57.
29. **Grüner, G. and Zawadowski, A.,** Magnetic impurities in non-magnetic metals, *Rep. Prog. Phys.,* 37, 1497, 1974; **Rizzuto, C.,** Formation of localized moments in metals: experimental bulk properties, *Rep. Prog. Phys.,* 37, 147, 1974.
30. **Fischer, K. H.,** Dilute magnetic alloys with transition metals as hosts, *Phys. Rep.,* 47, 225, 1978.
31. **Jaccarino, V. and Walker, L. R.,** Discontinuous occurrence of localized moments in metals, *Phys. Rev. Lett.,* 15, 258, 1965.
32. For a review, see **Perrier, J. P. and Tholence, J. L.,** Inhomogeneous character of the transition non magnetism-magnetism in alloys through magnetic measurements, *J. Phys. (Paris),* 35(Colloq. C4-163), 1974.
33. For example, **Brouers, F., Gautier, F., and Van der Rest, J.,** Local environment and magnetic properties in transitional binary alloys, *J. Phys. F,* 5, 975, 1975.
34. **Cyrot, M.,** Occurrence of local magnetic moments in disordered materials, in *Amorphous Magnetism,* Hooper, H. O. and de Graaf, A. M., Eds., Plenum Press, New York, 1973, 161.
35. **Massalski, T. B., Mizutani, U., Hartwig, K. T., and Hooper, R. W.,** Low temperature specific heat measurements on amorphous alloy ribbons, *Phys. Rev. Lett.,* 41, 661, 1978.
36. **Bagley, B. G. and Disalvo, F. J.,** The low-temperature magnetic properties of glassy and crystalline $Pd_{77.5}Cu_6Si_{16.5}$ in *Amorphous Magnetism I,* Hooper, H. O. and de Graaf, A. M., Eds., Plenum Press, New York, 1973, 143; **Müller, M., Beck, H., and Güntherodt, H. J.,** Magnetic properties of liquid Pd, Si and PdSi alloys, *Phys. Rev. Lett.,* 41, 983, 1978.
37. **Garoche, P., Calvayrac, Y., and Veyssié, J. J.,** Electronic properties of amorphous $Cu_{60}Zr_{40}$ alloy, *J. Phys. (Paris),* 41(Colloq. C8-766), 1980.
38. **Amamou, A. and Krill, G.,** Photoemission in a CuZr glassy metal, *Solid State Commun.,* 28, 957, 1978.
39. **Gruzalski, G. R., Weymouth, J. W., Sellmyer, D. J., and Giessen, B. C.,** Magnetism in amorphous transition-metal alloys, in *Amorphous Magnetism II,* Levy, R. A. and Hasegawa, R., Eds., Plenum Press, New York, 1977, 235.
40. **Amamou, A. and Durand, J.,** Giant moments and short range order in amorphous NiPB alloys, *Comm. Phys.,* 1, 191, 1976.
41. **Amamou, A., Aliaga-Guerra, D., Panissod, P., Krill, G., and Kuentzler, R.,** Electronic structure in amorphous and crystalline nickel phosphoborides through photoemission, NMR and specific heat, *J. Phys. (Paris),* 41(Colloq. C8-396), 1980.

42. **Tsuei, C. C. and Hasegawa, R.,** Kondo effect in amorphous palladium-silicon alloys containing transition metals, *Solid State Commun.*, 7, 1581, 1969; **Hasegawa, R. and Tsuei, C. C.,** s-d exchange interaction in amorphous CrPdSi and MnPdSi alloys, *Phys. Rev. B*, 2, 1631, 1970; **Hasegawa, R. and Tsuei, C. C.,** Kondo effect in amorphous FePdSi and CoPdSi alloys, *Phys. Rev. B*, 3, 214, 1971.
43. **Hasegawa, R.,** Magnetic states in amorphous FePdSi alloys, *J. Phys. Chem. Solids*, 32, 2487, 1971.
44. **Zentko, A., Duhaj, P., Potocky, L., Tima, T., and Bansky, J.,** Low-field magnetic susceptibility of amorphous CoPdSi and FePdSi alloys, *Phys. Status Solidi A*, 31 K 41, 1975; **Zentkova, A., Duhaj, P., Zentko, A., and Tima, T.,** Low-field magnetic susceptibility of amorphous CoPdSi and FePdSi alloys in the vicinity of the magnetic transition temperature, *Physica (Utrecht)*, 86—88 B, 787, 1977.
45. **Zentko, A., De-cong-Vinh, Zentkova, A., and Duhaj, P.,** Magnetic properties of amorphous NiPdSi alloys, *J. Phys. (Paris)*, 39(Colloq. C6-951), 1978.
46. **Dublon, G., Lin, C. H., and Bevk, J.,** The magnetic phase diagram of amorphous $Fe_xPd_{82-x}Si_{18}$, $2 \leq x \leq 25$, *Phys. Lett. A*, 76, 92, 1980 and references therein.
47. **Sharon, T. E. and Tsuei, C. C.,** Magnetism in amorphous FePdP alloys, *Phys. Rev. B*, 5, 1047, 1972.
48. **Marzwell, N. I.,** Magnetic and electrical properties of amorphous PdCoP alloys, *J. Magn. Magn. Mater.*, 5, 67, 1977; **Marzwell, N. I.,** Electrical and Magnetic Properties of PdMnP Amorphous Alloys, Report 822-S5-Nov. 1973, California Institute of Technology, Pasadena, 1973, unpublished.
49. **Hasegawa, R.,** Evidence for the $T^{1/2}$ singularity in Kondo alloys, *Phys. Rev. Lett.*, 28, 1376, 1972; **Hasegawa, R.,** Further evidence for the $T^{1/2}$ singularity in amorphous Kondo alloys, *Mater. Sci. Eng.*, 23, 293, 1976; **Hasegawa, R.,** Anomalous Electrical Resistivities and Magnetic Properties of Amorphous Metallic Alloys, California Institute of Technology, Pasadena, 1973, unpublished.
50. **Liang, V. K. C.,** Magnetic States in Amorphous $Pd_{41}Ni_{41}B_{18}$ Alloys Containing Chromium and Iron, Ph.D Thesis, California Institute of Technology, Pasadena, 1971, unpublished; **Liang, V. K. C.,** Concentration dependence of Kondo temperature and unitarity limit in nondilute magnetic amorphous alloys, *Solid State Commun.*, 9, 579, 1971; **Liang, V. K. C.,** Kondo effect in an amorphous $Ni_{41}Pd_{41}B_{18}$ alloy containing Cr, *Phys. Rev. B*, 7, 3215, 1973.
51. **Sinha, A. K.,** Electrical resistivity and magnetic susceptibility of amorphous CrNiPtP alloys, *J. Appl. Phys.*, 42, 5184, 1971.
52. **Amamou, A.,** Magnetism and local environment model in $(Ni_{1-x}Co_x)_{78}P_{14}B_8$ amorphous alloys, *IEEE Trans. Magn.*, MAG-12, 948, 1976.
53. **Hassanain, N. and Durand, J.,** unpublished results.
54. **Durand, J. and Poon, S. J.,** Spin-glasses in amorphous $(Ni,Fe)_{79}P_{13}B_8$ alloys, *J. Phys. (Paris)*, 38(Colloq. C6-953), 1978.
55. **Amamou, A.,** Ferromagnetic and antiferromagnetic coupling in amorphous $(Ni_{100-x}Mn_x)_{78}P_{14}B_8$, in *Amorphous Magnetism II*, Levy, R. A. and Hasegawa, R., Eds., Plenum Press, New York, 1977, 265.
56. **Mizoguchi, T. and Kubo, T.,** The magnetic properties of 3d impurities in amorphous CuZr alloys, *AIP Conf. Proc.*, 29, 167, 1976.
57. **Gruzalski, G. R. and Sellmyer, D. J.,** Magnetically dilute metallic glasses I — 3d moments, *Phys. Rev. B*, 20, 184, 1979.
58. **Amamou, A.,** Manganese impurities and spin-glass like properties in CuZr glassy metals, *J. Phys. (Paris)*, 41(Colloq. C8-670), 1980.
59. **Souletie, J. and Tournier, R.,** Specific heat and magnetization in dilute magnetic alloys, *J. Low Temp. Phys.*, 1, 95, 1969.
60. **Domb, E., Mac Donald, C. A., and Johnson, W. L.,** Magnetic impurities in amorphous superconductors: $T_{Kondo} > T_c$, *Solid State Commun.*, 30, 775, 1979.
61. **Matthias, B. T., Peter, M., Williams, H. J., Clogston, A. M., Corenzwit, E., and Sherwood, R. C.,** Local atomic moment associated with an iron atom dissolved in various transition metals alloys, *Phys. Rev.*, 125, 541, 1962.
62. For an experimental introduction to the Kondo problem, see **Daybell, M. D.,** The s-d model and the Kondo effect: thermal and transport properties, in *Magnetism V*, Suhl, H., Ed., Academic Press, New York, 1973, 121.
63. For a review on the Kondo behavior of the resistivity in amorphous metallic alloys, see **Tsuei, C. C.,** Some controversial aspects of electronic and magnetic interactions in the amorphous metallic state, in *Amorphous Magnetism II*, Levy, R. A. and Hasegawa, R., Eds., Plenum Press, New York, 1977, 181.
64. **Dierker, S. B., Gudmundsson, H., and Anderson, A. C.,** Resistance minima in amorphous PdSi alloy, *Solid State Commun.*, 29, 767, 1979.
65. **Kästner, J., Schink, H. J., and Wassermann, E. F.,** Electrical resistance and magnetoresistance of amorphous alloys $Gd_{67}Co_{33}$ and $Pd_{80}Si_{20}$, *Solid State Commun.*, 33, 527, 1980.
66. **Hamann, D. R.,** New solution for exchange scattering in dilute alloys, *Phys. Rev.*, 158, 570, 1967.
67. **Costa-Ribeiro, J., Souletie, J., Thoulouze, D., and Tournier, R.,** Determination of the number of magnetic Co atoms in dilute Au-Co alloys, *J. Phys. (Paris)*, 32(Colloq. C1-753), 1971.

68. **Buschow, K. H. J., Brouha, M., Biesterbos, J. W. M., and Dirks, A. G.,** Crystalline and amorphous rare-earth transition metal alloys, *Physica (Utrecht),* 91 B, 261, 1977.
69. **Mangin, Ph. and Marchal, G.,** Structural and magnetic properties in Fe_xSi_{1-x} amorphous alloys, *J. Appl. Phys.,* 49, 1709, 1978.
70. **Blandin, A.,** Theory versus experiments in the spin-glass systems, *J. Phys. (Paris),* 39(Colloq. C6-1499), 1978.
71. **Mydosh, J. A. and Nieuwenhuys, G. J.,** Dilute transition metal alloys. Spin glasses, in *Ferromagnetic Materials,* Vol. 1, Wohlfarth, E. P., Ed., North Holland, Amsterdam, 1980, 71.
72. **Poon, S. J. and Durand, J.,** Magnetic-cluster description of spin glasses in amorphous La-Gd-Au alloys, *Phys. Rev. B,* 18, 6253, 1978; **Durand, J. and Poon, S. J.,** Spin glasses in amorphous and crystalline RE alloys, *J. Phys. (Paris),* 40(Colloq. C5-231), 1979.
73. **Larkin, A. I. and Khmel'nitskii, D. E.,** Virial expansion for magnetic impurities in metals, *Soc. Phys. JETP,* 31, 958, 1970; **Matho, K.,** Thermodynamics of dilute magnetic alloys, *Physica (Utrecht),* 86-88B, 854, 1977.
74. **Beck, P. A.,** Properties of mictomagnets (spin glasses), *Prog. Mater. Sci.,* 23, 1, 1978.
75. **Ododo, J. C. and Coles, B. R.,** The critical concentration for the onset of ferromagnetism in CuNi alloys, *J. Phys. F,* 7, 2393, 1977.
76. **Chien, C. L.,** Hyperfine interactions and magnetic ordering temperatures of amorphous $Fe_xPd_{80-x}Si_{20}$ with low Fe concentrations, *Phys. Lett. A,* 68, 394, 1978.
77. **Durand, J.,** Magnetic regimes in amorphous NiFePB alloys, in *Amorphous Magnetism II,* Levy, R. A. and Hasegawa, R., Plenum Press, New York, 1977, 305.
78. **Chen, H. S., Krause, J. T., Shirakawa, K., and Masumoto, T.,** On the anomalies in density, Young's modulus and glass temperature of PdSi glasses, *J. Non-Cryst. Solids,* 41, 79, 1980. See the review of **Chen, H. S.,** Glassy metals, *Rep. Prog. Phys.,* 43, 353, 1980.
79. **Zaveta, K., Schneider, J., Handstein, A., and Zounova, F.,** Superparamagnetic behaviour of $Fe_{1.6}Ni_{78.4}P_{10}B_{10}$ metallic glass, in *Proc. 3rd Conf. Rapidly Quenched Metals,* Vol. 2, Cantor, B., Ed., The Metals Society, London, 1978, 156.
80. **Prater, J. T. and Egami, T.,** Annealing effects in Ni based amorphous alloys with dilute concentrations of Fe, *J. Appl. Phys.,* 50, 1706, 1979.
81. **Krause, J. K., Long, T. C., Egami, T., and Onn, D. G.,** Low-temperature specific heat of the metallic glasses $Fe_xNi_{80-x}P_{14}B_6$ and $(Fe_yNi_{100-y})_{79}P_{13}B_8$ for the spin-glass and spin-cluster-glass regimes, *Phys. Rev. B,* 21, 2886, 1980.
82. **De Gennes, P. G.,** Polarisation de charge (ou de spin) au voisinage d'une impureté dans un alliage, *J. Phys. Rad.,* 23, 630, 1962.
83. **Poon, S. J.,** Attenuation of magnetic interactions in amorphous metals, *Phys. Rev. B,* 21, 343, 1980.
84. **De Chatel, P. F.,** The RKKY interaction in disordered systems, *J. Magn. Magn. Mater.,* 23, 28, 1981.
85. **Hauser, J. J., Hsu, F. S. L., Kammlott, G. W., and Waszczak, J. V.,** Magnetic, electrical, structural, and thermal properties of amorphous MnSi, *Phys. Rev. B,* 20, 3391, 1979; **Cochrane, R. W., Strom-Olsen, J. O., and Rebouillat, J. P.,** A magnetization study of amorphous MnSi, *J. Appl. Phys.,* 50, 7348, 1979.
86. See the recent review of **Forester, D. W., Koon, N. C., Schelleng, J. H., and Rhyne, J. J.,** Amorphous YFe_2 — A concentrated spin-glass, *J. Appl. Phys.,* 50, 7336, 1979.
87. **Bhagat, S. M., Lloyd, J. N., and Pual, D. K.,** Spin resonances in amorphous YFe alloys, *J. Non-Cryst. Solids,* 10, 65, 1979; **Lloyd, J. N., Bhagat, S. M., Knudson, A. R., and Gossett, C. R.,** Evidence for a surface ferromagnetic layer in amorphous YFe_2, *J. Appl. Phys.,* 50, 1614, 1979.
88. **Berrada, A., Lapierre, M. F., Loegel, B., Panissod, P., and Robert, C.,** A study of the magnetic and transport properties of amorphous electrodeposited nickel-phosphorus alloys, *J. Phys. F,* 8, 845, 1978.
89. **Hauser, J. J.,** Amorphous ferromagnetic AgX (X = Ni, Co, Gd) alloys, *Phys. Rev. B,* 12, 5160, 1975.
90. **Liénart, A., Rebouillat, J. P., Garoche, P., and Veyssié, J. J.,** Magnetic properties of amorphous YNi alloys, *J. Phys. (Paris),* 41(Colloq. C8-658), 1980; **Garoche, P., Liénart, A., Rebouillat, J. P., and Veyssié, J. J.,** Low-temperature specific heat of amorphous $Ni_{78}Y_{22}$, *J. Phys. F,* 9, L 129, 1979.
91. **Pan, D. and Turnbull, D.,** Magnetic properties of amorphous CoP alloys, *J. Appl. Phys.,* 45, 1406, 1974.
92. **Marchal, G., Teirlinck, D., Mangin, Ph., Janot, Ch., and Hübsch, J.,** Magnetic behaviour of vapour quenched Co_xSn_{1-x} alloys, *J. Phys. (Paris),* 41(Colloq. C8-662), 1980.
93. **Heiman, N. and Kazama, N.,** Concentration dependence of the Co moment in amorphous alloys with Y, La and Zr, *Phys. Rev. B,* 17, 2215, 1978; **Heiman, N. and Kazama, N.,** Magnetic properties of amorphous alloys of Fe with La, Ln, Y and Zr, *Phys. Rev. B,* 19, 1623, 1979.
94. **Suran, G., Sztern, J., Aboaf, J. A., and McGuire, T. R.,** Magnetic properties of amorphous CoTi thin films, *IEEE Trans. Magn.,* MAG-17, 3065, 1981.
95. **Massenet, O., Daver, H., and Geneste, J.,** Magnetic ordering and magnetization in amorphous FeGe films, *J. Phys.,* 35(Colloq. C4-279), 1974.

96. **Rodmacq, B., Piecuch, M., Janot, Ch., Marchal, G., and Mangin, Ph.**, Structure and magnetic properties of amorphous Fe_xSn_{1-x} alloys, *Phys. Rev. B,* 21, 1911, 1980; **Teirlinck, D., Piecuch, M., Geny, J. F., Marchal, G., Mangin, Ph., and Janot, Ch.**, Magnetic phase diagram in Fe_xSn_{1-x} amorphous alloys, *IEEE Trans. Magn.*, MAG-17, 3079, 1981.
97. **Buschow, K. H. J. and Van der Kraan, A. M.**, Magnetic properties of amorphous $Th_{1-x}Fe_x$ alloys, *Phys. Status Solidi A,* 53, 665, 1979.
98. **Coey, J. M. D., Givord, D., Liénart, A., and Rebouillat, J. P.**, Amorphous Yttrium-iron alloys, *J. Phys. F,* 11, 2707, 1981.
99. **Chien, C. L. and Unruh, K. M.**, Magnetic properties of amorphous Fe_xB_{100-x} ($0 \leq x \leq 90$), *Phys. Rev. B,* 24, 1556, 1981.
100. For recent reviews, see **Yeshurun, Y., Salamon, M. B., Rao, V. K., and Chen, H. S.**, Critical phenomena in amorphous ferromagnetic and spin-glass alloys, *Phys. Rev.*, 24, 1536, 1981; **Geohegan, J. A. and Bhagat, S. M.**, Magnetic phases of amorphous transition-metal-metalloid alloys, *J. Magn. Magn. Mater.*, 25, 17, 1981; **Rao, K. V. and Chen, H. S.**, Manifestations of para-ferro-spin glass transitions in glassy alloys, in *Rapidly Quenched Metals IV,* Masumoto, T. and Suzuki, K., Eds., The Japan Institute of Metals, Sendai, 1982, 1073.
101. **Donnelly, T. A., Egami, T., and Onn, D. G.**, Low-temperature specific heat of the metallic glasses $Fe_xNi_{80-x}P_{14}B_6$ with $x = 0, 20, 40, 60, 80$, *Phys. Rev. B,* 20, 1211, 1979.
102. For a review, see **Nieuwenhuys, G. J., Verbeeck, B. H., and Mydosh, J. A.**, Towards an uniform magnetic phase diagram for magnetic alloys with mixed types of order, *J. Appl. Phys.*, 50, 1685, 1979.
103. **Cochrane, R. W., Harris, R., and Zuckermann, M. J.**, The role of structure in the magnetic properties of amorphous alloys, *Phys. Rep.*, 48, 1, 1978.
104. **Moorjani, K., Ghatak, S. K., Rao, K. V., Kramer, B., and Chen, H. S.**, Spin-glass-paramagnetic phase boundary in amorphous metallic alloys, *J. Phys.*, 41(Colloq. C8-718), 1980.
105. **Lin, S. C. H.**, Resistivity minimum in an amorphous ferromagnetic alloy, *J. Appl. Phys.*, 40, 2173, 1969.
106. For a recent review, see **Babic, E.**, The resistivities of amorphous ferro-magnets, in *Proc. Conf. Amorphous Metallic Materials,* Smolenice (Czechoslovakia), Duhaj, P. and Mrafko, P., Eds., Veda, Bratislava, 1980, 15.
107. **Cochrane, R. W., Harris, R., Strom-Olsen, J. O., and Zuckermann, M. J.**, Structural manifestations in amorphous alloys: resistance minima, *Phys. Rev. Lett.*, 35, 676, 1975.
108. References for these crystalline systems were collected in Durand, J., Concentration effects in temperature dependence coefficients of electrical resistivity of amorphous alloys containing Fe, *Int. Conf. Metallic Glasses, Science and Technology,* Vol. 1, Hargitai, C., Bakonyi, I., and Kemeny, T., Eds., Central Research Institute for Physics, Budapest, 1981, 399.
109. **Poon, S. J. and Durand, J.**, From superconductivity to ferromagnetism in amorphous La-Gd-Au alloys, in *Amorphous Magnetism II,* Levy, R. A. and Hasegawa, R., Eds., Plenum Press, New York, 1977, 245.
110. **Rossiter, P. L.**, A comment on the electrical resistivity during phase separation in alloys, *Philos. Mag. B,* 42, 561, 1980, and references therein.
111. **Rao, K. V., Egami, T., Gudmundsson, H., Aström, H. U., Chen, H. S., and Nägele, W. A.**, A universal plot for the temperature dependence below the minimum in the resistivity of amorphous Fe-Ni alloys, *J. Appl. Phys.*, 52, 2187, 1981, and references therein.
112. **Fert, A., Asomoza, R., Campbell, I. A., and Jouve, H.**, Contribution de l'ordre magnétique à la résistivité d'alliages amorphes, *C.R. Acad. Sci. Ser. B,* 285, 113, 1977.
113. **Cochrane, R. W. and Ström-Olsen, J. O.**, Temperature and magnetic field dependences of the resistivity of amorphous alloys, *J. Phys. F,* 7, 1799, 1977; **Stewart, A. M. and Philipps, W. A.**, Electrical and thermal properties of some amorphous metals, *Philos. Mag.*, 37, 561, 1978.
114. **Durand, J., Thompson, C., and Amamou, A.**, Magnetic and transport properties of amorphous ferromagnetic $(Fe_{100-x}M_x)_{79}P_{13}B_8$ obtained by splat-cooling (M = Mn, Cr, V) ($1 \leq x \leq 4$), in *Proc. 3rd Conf. Rapidly Quenched Metals,* Vol. 2, Cantor, B., Ed., The Metals Society, London, 1978, 109; **Rao, K. V., Gudmundsson, H., Amström, H. U., and Chen, H. S.**, Influence of metal-metalloid elements on the electrical resistance of amorphous materials, *J. Appl. Phys.*, 50, 1592, 1979.
115. **Bergmann, G. and Marquardt, P.**, Temperature dependence of the resistivity in amorphous ferromagnetic alloys. Spin-wave contribution and resistivity minimum, *Phys. Rev. B,* 17, 1355, 1978; **Bergmann, G. and Marquardt, P.**, Resistivity of amorphous ferromagnetic Fe_xAu_{1-x} alloy. Anisotropy and field dependence, *Phys. Rev. B,* 18, 326, 1978.
116. **Grest, G. S. and Nagel, S. R.**, Effective-field distributions and resistivity minima in amorphous ferromagnets, *Phys. Rev. B,* 19, 3571, 1979.
117. **Continentino, M. A. and Rivier, N.**, On the resistivity of amorphous ferromagnets, *J. Phys. F,* 8, 1187, 1978.

118. **Béal-Monod, M. T.,** Kondo-resistivity due to a pair of interacting impurities, *Phys. Rev.*, 178, 874, 1969; **Matho, K. and Béal-Monod., M. T.,** Kondo temperature of interacting magnetic impurities, *J. Phys. F*, 3, 136, 1973; **Levin, K. and Mills, D. L.,** Contribution of giant clusters to the resistivity, Application to Ni-Cu, *Phys. Rev. B*, 9, 2354, 1974.
119. **Kok, W. C. and Anderson, P. W.,** Comments on the paramagnetic Curie temperature in amorphous magnetic alloys, *Philos. Mag.*, 24, 1141, 1971.
120. **Felsch, W.,** Schichten aus amorphem Eisen, *Z. Phys.*, 195, 201, 1966; **Felsch, W.,** Ferromagnetismus von amorphem Eisen, *Z. Phys.*, 219, 280, 1969.
121. **Kuz'menko, V. M., Mel'nikov, V. I., and Sudovtsov, A. I.,** Galvanometric properties and magnetization of amorphous iron films, *Sov. J. Low Temp. Phys.*, 2, 723, 1976.
122. **Wohlfarth, E.,** The possibility that ϵ-Fe is a low-temperature superconductor, *Phys. Lett. A*, 75, 141, 1979; **Kümmerle, W. and Gradmann, U.,** Ferromagnetism in γ-Fe, *Solid State Commun.*, 24, 33, 1977.
123. See, for example, **Friedel, J.,** Metallic alloys, *Nuovo Cimento,* Suppl. 7, 287, 1958.
124. **Albert, P. A., Kovac, Z., Lilienthal, H. R., McGuire, T. R., and Nakamura, Y.,** Effect of phosphorus on the magnetization of nickel, *J. Appl. Phys.*, 38, 1258, 1967.
125. **Takahashi, M., Kim, C. O., Koshimura, M., and Suzuki, T.,** Temperature dependence of the magnetic moment in rapidly quenched NiB alloys, *Jpn. J. Appl. Phys.*, 17, 741, 1978.
126. **Nose, M. and Masumoto, T.,** Characteristics of (Fe, Co, Ni)-Zr amorphous alloys, *Sci. Rep. Res. Inst. Tohoku Univ. Ser. A*, 222(Suppl.), 1980; **Masumoto, T., Ohnuma, S., Shirakawa, K., Nose, M., and Kobayashi, K.,** Magnetic properties of metal-metal amorphous alloys, *J. Phys. (Paris),* 41(Colloq. C8-686), 1980.
127. **Hasegawa, R. and Ray, R.,** Magnetization of glassy CoB alloys, *J. Appl. Phys.*, 50, 1586, 1979; **Watanabe, H., Morita, H., and Yamauchi, H.,** Magnetic properties of amorphous CoB alloys, *IEEE Trans. Magn.*, MAG-14, 944, 1978.
128. **Hasegawa, R. and Ray, R.,** Iron-boron metallic glasses, *J. Appl. Phys.*, 49, 4174, 1978; **Durand, J. and Yung, M.,** Electronic and magnetic properties of amorphous FeP, FeB and FePB alloys, in *Amorphous Magnetism II,* Levy, R. A. and Hasegawa, R., Eds., Plenum Press, New York, 1977, 275.
129. **Ray, R. and Hasegawa, R.,** Rapidly quenched metastable iron-boron solid solutions, *Solid State Commun.*, 27, 471, 1978.
130. **Maurer, M., Cadeville, M. C., and Sanchez, J. P.,** Electronic structure of dilute $Fe_{1-x}P_x$ alloys, *J. Phys. F*, 9, 271, 1979.
131. **Kazama, N. and Heiman, N.,** Magnetic properties of amorphous FeC thin films, *J. Appl. Phys.*, 49, 1706, 1978.
132. **Shigematsu, T., Shinjo, T., Bando, Y., and Takada, T.,** Mössbauer effect of amorphous FeSb thin films, *J. Magn. Magn. Mater.*, 15—18, 1367, 1980.
133. **Wiesner, H. and Schneider, J.,** Magnetic properties of amorphous FeP alloys containing Ga, Ge and As, *Phys. Status Solidi A*, 26, 71, 1974.
134. **Durand, J.,** Concentration dependence of the magnetic properties in amorphous FePB alloys, *IEEE Trans. Magn.*, MAG-12, 945, 1976.
135. **Kazama, N., Masumoto, T., and Mitera, M.,** Effects of metalloids on the magnetic properties of Fe and Co based amorphous alloys, *J. Magn. Magn. Mater.*, 15—18, 1331, 1980.
136. **Narita, K., Fukunaga, H., and Yamasaki, J.,** Effect of metalloid content on Curie-temperature and magnetic moment of amorphous FeSiB alloys, *Jpn. J. Appl. Phys.*, 16, 2063, 1977; **Narita, K., Fukunaga, H., Yamasaki, J., and Hara, K.,** Compositional effects on magnetic properties of FeSiB glassy alloys, *J. Magn. Magn. Mater.*, 19, 145, 1980.
137. **Luborsky, F. E., Walter, J. L., and Wohlfarth, E. P.,** The saturation magnetization, Curie temperature and size effect in amorphous iron alloys, *J. Phys. F*, 10, 959, 1980; **Luborsky, F. E.,** Composition dependence of the Curie temperatures of amorphous alloys, *J. Appl. Phys.*, 51, 2808, 1980.
138. **Johnson, W. L. and Williams, A. R.,** Structure and properties of transition-metal-metelloid glasses based on refractory metals, *Phys. Rev. B*, 20, 1640, 1979.
139. **Mizoguchi, T., Yamauchi, K., and Miyajima, H.,** Ferromagnetism of amorphous iron alloys, in *Amorphous Magnetism I,* Hooper, H. O. and de Graaf, A. M., Eds., Plenum Press, New York, 1973, 325.
140. **Cadeville, M. C. and Daniel, E.,** Structure électronique de borures à base de métaux de transition, *J. Phys. (Paris)*, 27, 449, 1966.
141. **Hasegawa, R.,** Magnetic properties of glassy $(FeNi)_{86}B_{14}$ alloys, *J. Phys. (Paris),* 41(Colloq. C8-701), 1980.
142. **Sinha, A. K.,** Temperature and field dependence of magnetization of amorphous (Fe,Mn) alloys, *J. Appl. Phys.*, 42, 338, 1971; **Hasegawa, R.,** Magnetic properties of an amorphous MnPC alloy, *Phys. Rev. B*, 3, 1631, 1971.
143. **Senno, H., Sakakima, H., and Hirota, E.,** Effect of Mn substitution on magnetic properties of CoSiB and CoFeSiB amorphous alloys, *Sci. Rep. Res. Inst. Tohoku Univ. Ser. A*, 276(Suppl.), 1980.

144. **Aso, K., Makino, Y., Uedaira, S., Ito, S., and Hayakawa, M.,** Effect of Cr substitution on magnetic properties in Fe base amorphous alloys, *Sci. Rep. Res. Inst. Tohoku Univ. Ser. A,* 282(Suppl.), 1980.
145. **Chien, C. L. and Hasegawa, R.,** Mössbauer study of glassy alloys $(Fe,Mo)_{80}B_{20}$, *J. Appl. Phys.,* 49, 1721, 1978; **Chien, C. L. and Chen, H. S.,** Effects of Mo on the magnetic properties of amorphous $(Fe_xMo_{1-x})_{75}P_{16}B_6Al_3$, *J. Appl. Phys.,* 50, 1574, 1979.
146. **Wright, A. C.,** Neutron magnetic scattering studies of amorphous solids, *J. Non-Cryst. Solids,* 40, 325, 1980.
147. For short reviews, see **Durand, J., Panissod, P., Aliaga Guerra, D., and Qachaou, A.,** NMR and Mössbauer spectroscopy in amorphous alloys, *Int. Conf. Metallic Glasses, Science and Technology,* Vol. 1, Hargitai, C., Bakonyi, I., and Kemeny, T. E., Eds., Central Research Institute for Physics, Budapest, 1981, 41; **Durand, J. and Panissod, P.,** Investigation of the atomic scale structure of amorphous metallic alloys by means of bulk and local magnetic measurements, *IEEE Trans. Magn.,* MAG-17, 2595, 1981; **Panissod, P., Durand, J., and Budnick, J. I.,** Hyperfine fields in metallic glasses, *Nucl. Instrum. Methods,* 199, 99, 1982.
148. **Cowlan, N., Davies, H. A., and Dini, K.,** Neutron measurements of magnetic correlations in metallic glasses, *J. Non-Cryst. Solids,* 40, 377, 1980.
149. **Durand, J. and Lapierre, M. F.,** NMR of ^{59}Co in amorphous CoP alloys, *J. Phys. F,* 6, 1185, 1976.
150. **Vincze, I., Kemeny, T., and Arajs, S.,** Short-range order in transition metal-metalloid glasses, *Phys. Rev. B,* 21, 937, 1980.
151. **Mc Coll, J. R., Murphey, D., Cargill, G. S., III, and Mizoguchi, T.,** Spin-wave dispersion and temperature dependence of magnetization in an amorphous CoP alloy, *AIP Conf. Proc.,* 29, 172, 1976.
152. **Birgeneau, R. J., Tarvin, J. A., Shirane, G., Gyorgy, E. M., Sherwood, R. C., Chen, H. S., and Chien, C. L.,** Spin-wave excitations and low-temperature magnetization in the amorphous metallic ferromagnetic $(Fe_xNi_{1-x})_{75}P_{16}B_6Al_3$, *Phys. Rev. B,* 18, 2192, 1978.
153. **Bhagat, S. M., Spano, M. L., Chen, H. S., and Rao, K. V.,** Magnetization of amorphous alloys. Deviations from spin-wave theory at low temperature, *Solid State Commun.,* 33, 303, 1980.
154. **Mook, H. A.,** Magnetic excitations in amorphous ferromagnets, *J. Appl. Phys.,* 49, 1665, 1978.
155. **Rhyne, J. J., Lynn, J. W., Luborsky, F. E., and Walter, J. L.,** Spin excitations in amorphous transition-metal boron glasses, *J. Appl. Phys.,* 50, 1583, 1979; **Bhagat, S. M., Spano, M. L., and Rao, K. V.,** Temperature dependence of the magnetization in amorphous alloys, *J. Appl. Phys.,* 50, 1580, 1979.
156. **Suran, G. and Gambino, R. J.,** Spin-wave spectra of non-magnetostrictive amorphous alloy film, *J. Appl. Phys.,* 50, 7671, 1979; **Krishnan, R., Pravad, S., Sztern, J., Battarel, C., and Morillo, R.,** Spin-wave excitations studies in amorphous $Ni_{50}Co_{40}P_{10}$ films, *Phys. Rev. B,* 21, 1246, 1980.
157. **Ishikawa, Y., Xyanyu, Z., Onodera, S., Ishio, S., and Takahashi, M.,** Spin-wave excitation in amorphous ferromagnetic alloys, in *Rapidly Quenched Metals IV;* Masumoto, T. and Suzuki, K., Eds., The Japan Institute of Metals, Sendai, 1982, 1093.
158. **Balogh, J. and Vincze, I.,** Temperature dependence of the hyperfine field distribution in amorphous ferromagnet, *Solid State Commun.,* 25, 695, 1978; **Schurer, P. J. and Morrish, A. H.,** Long and short-range interactions in ferromagnetic amorphous alloys, *Solid State Commun.,* 30, 21, 1979.
159. **Tsuei, C. C., Longworth, G., and Lin, S. C. H.,** Temperature dependence of the magnetization of an amorphous ferromagnet, *Phys. Rev. B,* 170, 603, 1968.
160. **Mizoguchi, T., Ueda, N., Yamauchi, K., and Miyajima, H.,** Critical behaviour of an amorphous ferromagnet near the Curie point, *J. Phys. Soc. Jpn.,* 34, 1691, 1973.
161. **Kaul, S. N.,** Anomalous magnetic behaviour in the critical region of an amorphous ferromagnetic alloy containing chromium, *Phys. Rev. B,* 22, 278, 1980; **Gaunt, P., Ho, S. C., Williams, G., and Cochrane, R. W.,** Critical behaviour of an amorphous ferromagnet, *Phys. Rev. B,* 23, 251, 1981.
162. **Kaul, S. N.,** Critical phenomena and giant superparamagnetic moments above Curie point of the amorphous $Ni_{60}Fe_{20}P_{14}B_6$ alloys, *Phys. Rev. B,* 23, 1205, 1981; **Kaul, S. N.,** Detailed magnetization study of an amorphous ferromagnet, *Phys. Rev. B,* 24, 6550, 1981.
163. **Malmhäll, R., Bäckström, G., Rao, K. V., and Egami, T.,** Compositional short-range order effect in the Hall resistivity of amorphous $Fe_{27}Ni_{53}P_{14}B_6$, *J. Appl. Phys.,* 50, 7656, 1979; **Malmhäll, R., Bhagat, S. M., Rao, K. V., and Bäckström, G.,** Transport properties of amorphous ferromagnets, *Phys. Status Solidi A,* 53, 641, 1979.
164. **Durand, J., Raj, K., Poon, S. J., and Budnick, J. I.,** Local and bulk magnetic properties of amorphous ferromagnetic Gd_2Co, *IEEE Trans. Magn.,* MAG-14, 722, 1978.
165. **Harris, A. B.,** Effect of random defects on the critical behaviour of Ising models, *J. Phys. C,* 7-1671, 1974; **Lubensky, T. C.,** Critical properties of random-spin models from the ϵ-expansion, *Phys. Rev. B,* 11, 3573, 1975.
166. **Schowalter, L. J., Salamon, M. B., Tsuei, C. C., and Craven, R. A.,** The critical specific heat of an amorphous ferromagnet, *Solid State Commun.,* 24, 525, 1977.

167. **Craven, R. A., Tsuei, C. C., and Stephens, R.**, Specific heat and resistivity of an amorphous alloy, $Fe_{34}Pd_{46}P_{20}$, near its ferromagnetic phase transition, *Phys. Rev. B,* 17, 2206, 1978; **Babic, E., Marohnic, Z., Saub, K., Fogarassy, B., and Kemeny, T.**, Curie point anomalies in metallic glasses, *J. Magn. Magn. Mater.,* 15—18, 249, 1980.
168. **Tarvin, J. A., Shirane, G., Birgeneau, R. J., and Chen, H. S.**, High temperature spin dynamics in an amorphous ferromagnet, *Phys. Rev. B,* 17, 241, 1978.
169. **Kronmuller, H. and Fahnle, M.**, Micromagnetism theory of phase transitions in inhomogeneous ferromagnets, *Phys. Status Solidi B,* 97, 513, 1980; **Kronmuller, H. and Fahnle, M.**, *Phys. Status Solidi B,* 98, 219, 1980; **Kronmuller, H. and Fahnle, M.**, *Phys. Status Solidi B,* 101, 713, 1980.
170. **Chen, H. S.**, Effects of composition and structure on magnetic properties of amorphous PePC alloys, *Phys. Status Solidi A,* 17, 561, 1973; **Chen, H. S., Sherwood, R. C., and Gyorgy, E. M.**, The influence of composition and aging on the Curie temperature of metallic glasses, *IEEE Trans. Magn.,* MAG-13, 1538, 1977.
171. **Panissod, P., Aliaga-Guerra, D., Amamou, A., Durand, J., Johnson, W. L., Carter, W. L., and Poon, S. J.**, Local symmetry around the glass-former sites in amorphous metallic alloys through electric quadrupole effects, *Phys. Rev. Lett.,* 44, 1465, 1980.
172. **Vincze, I. and Van der Woude, F.**, Correlation between amorphous and metastable crystalline alloys, *J. Non-Cryst. Solids,* 42, 499, 1980.
173. **Panissod, P., Qachaou, A., Durand, J., and Hasegawa, R.**, Hyperfine field, concentration and magnetic anisotropy distributions in amorphous CoB and CoP alloys, *Int. Conf. Amorphous Systems Investigated by Nuclear Methods,* Vol. 1, Kajcsos, Zs. et al., Eds., Central Research Institute for Physics, Budapest, 1981, 543.
174. **Durand, J., Aliaga-Guerra, D., Panissod, P., and Hasegawa, R.**, Hyperfine field distribution at ^{59}Co in amorphous Fe(Co)PB alloys, *J. Appl. Phys.,* 50, 7668, 1979.
175. **Goto, M., Tange, H., Tokunaga, T., Fujii, H., and Okamoto, T.**, Magnetic properties of the $(Fe_{1-x}M_x)_3P$ compounds, *Jpn. J. Appl. Phys.,* 16, 2175, 1977.
176. **Kullmann, U. and Dietz, G.**, Compositional inhomogeneities in electrodeposited CoP alloys, *Appl. Phys.,* 24, 239, 1981.
177. **Chi, G. C. and Cargill, G. S., III**, Annealing effects in amorphous CoP alloys, *Mater. Sci. Eng.,* 23, 155, 1976; **Chi, G. C. and Cargill, G. S., III**, Structural characterization of amorphous electrodeposited cobalt-phosphorous alloys, *J. Appl. Phys.,* 50, 2713, 1979.
178. **Mizutani, U. and Massalski, T. B.**, Hall-effect measurements and the electronic structure of amorphous PdSi(Cu) alloys, *Phys. Rev. B,* 21, 3180, 1980.
179. **Inoh, H. and Nanao, S.**, A study of local atomic and electronic structure of amorphous alloys by Mössbauer spectroscopy, *Sci. Rep. Res. Inst. Tohoku Univ. Ser. A,* 105(Suppl.), 1978.
180. **Johnson, W. L.**, Superconductivity in metallic glasses, *J. Phys.,* 41(Colloq. C8-731), 1980.
181. **Logan, J. and Yung, M.**, The low-temperature electrical resistivity of amorphous FeP alloys, *J. Non-Cryst. Solids,* 21, 151, 1976; **Logan, J.**, The RDF of amorphous FeP alloys, *Phys. Status Solidi A,* 32, 361, 1975.
182. **Takahashi, M. and Koshimura, M.**, Mössbauer effect of amorphous Fe-P alloys, *Jpn. J. Appl. Phys.,* 18, 685, 1979.
183. **Tenhover, M.**, Fe_x^{57} Mössbauer effect study of the metallic glasses $Y_{1-x}Fe_s$, *J. Phys. Chem. Solids,* 42, 325, 1981.
184. **Predel, B.**, T[illegible]namic investigations on the formation and decomposition of metallic glasses, *Physica,* 103 B, 11[illegible]1.
185. **Mangin, Ph., Marchal, G., Mourey, C., and Janot, Ch.**, Physical studies of $AuSi_{1-x}$ amorphous alloys, *Phys. Rev. B,* 21, 3047, 1980; **Marchal, G., Mangin, Ph., and Janot, Ch.**, Crystallization of amorphous Au_xSi_{1-x} alloys, *Philos. Mag.,* B 42, 81, 1980.
186. **Kim, O. C. and Johnson, W. L.**, Amorphous phase separation in the metallic glasses $(Pb_{1-y}Sb_y)_{1-x}Au_x$, *Phys. Rev. B,* 23, 143, 1981.
187. **Tanner, L. E. and Ray, R.**, Phase separation in ZrTiBe metallic glasses, *Scr. Metall.,* 14, 657, 1980.
188. For a review, see **Fukamichi, K., Kikuchi, M., Hiroyoshi, H., and Masumoto, T.**, Invar and Elinvar characteristics in amorphous Fe base alloys, *Sci. Rep. Res. Inst. Tohoku Univ. Ser. A,* 199(Suppl.), 1978; **Fukamichi, K., Kikuchi, M., Hiroyoshi, H., and Masumoto, T.**, Invar and Elinvar amorphous alloys, *IEEE Trans. Magn.,* MAG-15, 1404, 1979.
189. **Walter, J. L. and Bartram, S. F.**, Crystallization of some amorphous alloys, in *Proc. 3rd Conf. Rapidly Quenched Metals,* Vol. 1, Cantor, B., Ed., The Metals Society, London, 1980, 307; see also some critical comments, ibid., 441, 442.
190. For a recent discussion of the Invar problem, see **Ododo, J. C.**, The onset of ferromagnetism in transition metal alloys, *J. Phys. F,* 10, 2515, 1980.

Chapter 5

APPLICATION OF METALLIC GLASSES IN LOW-FREQUENCY MAGNETIC DEVICES

Ryusuke Hasegawa

TABLE OF CONTENTS

I. INTRODUCTION

The most widely used magnetic devices in the low frequency (50/60 Hz) regime are power transformers and electric motors. The magnetic cores used in these applications are usually silicon and carbon steels of various grades. The most advanced crystalline core materials are grain-oriented iron-silicon alloys which meet the requirements of high saturation magnetic induction (B_s) of about 2 tesla, low ac core loss of about 1.3 W/kg at the induction level of 1.6 T (16 kG) and economic feasibility. The relationship between microstructure such as grain orientations and magnetic properties of silicon steel has been well established.[1] Based on the understanding of this relationship, steady improvements are being made.[2]

Possible use of metallic glasses in the above-mentioned area of magnetic applications was realized with the synthesis of a noncrystalline $Fe_{80}B_{20}$ alloy in ribbon form by a rapid quenching technique. This material was shown to have a B_s value at room temperature of about 1.6 T and ac core loss lower than that of a crystalline 50Fe–50Ni alloy having the same B_s value.[3] Considerable efforts have been since directed at increasing the saturation induction and reducing the core loss (L) even further. A number of systematic approaches adopted in these efforts are based on known metallurigcal facts and limited understanding of magnetism in noncrystalline solids. This is because the level of understanding of the relationship between physical properties and local atomic short range order characteristic of metallic glasses is lower than in the crystalline case. Attempts, however, are made in the following sections to correlate the magnetic properties and structure of the materials relevant to the present section, whenever possible. More importantly stressed is the fact that considerably lower ac core loss and exciting power can be achieved in metallic glasses than in conventional silicon steels. This is quite significant in light of diminishing energy sources and increasing energy costs.

II. MAGNETIC PROPERTIES

A. Saturation Magnetization

1. Iron-Transition Metal-Boron Series

One approach in designing a glassy alloy with a high saturation induction $B_s(RT)$ at room temperature may be guided by the curves shown in Figure 1.[3,4] In the case of $(Fe_{1-x}Co_x)_{80}B_{20}$, it is seen that saturation moment (μ_s peaks around x ~0.15 and the Curie temperature, T_c, increases with x up to about x = 0.5. This means that $B_s(RT) = 4\pi M_s(RT)d$ could be maximized near x = 0.15, where $M_s(RT)$ is the saturation magnetization at room temperature and *d* is the mass density. The effect of the boron content on the magnetic properties include the following: Decreasing the boron content results in increase of μ_s per transition metal atom,[5,6] decrease of T_c for Fe–B (see Figure 4)[5] and increase of T_c for Co–B system.[6] Combination of these effects leads to maximizing the $M_s(RT)$ value in the ternary Fe–Co–B system which is summarized in Figure 2.[7] Although an induction level of as high as 1.91 T has been observed in $Fe_{69}Co_{18}B_{13}$, its low crystallization temperature (T_x ~ 630 K) and closeness to the glass-crystalline composition boundary make this alloy impractical from technological standpoints. One of the more practical alloys, $Fe_{67}Co_{18}B_{15}$, has $B_s(RT)$ = 1.81 T and is compared in Section IV with other Fe-rich alloys in light of ac core losses.

With Ni as the transition metal (*TM*) and with the boron content of 20 at.%, the rate of decrease of μ_s with Ni content (see Figure 1) is larger than the rates of increases of T_c and d, resulting in decrease of B_s with Ni content. For example, $B_s(RT)$ = 1.51 T is obtained for $Fe_{70}Ni_{10}B_{20}$. The situation is different when the boron content is decreased from 20 at.%, an example of which is summarized in Figure 3. The general trend of μ_s vs. *TM* composition for the 14 at.% B case resembles that of the crystalline case. Furthermore, the rapid increase[8] of T_c and *d* with Ni content result in a slight increase of $B_s(RT)$ up to Ni/(Ni + Fe)~0.1. A B_s value of 1.63 T has been obtained for $Fe_{78}Ni_8B_{14}$.

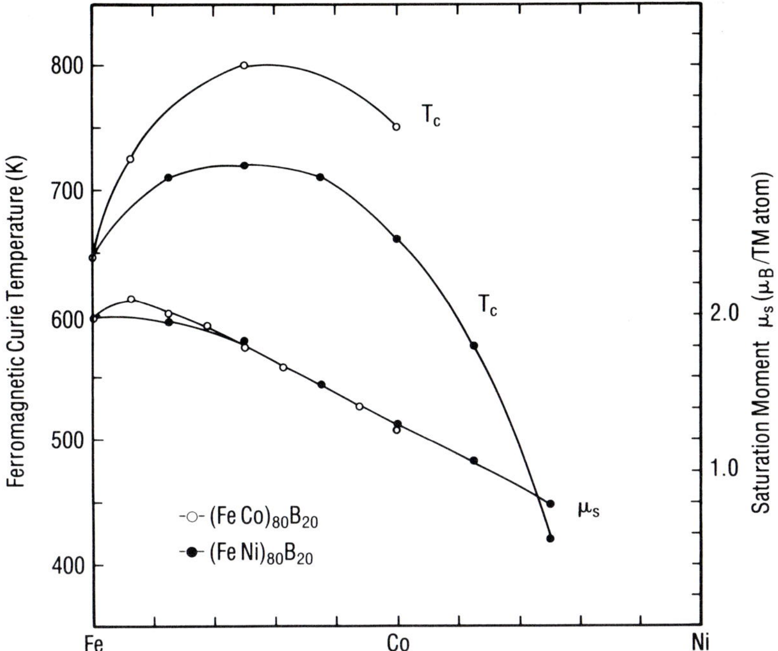

FIGURE 1. Saturation moment μ_s at T = 4.2 K and ferromagnetic Curie temperature T_c as a function of transition metal (*TM*) composition for $TM_{80}B_{20}$ metallic glasses.[3,4]

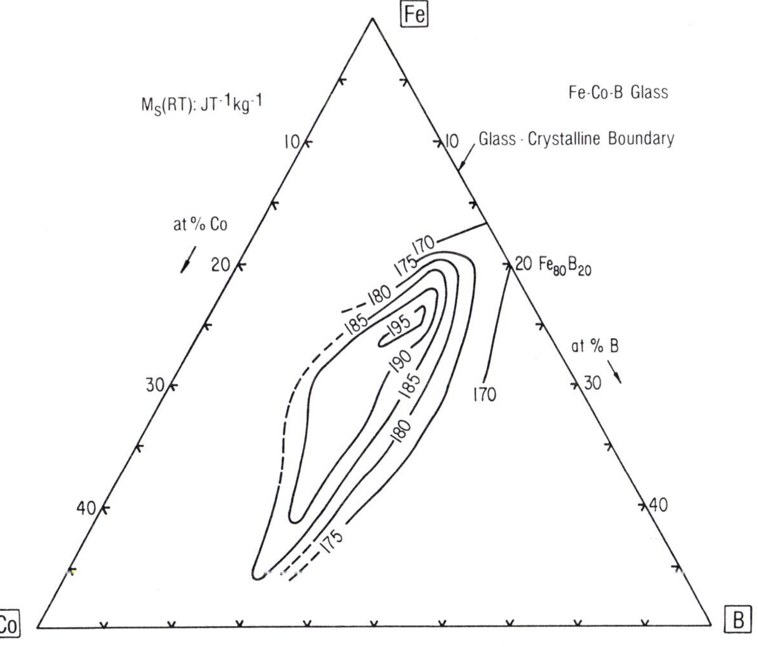

FIGURE 2. Partial Fe–Co–B ternary diagram showing contours of equal magnetization in 640 kA/m at room temperature. (From O'Handley, R. C., Chou, C.-P., and DeCristofaro, N., *J. Appl. Phys.*, 50, 3603, 1979. With permission.)

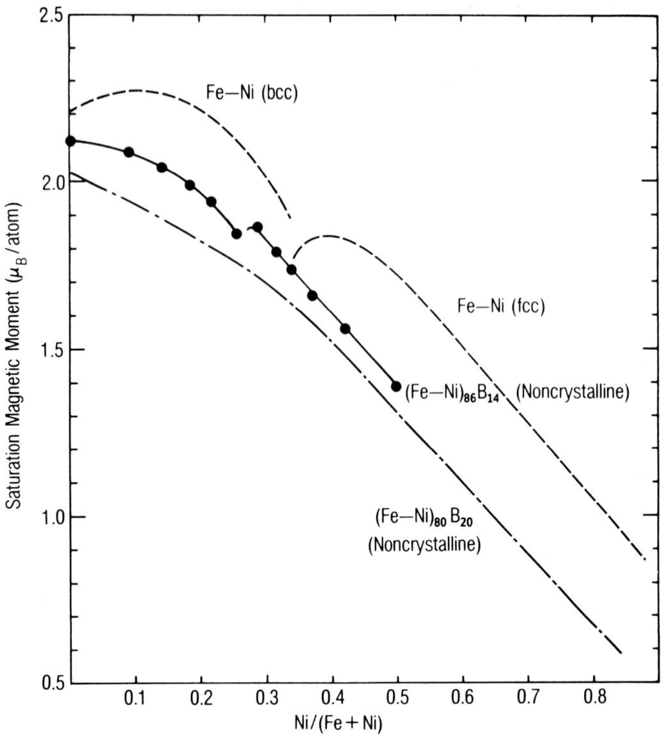

FIGURE 3. Saturation magnetic moment per *TM* atom at 4.2 K vs. Ni content for glassy and crystalline alloys. (From Hasegawa, R., *J. Phys. (Paris)*, 41, C8-701, 1980. With permission.)

Transition metals other than Co and Ni interact with Fe either antiferromagnetically (e.g., *TM* = V, Cr, and Mn) or nonmagnetically (e.g., *TM* = Zr, Nb, Mo, etc.). These cases, therefore, are not relevant to the context of this chapter.

2. *Iron-Boron-Metalloids Series*

Efforts were made to increase the B_s value by varying the composition of Fe–B system immediately after the successful synthesis of the glassy $Fe_{80}B_{20}$ alloy. It was found, however, that with decreasing B content T_c decreases rapidly while μ_s increases slightly.[5] This is shown in Figure 4. Two different explanations have been proposed for the observation. One is based on the hypothesis that the local atomic arrangement for Fe atoms tends to take an fcc-like structure as B content approaches zero.[9] The local structure in this hypothesis is not expected to change considerably by varying B content. The mass density data taken on the binary system, however, indicate clearly a considerable difference in the composition dependence of *d* above and below about 20 at.% B, suggesting a change in the atomic structure near this composition. This has led to another notion that the basis of the local atomic arrangement changes from nearly 12-fold coordinated random packing structure in high B-containing alloys to a bcc-like one in the low B region.[5] Currently available information on the local structure, however, is not precise enough to determine the type of local atomic arrangement which is responsible for the observed properties. The efforts to obtain high values of B_s by modifying the metalloid (*M*) part of the composition, therefore, is empirical. Some examples are described below.

The magnetization M_s at 295 K for various substitutions (*M* for boron is summarized in Figure 5. Aluminum substitution reduces M_s(RT) considerably and therefore is of little

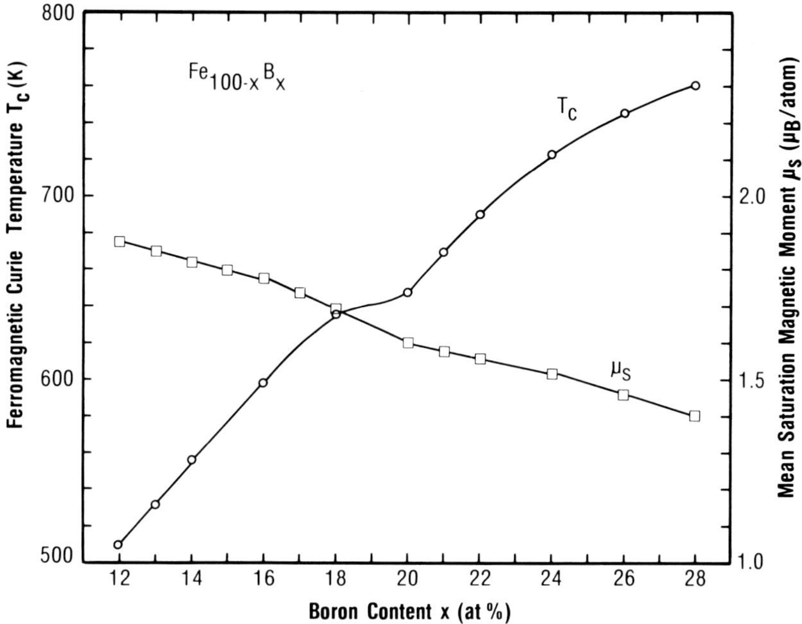

FIGURE 4. Mean saturation moment $\bar{\mu}_s$ and Curie temperature T_c for binary Fe–B glasses. (From Hasegawa, R. and Ray, R., *J. Appl. Phys.*, 50, 1586, 1979. With permission.)

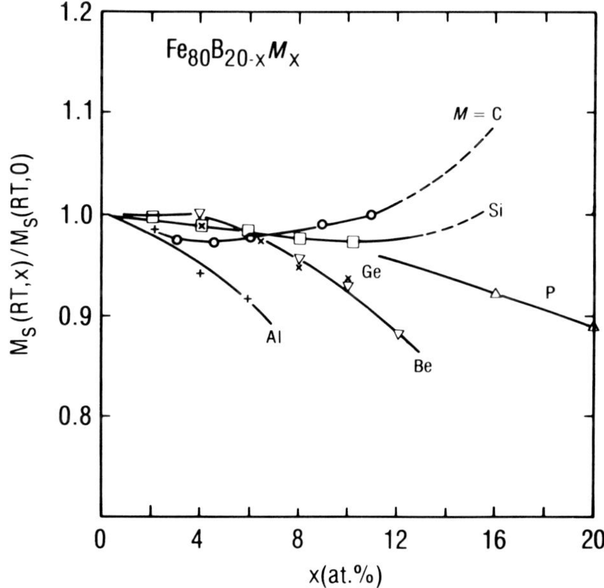

FIGURE 5. Saturation magnetization at 295 K relative to that of $Fe_{80}B_{20}$ [M_s(RT) = 170 JT^{-1} kg^{-1}] as a function of metalloid substitution.[7,12]

interest here. It is noted that M_s(RT) can be increased significantly when M = C. For example, a value of M_s(RT) = 178 $JT^{-1}kg^{-1}$ corresponding to B_s(RT) = 1.75 T has been reported for $Fe_{85}B_8C_7$.[10] However, it has been found that the carbon-containing alloys are thermally and magnetically unstable.[11] Although phosphorus tends to ease the glass form-

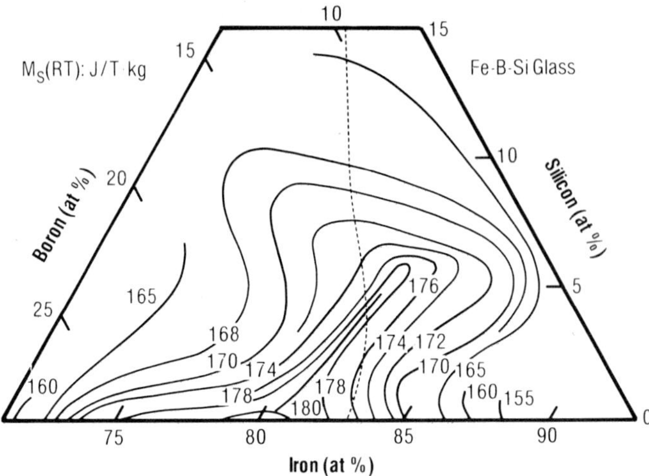

FIGURE 6. Saturation magnetization at room temperature for Fe–B–Si alloys.[14] The dashed line represents the eutectic trough.[15]

ability of iron-base alloys, it also reduces the thermal stability of the alloys.[3] Beryllium up to 4 at.% is an attractive candidate because it does not decrease M_s(RT) while decreasing the magnetostriction appreciably (see Section C).[12] However, due to toxic nature of Be, technological use of this alloy system is probably not realistic. The change of M_s(RT) with increasing Ge content is somewhat similar to the case with Be. Larger mass density of Ge, however, makes the material denser, resulting in higher induction values. It is reported that B_s(RT) = 1.72 T has been achieved in $Fe_{82}B_{13}Ge_5$.[13] This B_s value is among the highest attained in Fe–B base alloys. The relatively low thermal stability and the higher material cost of this alloy, however, make it less feasible in commercial frequency applications. Thus, considering both technical and economical aspects of the present application of these materials, one is likely led to choose Si as a replacement for B.

The equimagnetization lines at room temperature for glassy Fe–B–Si alloys are shown in Figure 6.[14] Also indicated in the figure is the eutectic trough (melting point minimum).[15] Thus the composition at which M_s(RT) has a broad maximum and the melting point is the lowest is near $Fe_{82}B_{14}Si_4$. More detailed data on M_s(RT), T_c and T_x near this composition are shown in Figure 7, where considerable increases of T_c and T_x are noticed in the Si-containing alloys.[7] It appears on the basis of Figures 6 and 7 that the optimum composition would be $Fe_{82}B_{18-x}Si_x$ with $4 < x < 10$. The upper limit for x is determined by the glass-crystalline phase boundary which is near x = 12.5 for liquid quenching rates (~10^6 K/sec) currently accessible. Although much improvement in the value of B_s(RT) over that for $Fe_{80}B_{20}$ has not been made, increases of T_c and T_x attained by replacing B by Si in the Fe–B system should be of practical importance.

The alloys near the optimal composition are hypoeutectic (low in metalloid content) and bcc Fe(B,Si) is their primary phase on equilibrium solidification as expected from previous work.[5] This means that these glassy alloys could be prone to local crystallization during rapid solidification. This appears to be the case, since reproducibility of various soft magnetic properties, which are sensitive to homogeneities of the materials, has been found to be poor in such alloys as $Fe_{82}B_{12}Si_6$. This has led to the development of Fe–B–Si–C alloys, which has resulted in an optimal composition close to $Fe_{82}B_{12}Si_4C_2$.[11,16] The ratio of Si/C near 2 seems to inhibit the local crystallization mentioned above.[16] The tendency of the local crystallization is expected to be much reduced when the metalloid content is increased to hypereutectic levels because of the higher free energy for the equilibrium Fe_3B or Fe_2B state

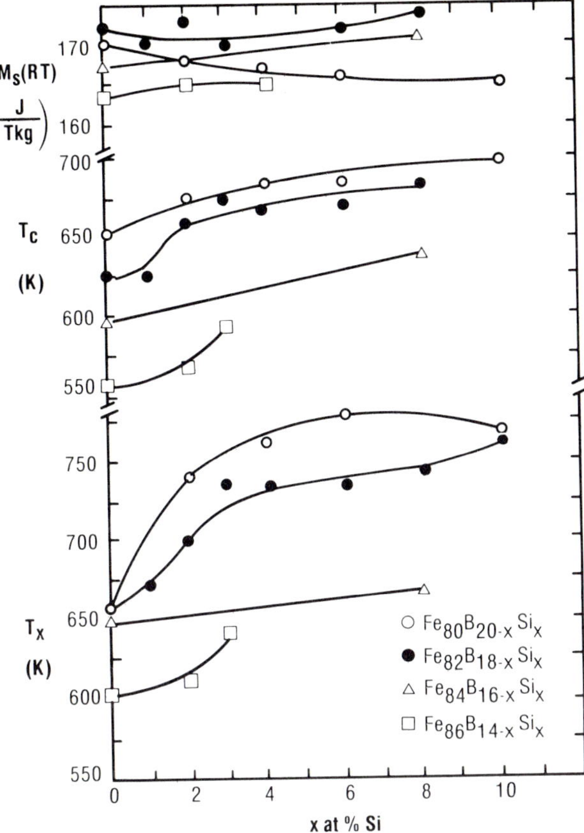

FIGURE 7. Room temperature saturation magnetization M_s(RT), Curie temperature T_c and primary crystallization temperature T_x for Fe–B–Si glasses. (From O'Handley, R. C., Chou, C.-P., and DeCristofaro, N., *J. Appl. Phys.*, 50, 3603, 1979. With permission.)

than that for bcc Fe.[17] When the metalloid content increases, Curie temperature increases while saturation moment and mass density decrease. This leads to a reduction of B_s(RT) but also to a smaller rate of decrease of B_s with temperature, which is generally favorable in magnetic device applications. Based on these facts, a new metallic glass, $Fe_{78}B_{13}Si_9$, has been developed for low frequency magnetic applications (see Section IV).[18] Comparison of the temperature dependence of B_s for $Fe_{78}B_{13}Si_9$ and $Fe_{81}B_{13.5}Si_{3.5}C_2$ is shown in Figure 8.

B. Magnetic Anisotropy

One of the dominant contributions to the bulk magnetic anisotropy resulting in a magnetically easy axis of a crystalline ferromagnet is the magnetocrystalline anisotropy which arises from magnetic interactions between atoms situated at certain lattice sites. In a solid where atomic arrangement is completely random, the magnetocrystalline anisotropy averaged over a certain volume vanishes. This situation may be realized in disk-shaped ferromagnetic noncrystalline foils prepared by a piston and anvil technique.[19] For example, such a foil of $Fe_{75}P_{15}C_{10}$ exhibits an in-plane uniaxial anisotropy energy, K_u, of less than 50 J/m^3.[20] Local anisotropy energy, however, must be substantially larger than this value, since magnetic domains can be observed in this foil.[21]

Chill casting techniques to produce glassy alloys in ribbon form appear to introduce a uniaxial anisotropy along the ribbon axis.[22] Since the temperature dependence of K_u is best

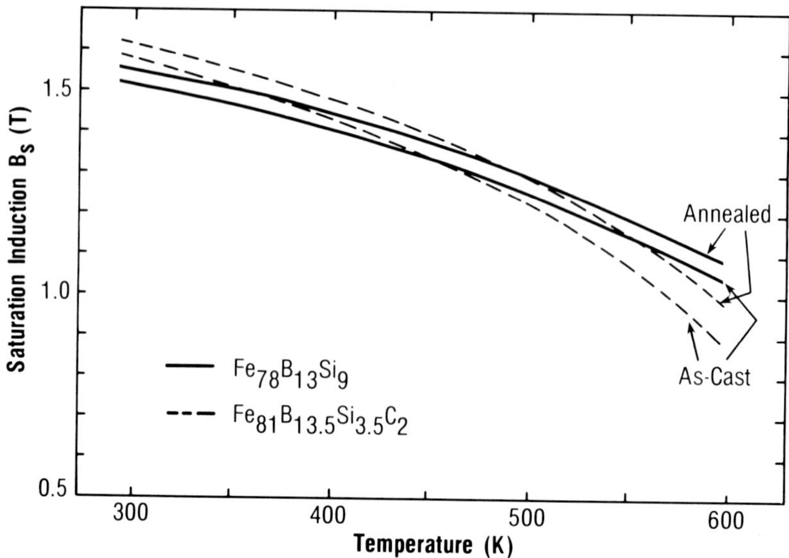

FIGURE 8. Saturation induction as a function of temperature for as-cast and annealed (at T_c) Fe–B–Si and Fe-B–Si–C alloys.[18]

described by $K_u(T) \propto [M_s(T)]^2$, the origin of this anisotropy ($K_u = 10^2 - 10^3$ J/m^3) is probably due to atomic pair ordering[22] which may be introduced during rapid solidification. It has also been found that magnetic fields applied at elevated temperatures near the Curie temperature but below the crystallization temperature of the material modify the magnitude and the direction of the bulk magnetic anisotropy. The field-induced uniaxial anisotropy energy can reach the order of 10^3 J/m^3 in liquid-quenched *TM*-base glassy alloys. Figures 9 and 10 represent the maximum induced K_u values for the Fe–B and Fe–B–Si systems.[23]

It has been speculated that the field-induced anisotropy also originates from atomic pair ordering.[24] The notion that atomic pair ordering is the origin of the as-quenched and field-induced anisotropy seems incompatible with the random atomic arrangement implied for noncrystalline solids but is plausible for the following reason. As pointed out previously,[25] the anisotropy energy which is dipolar in origin, is of the order of 10^{-22} J per atomic pair. The observed value of $K_u \sim 10^3$ J/m^3 then corresponds to about 10^{25} atom pairs/m^3 which is considerably smaller than the number of magnetic atoms ($\sim 8 \times 10^{28}$ atoms/m^3), for example, in glassy $Fe_{80}B_{20}$ alloy. In terms of Néel-Taniguchi theory of directional ordering,[25] the induced uniaxial anisotropy is proportional to $-N\ell\ell_a/k_BT_a$ where N is the total number of atomic pairs, ℓ and ℓ_a are the average energies for the dipole interactions at measuring and annealing temperatures (T_a), respectively, and k_B is the Boltzmann constant. This implies that $K_u \propto [M(T)]^2$ for a given composition and $K_u \propto c_i A^2 c_B{}^2$ for a given temperature when two atoms A and B with the relative concentrations of c_A and c_B are responsible for the magnetic interactions. The first prediction has been confirmed in a number of metallic glasses including $(Fe\text{–}Ni)_{80}P_{14}B_6$[22,26] and Fe–B.[27] The second prediction, however, has not been widely verified.[24,26] More investigation is clearly needed to clarify the origin of the magnetic anisotropy.

Magnetic anisotropy is needed in many magnetic applications of materials. In power transformers, for example, the anisotropy must be high enough to keep the remanence to saturation ratio close to 1, but sufficiently low to maintain relatively low coercivity so that ac core loss and exciting power can be as low as possible. It appears that $K_u = 500$ to 1000 J/m^3 is optimal for low-frequency applications as discussed in Section IV.

Once a desired level of the anisotropy is achieved, e.g., by field-annealing, it must be thermally stable in the temperature range in which the devices are operated. The magnetic

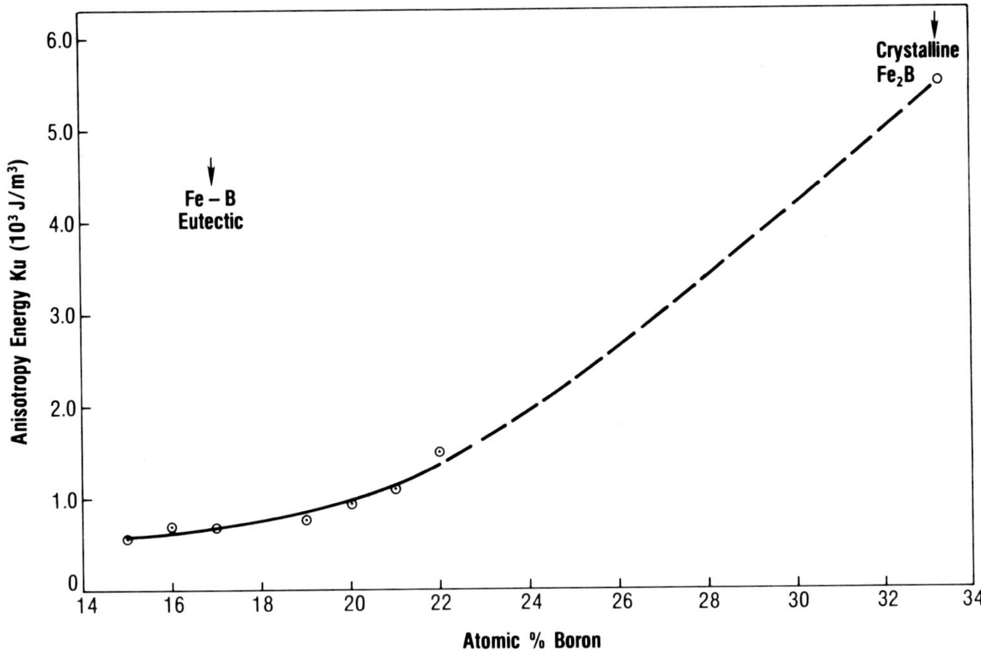

FIGURE 9. Variation of magnetic anisotropy vs. boron content for glassy Fe– B system. (From Nesbitt, E. A., Hasegawa, R., O'Handley, R. C., and Sullivan, M. O., *J. Appl. Phys.*, 51, 3328, 1980. With permission.)

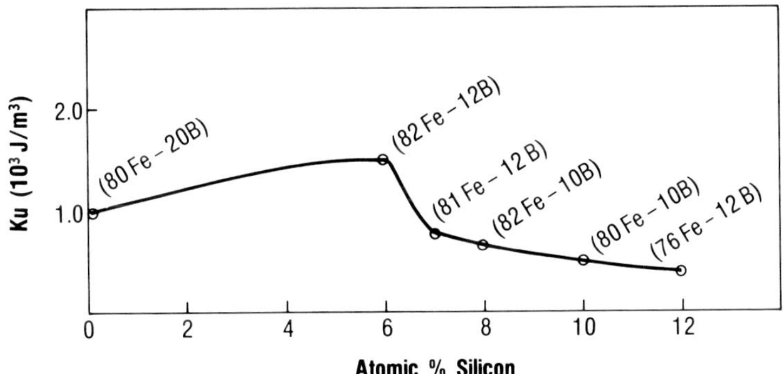

FIGURE 10. Effect of replacing boron with silicon on the magnetic anisotropy. Samples were cooled from above their Curie temperatures at a rate of 1 to 2 K/min. (From Nesbitt, E. A., Hasegawa, R., O'Handley, R. C., and Sullivan, M. O., *J. Appl. Phys.*, 51, 3327, 1980. With permission.)

stability thus defined may be inferred from the kinetics of magnetic anisotropy reorientation. The thermally activated process can be described by $\tau = \tau_0 \exp (E_a/k_BT)$ where τ is the time required for a given degree of anisotropy reorientation at a given temperature T, E_a is the activation energy for the process, τ_0 is the time constant and k_B is the Boltzmann constant. Some examples of the above relationship are shown in Figure 11 for some representative glassy alloys. The time constant τ_0 is important in determining the stability of the system. It, however, depends on various factors including the entropy of the system and is not well characterized at the present moment. The data in Figure 11 were taken on samples with different initial magnetic states, which makes the comparison of τ_0 less meaningful. The activation energy, on the other hand, indicates the relative ease or difficulty for the thermal

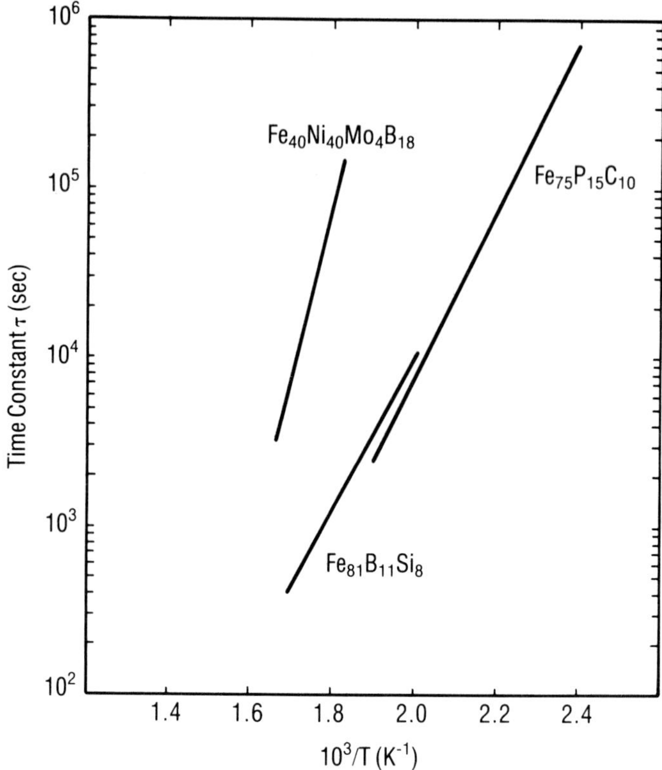

FIGURE 11. Time constant for the magnetization reorientation as a function of the inverse of temperature for various metallic glasses.[28,29]

process. The quantity E_a ranges from about 1.6×10^{-19} J (1 eV) for $Fe_{75}P_{15}C_{10}$[28] and $Fe_{81}B_{11}Si_8$[23] to about 3.2×10^{-19} J (2 eV) for $Fe_{40}Ni_{38}Mo_4B_{18}$.[29] These values for E_a correspond to those for atomic diffusion processes in metallic solids and are relatively low. However, they seem to be sufficiently high for such applications as power transformers in which the core materials are constantly magnetized along a certain direction.

C. Magnetostriction

When a magnetic material is placed in magnetic fields, its physical dimension generally changes slightly due to magnetoelastic effects. The effects may be described by the Hamiltonian[30]

$$H = \vec{S}_i \cdot \overleftrightarrow{J}_{ij} \cdot \vec{S}_j + \vec{S}_i \cdot \overleftrightarrow{D} \cdot \vec{S}_i + c\epsilon^2/2$$

containing exchange, crystal field and elastic terms, where $\mathbf{S}_{i(j)}$ is the spin operator for atom i (j). The exchange ($\vec{\mathbf{J}}_{ij}$) and the crystal field ($\vec{\mathbf{D}}$) parameters can be considered as a function of strain ϵ, and the above Hamiltonian can be expanded by using strain derivatives of $\vec{\mathbf{J}}_{ij}$ and $\vec{\mathbf{D}}$. One can, for example, minimize the free energy with respect to ϵ to obtain the equilibrium strain. The coefficients of the resulting spin correlation functions such as $<\vec{S}_i \cdot \vec{S}_j>$ or $<S_{iz} S_{jz}>$ and $<S_i^2>$ or $<(S_{iz})^2>$ contain $\partial\vec{\mathbf{J}}_{ij}/\partial\epsilon$ and $\partial\vec{\mathbf{D}}/\partial\epsilon$, respectively. These terms are responsible for the magnetoelastic effects considered here. Among the effects, volume and linear magnetostrictions are discussed below because they are important quantities affecting magnetic properties of metallic glasses.

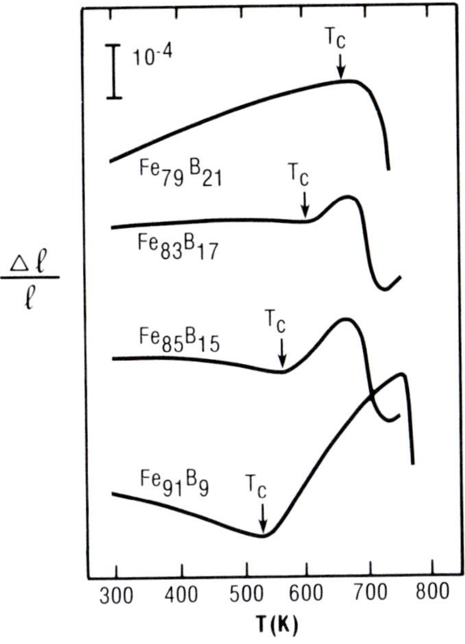

FIGURE 12. Thermal expansion as a function of temperature for Fe–B glasses.[34]

1. Volume Magnetostriction

The isotropic terms discussed above containing $>\vec{S}_i \cdot \vec{S}_j>$ and $<S_i^2>$ originate from isotropic exchange and Coulomb interactions. These interactions can be observed as volume magnetostriction, a change of volume (V) of a magnetic material in applied fields of $\partial\omega/\partial H = (V^{-1})\ \partial V/\partial H$. For metallic glasses the quantity $\partial\omega/\partial H$ becomes as large as $3 \times 10^{-11}\ (A/m)^{-1}$ for $Fe_{78}B_{14}Si_8$, which is about ten times as large as that in crystalline state.[32] This effect, in turn leads to an anomalous thermal expansion coefficient;[33]

$$\alpha_m(T) = (1/3)\partial\omega_s(T)/\partial T = (1/3)(\partial\omega/\partial H)(\partial M_s/\partial T)/(\partial M_s/\partial H)$$

which is magnetic in origin. Since $\partial M_d/\partial H > 0$ and $\partial M_s/\partial T < 0$, the sign of $\alpha_m(T)$ is negative when $\partial\omega/\partial H > 0$. The thermal expansion coefficient in the paramagnetic state of the material (α_p) is, on the other hand, positive. The resultant thermal expansion coefficient $\alpha = \alpha_m + \alpha_p$ could then become zero if α_m becomes significant. This situation has been experimentally observed in binary $Fe_{100-x}B_x$ glassy alloys[34] which is shown in Figure 12. The coefficient α becomes zero at about $x = 17$. Many metallic glasses designed for low frequency applications have compositions near $Fe_{83}B_{17}$ and therefore α values of these technologically important materials have relatively low thermal expansion coefficient. For example, one has $\alpha = 4 \times 10^{-6}$ and 3×10^{-6} for $Fe_{82}B_{12}Si_6$ and $Fe_{78}B_{13}Si_9$, respectively, low frequency properties of which alloys are discussed in Section IV.

2. Linear Magnetostriction

The anisotropic effects involving the spin correlation functions, $<S_{iz}\ S_{jz}>$ and $<(S_{iz})^2>$ mentioned earlier, arise mainly from spin-orbit interactions which are, in general, weaker than interatomic exchange interactions in transition metals. The terms responsible for the linear magnetostriction (λ) to first order are $(\partial J_{ij}/\partial\epsilon)<S_{iz}\ S_{jz}>$ and $(\partial D/\partial\epsilon)<(S_{iz})^2>$, representing two-ion and single-ion exchange anisotropy, respectively. Experimentally, the quantity λ is measured as a rate ($\Delta\ell/\ell$) of change in the linear dimension (ℓ) of a magnetic

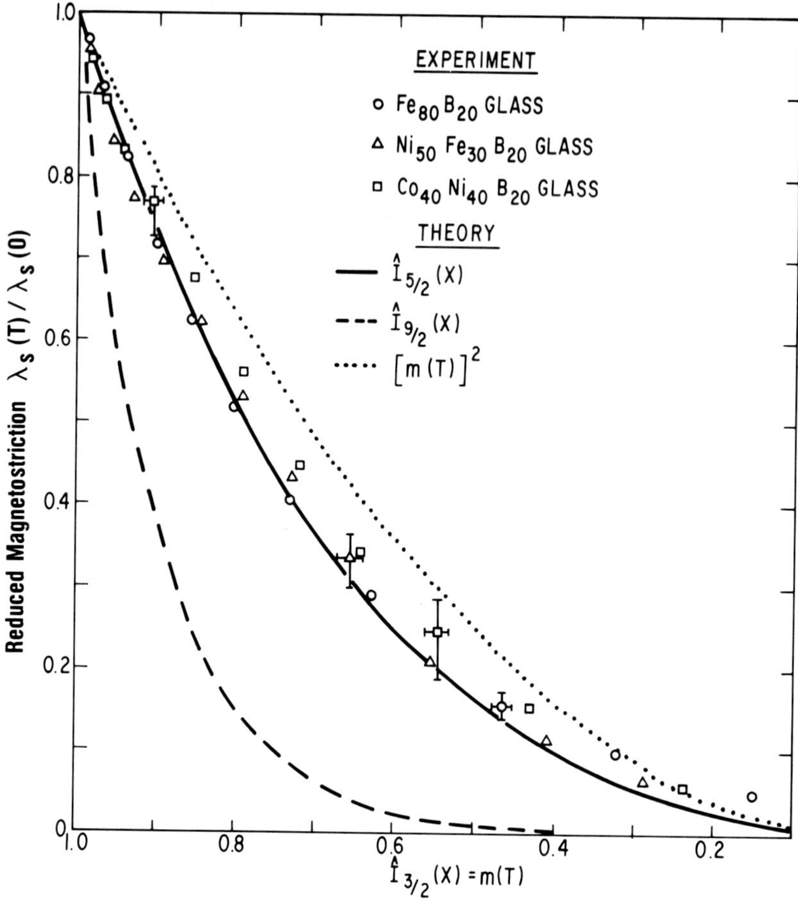

FIGURE 13. Reduced magnetostriction versus reduced magnetization for three representative metallic glasses. The theoretical curves, $\hat{I}_{5/2}(X)$, $\hat{I}_{9/2}(X)$, and $[m(T)]^2$ are for single-ion anisotropy of uniaxial symmetry, cubic symmetry and anisotropic two-ion exchange of uniaxial symmetry, respectively, where $\hat{I}_{\ell+1/2}(X) = I_{\ell+1/2}(X)/I_{1/2}(X)$, $I_{\ell+1/2}$ being the hyperbolic Bessel function of order $(\ell + 1/2)$. (From O'Handley, R. C., *Phys. Rev. B*, B18, 930, 1978. With permission.)

material in magnetic fields. The above expressions imply that λ varies as M^2 where M is the magnetization. This is indeed the case with many metallic glasses. For Fe and Fe–Ni base metallic glasses, an empirical relation $\lambda_s = 12.1 \times 10^{-10} M_s^2$ has been reported, where λ_s is the saturation magnetostriction and M_s is in $JT^{-1}kg^{-1}$ (emu/g), both being at room temperature.[35] Furthermore, temperature dependence of λ sheds some light on the nature of the magnetostriction, since the above-mentioned anisotropic spin correlation functions have characteristic temperature dependences. The reduced magnetostrictions, $\lambda_s(T)/\lambda_s(0)$, are shown in Figure 13 as a function of reduced magnetization $[M_s(T)/M_s(0) = m(T)]$ for three representative metallic glasses, $Fe_{80}B_{20}$, $Ni_{50}Fe_{30}B_{20}$, and $Co_{40}Ni_{40}B_{20}$ with $10^6\lambda_s$ (295 K) = +32, +8, and −7, respectively.[31] Also included in the figure are theoretical curves for single-ion and two-ion behavior. The experimental data seem to follow the theoretical curve for the single-ion uniaxial strain symmetry with a small contribution from two-ion uniaxial symmetry for the Co-containing alloys. Thus the linear magnetostriction of many metallic glasses appear to originate mainly from crystal field effects.

Although linear magnetostriction is a small quantity of the order of 10^{-5} for many soft ferromagnets, it affects the magnetic properties of the materials appreciably. Some of the major effects include magnetomechanical loss as evidenced by the familiar hum in transformers, reduction of initial permeability and contribution to the phase lag between excitation and resultant flux. Smaller values of λ_s, therefore, are desirable in many magnetic device applications including low frequency transformers. Many metallic glasses such as $Fe_{78}B_{13}Si_9$ considered for these applications, however, have relatively large values of λ_s of about 30 $\times$ 10^{-6}. Efforts to reduce this quantity without decreasing M_s in Fe-rich metallic glasses have been somewhat successful in low boron-containing Fe–Ni–B alloys,[8] which are unfortunately less economical because of the nickel content. One can, on the other hand, take advantage of the large magnetostriction of some metallic glasses to fabricate magnetomechanical transducers which are discussed in Chapter 6.

D. Magnetization Process and Domains

One of the unique magnetic properties of 3d transition metal base glassy ferromagnets is their magnetic softness. Especially noted are their extremely low values of coercivity; $H_c < 8$ A/m (0.1 Oe) can be easily achieved in as-cast materials. This reflects the magnetically homogeneous nature of metallic glasses with a relatively small number of domain wall pinning centers. A more quantitative argument can be made by using a formula for H_c[36]

$$H_c = 0.38\,(KW/M_s\delta)\,|A/A' - K'/K|$$

where A and K are exchange and anisotropy energy, respectively, of the matrix, A′ and K′ are the corresponding quantities of a region of size W imbedded in the matrix and δ is the domain wall width parameter, $(A/K)^{1/2}$. The low values of H_c in noncrystalline ferromagnets indicate that the pinning strength $W|A/A'-K'/K|$ is small in these materials, implying their homogeneous nature. The value of H_c can be further reduced by heat treatments. An example is shown in Figure 14, in which B–H loops of as-cast and field-annealed $Fe_{80}B_{20}$ are compared.[3] The decrease of H_c by a factor of 2 is normally attained in Fe-rich glassy alloys. The heat-treatment also introduces an increase of the slope of the resistivity vs. temperature curves for these materials. These findings suggest that annealing tends to make glassy alloys more homogeneous. The reduction of H_c is significant because it reduces the hysteresis loss when the material is used for cores in low-frequency magnetic devices. Field-annealing modifies the magnetic anisotropy as mentioned earlier (see Section B) and can be used to control the remanence, B_r. In the case of Figure 14, the quantity B_r measured along the ribbon direction increases because the annealing field was applied along the ribbon direction.

In addition to the magnetization processes exemplified in Figure 14, the behaviors of magnetic domains during excitation determine the performances of magnetic materials. Static domain configurations are determined by a balance of the exchange, anisotropy, magnetoelastic, domain wall, and magnetostatic energies. Dynamic domain properties are more complicated because various wall damping mechanisms are involved. An example of static domain pattern for a typical glassy ferromagnet, $Fe_{80}B_{20}$, is shown in Figure 15. It is noticed that the material has a small number of unusually large domains. This leads to a relatively large eddy current loss during ac excitation because flux reversal must be carried out by a small number of domains resulting in faster wall motions. For a given domain structure, however, the eddy current losses in glassy ferromagnets are smaller when compared to crystalline metallic ferromagnets, because the former have higher resistivities ($\rho \sim 150$ $\mu\Omega$-cm) than the latter ($\rho \sim 50$ $\mu\Omega$-cm).

III. THERMAL STABILITY

Upon heating, noncrystalline states of glassy alloys transform into crystalline states at

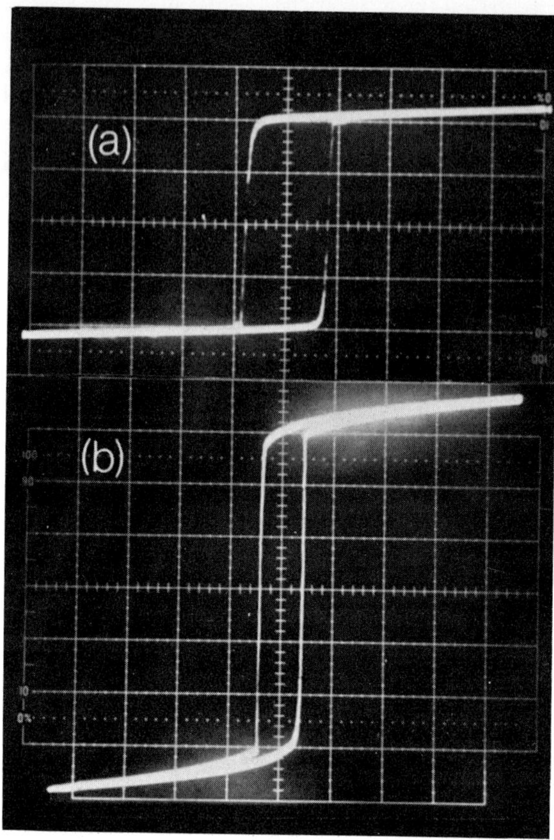

FIGURE 14. B–H loops for as-cast (a) and field-annealed (b) glassy $Fe_{80}B_{20}$ alloys. The vertical axis (B): 0.4 T/div.; the horizontal axis (H): 8 Am^{-1}/div.

temperatures defined as crystallization temperatures, T_x. In addition to the magnetic properties discussed in this chapter, glassy alloys for use in magnetic devices should, therefore, be designed to be thermally stable against crystallization. At T_x, an appreciable heat release takes place because of the exothermic nature of the phase transition, which makes the measurement of T_x relatively easy. The temperature T_x is also a function of the heating rate. For a given heating rate, then T_x can be considered as a measure of the thermal stability of a metallic glass. Since both time and temperature are involved in the crystallization process, a more realistic approach to the question of thermal stability would be the use of a crystallization rate equation.[37] Following Avrami,[38] the crystalline fraction x(t) at time t may be given by

$$x(t) = 1 - \exp(-Kt^n)$$

where n is a constant known as the Avrami exponent and depends on the nucleation and growth mechanisms. The quantity K is assumed to obey the Arrhenius law;

$$K(T) = K_0 \exp(-Q_a/k_BT)$$

where K_0 is a constant, Q_a is the apparent activation energy for crystallization, k_B is the Boltzmann constant, and T is the absolute temperature. When the crystallization rate, dx/

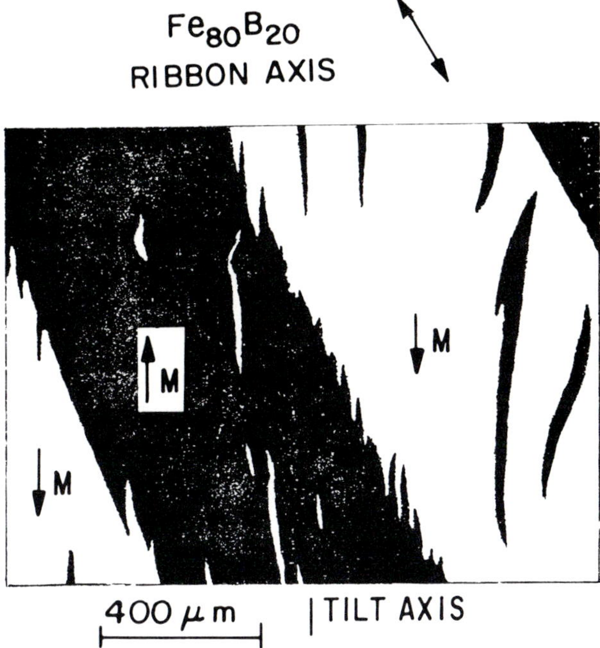

FIGURE 15. Domain pattern of as-cast $Fe_{80}B_{20}$ ribbon sample taken by scanning electron microscopy. The direction of the remnant magnetization is indicated by arrows. (From Hasegawa, R., O'Handley, R. C., and Mendelsohn, L. I., *AIP Conf. Proc.*, No. 34, 298, 1977. With permission.)

dt, is at a maximum for a given heating rate, one obtains $d\ell n t_m/dT^{-1} = Q_a/nk_B$ where t_m is the time to reach this maximum. This allows an experimental determination of $Q_a/n = Q_c$, which may be defined as an effective activation energy for crystallization as the above set of rate equations implies. Furthermore, the exponent n can be shown to be partitioned as[39]

$$n = a + bp$$

where a is related to the nucleation rate varying from 0 (quenched-in nuclei) to 1 (constant nucleation) and b has the values 1, 2, or 3 representing the dimensionality of the crystal growth. The quantity p refers to the mechanism of the crystal growth, i.e., p = 0.5 for diffusion and p = 1 for interface controlled growth. The analysis of many experimental data based on the above phenomenological treatment seems to describe the crystallization processes of many different metallic glasses relatively well.[40,41] One example[41] which is relevant to this chapter is illustrated in the following.

The data of Q_c taken from Fe–B–Si metallic glasses are shown in Figure 16, which may be summarized as follows:

1. When Si content is less than about 5 at.%, the activation energy Q_c is relatively independent of the boron content.
2. In the hypoeutectic (low metalloid) region, the value Q_c is relatively insensitive to Si content, but decreases rapidly with increasing Fe content.
3. In the hypereutectic region, the effect of Si on Q_c is very strong up to about 10 at.% Si, beyond which Q_c is somewhat independent of Si content.

The boron content does not seem to affect Q_c values. Based on these findings, one can

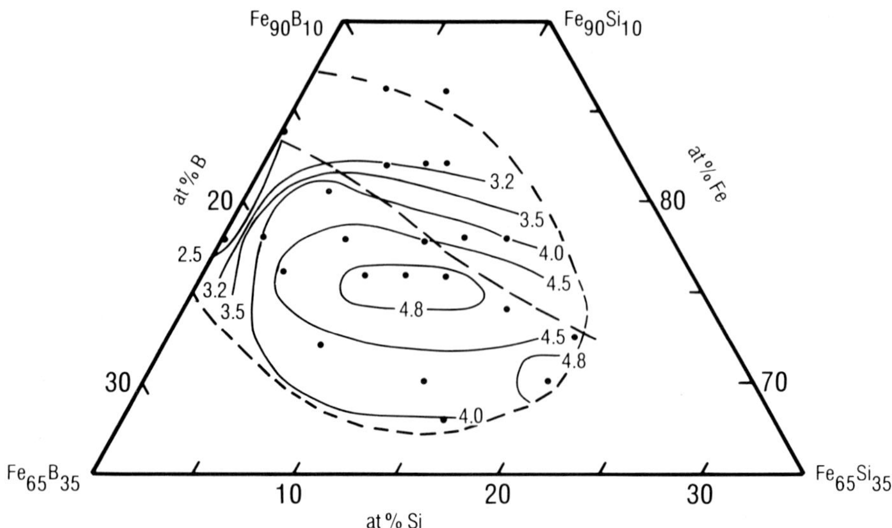

FIGURE 16. Contours of constant Q_c (in eV) in the Fe–B–Si glassy system. The dashed envelope defines the glass-crystalline boundary. The dashed line originating from $Fe_{83.5}B_{16.5}$ is the eutectic trough. (From Ramanan, V. R. V. and Fish, G. E., *J. Appl. Phys.*, 53, 2273, 1982. With permission.)

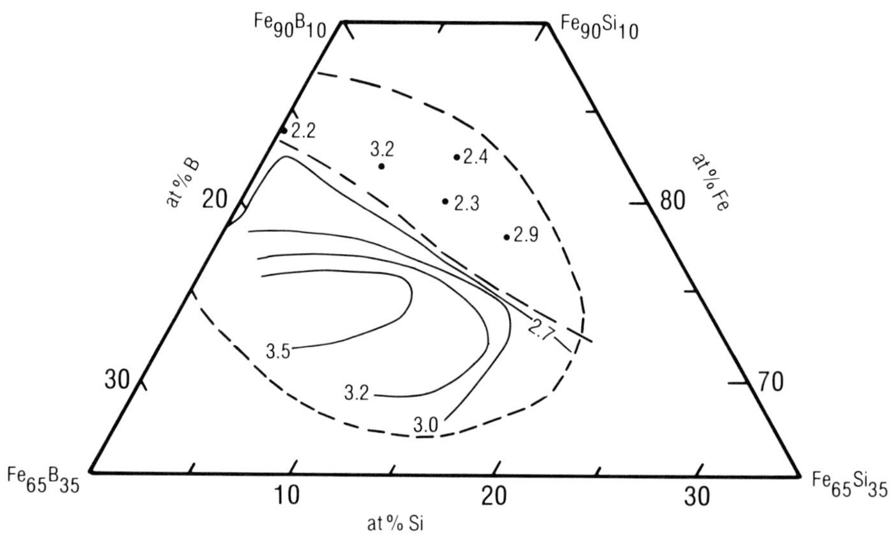

FIGURE 17. Contours of constant n in the hypereutectic Fe–B–Si glasses. The data points represent the n values in hypoeutectic region. (From Ramanan, V. R. V. and Fish, G. E., *J. Appl. Phys.*, 53, 2273, 1982. With permission.)

conclude that the most thermally stable Fe–B–Si glasses have 5 to 10 at.% Si and 75 to 78 at.% Fe. As shown in the next section, glassy alloys in this system with the improved low-frequency properties fall in this concentration region.

The behavior of n in the Fe–B–Si system is given in Figure 17. This figure indicates that n is more or less constant at about 2.5 along the eutectic trough and in most of the hypoeutectic region. Since the growth rate in this region is found to be diffusion controlled,[42] it is expected that a roughly linear nucleation rate results, i.e., $n = a + bp$ with $n = 2.5$, $a = 1$, $b = 3$, and $p = 0.5$. In the hypereutectic region, the exponent n is larger than 3 and increases slightly with the boron content. This indicates that growth is no longer purely diffusion controlled (p = 0.5) but probably tends to be interface-controlled (p = 1) accompanied by suppression of nucleation ($a<1$). This is probably the cause of the enhancement of thermal stability of metallic glasses containing Si.

IV. LOW FREQUENCY MAGNETIC PROPERTIES

When a magnetic material is placed in an alternating field H, the resultant induction B cannot follow H exactly, introducing a phase shift ϕ. This leads to an imaginary part in the permeability (B/H) which in turn introduces a magnetic loss in the material. The origin of ϕ includes the nonlinear nature of the material, eddy currents and magnetoelastic effects. Phenomenologically the magnetic loss at frequency f and induction B is given by $L(f,B) = \int BdH = P_e(f,B)\sin\phi$. Here P_e is the exciting power which is defined as the power required to maintain a given flux density. Obviously smaller values of L and P_e are desired in practical applications, especially in distribution and power transformers and electric motors because of the vast amount of energy involved. The major losses in these devices are hysteresis and eddy current losses. Metallic glasses with their unusually low coercivities and high resistivities offer advantages, in the two aspects of the magnetic losses, over the conventional Fe–Si electrical sheets.

Figure 18 summarizes the results of 60 Hz core loss measurements on representative glassy ferromagnets with high saturation induction.[7] For comparison, handbook data for M-4 grade 275 μm thick grain-oriented 3% Si/Fe sheets[43] and for a recently developed Orientcore HiB®[44] are also included in the figure. Glassy alloys represent significant improvements in the core loss over the crystalline materials widely used in transformers. Among the metallic glasses, $Fe_{82}B_{10}Si_8$ shows the lowest core loss with L(60 Hz, 1.6 T) = 0.24 W/kg which is about 1/5 of the corresponding core loss of the commercial materials. The importance of Si in reducing the core loss is clearly noticed in the figure. The substitution of silicon for some of the boron in Fe–B alloys results in the values of magnetic anisotropy energy K_u ranging from about 380 to 1500 J/m^3 as indicated in Figure 10. The superior soft magnetic properties obtained in the alloys having compositions near $Fe_{82}B_{10}Si_8$ indicate that K_u = 500 to 1000 J/m^3 seems to be optimal for the magnetic anisotropy.

It is noticed that the core loss for $Fe_{67}Co_{18}B_{15}$ with B_s(RT) = 1.8 T is significantly higher than those of Fe–B and Fe–B–Si alloys. This is partly due to the larger magnetostriction ($\lambda \sim 35 \times 10^{-6}$) of the Co-containing alloy as compared with the Fe–B–Si alloys ($\lambda \sim 30 \times 10^{-6}$). The presence of Co would undoubtedly increase the value of K_u beyond the optimal value of 10^3 J/m^3, resulting in increase of H_c and hence the hysteresis loss. In fact, optimally heat-treated $Fe_{82}B_{10}Si_8$ and $Fe_{67}Co_{18}B_{15}$ exhibit $H_c \sim 2.5$ and ~ 4 A/m, respectively. Although the core loss of the Co-containing alloys is higher than those of Fe–B–Si glasses, it is still less than 1/2 of that of the commercial materials. Such alloys, therefore, are of practical use where higher flux density is more important.

Since the upper limit for Si content in glassy $Fe_{82}(B,Si)_{18}$ is about 10 at.% by liquid-quenching techniques currently available,[7] easily fabricable $Fe_{82}B_{12}Si_6$ was chosen as a candidate for the low core-loss high induction glassy alloy. However, because of reasons described in Section I, alloys of the compositions $Fe_{81}B_{13.5}Si_{3.5}C_2$ and $Fe_{78}B_{13}Si_9$ have been proposed for the same purpose. The low frequency properties as a function of temperature of these glassy ferromagnets are shown in Figure 19.[18] This figure clearly indicates that the temperature dependence of L and P_e are greatly improved in $Fe_{78}B_{13}Si_9$ over $Fe_{81}B_{13.5}Si_{3.5}C_2$. Furthermore, both ferromagnetic Curie (T_c = 688 K) and crystallization temperature (T_x = 823 K) of the former are higher than those T_c = 648 K, T_x = 748 K of the latter, which makes $Fe_{78}B_{13}Si_9$ magnetically and thermally more stable.

Saturation induction greater than 1.7 T probably cannot be achieved in relatively inexpensive and stable Fe–B base metallic glasses as discussed in Section I. However, the current

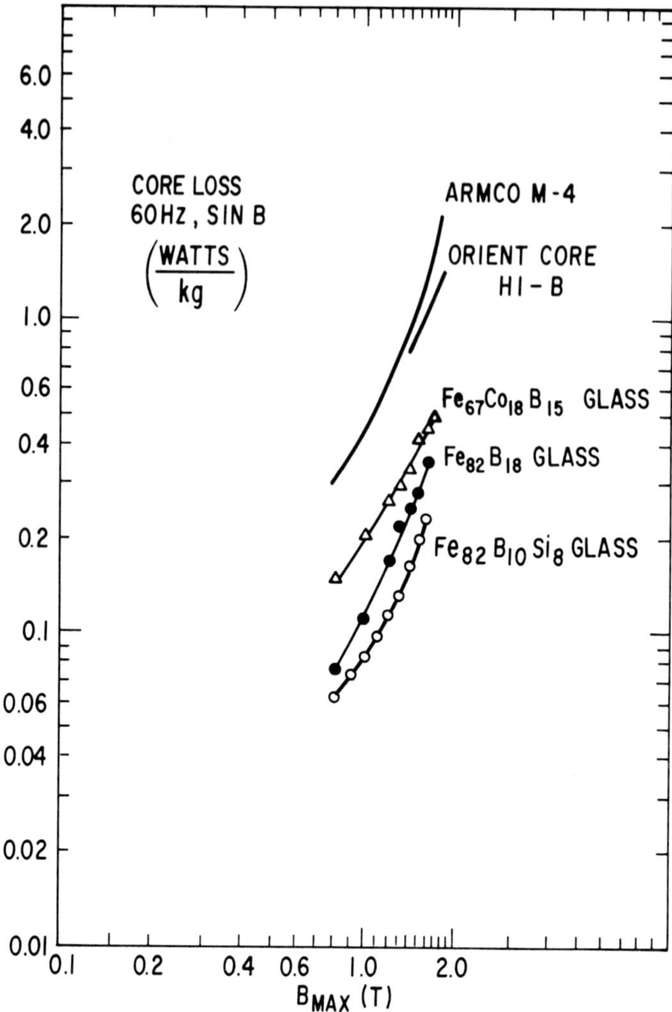

FIGURE 18. Core loss under sinusoidal flux conditions at 60 Hz vs. maximum induction for three representative metallic glasses with high induction. The data for the commercial grain-oriented silicon steels are also shown for comparison.[7,43,44]

trend is leaning toward operating transformers at lower induction levels to reduce losses. The lower saturation induction of metallic glasses is therefore no longer a drawback.[45] The presence of about 20 at.% nonmagnetic atoms leads to low Curie temperatures which could limit the operating induction level. Due to the exceptionally low core loss of glassy ferromagnets, however, the temperature rise during device operation is expected to be much lower than that (~65°C above ambient) for the conventional devices. Furthermore, development of new materials with higher Curie and crystallization temperatures is a continuing effort in many laboratories. With the improved properties and large scale production of high quality materials, metallic glasses will undoubtedly find their way in many energy-saving electrical power devices.

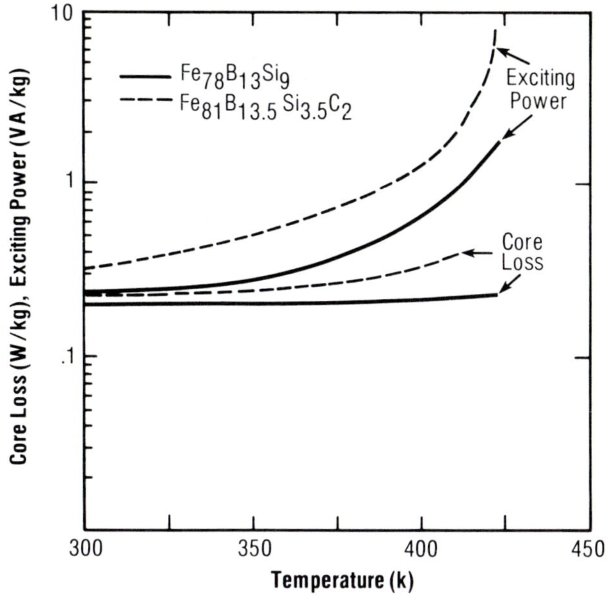

FIGURE 19. Core loss and exciting power as a function of temperature for two representative high induction metallic glasses. Data are for the maximum induction level of 1.4 tesla.[18]

REFERENCES

1. **Bozorth, R. M.,** *Ferromagnetism*, D. van Nostrand, New York, 1951, 67.
2. **Barisoni, M. and Candiotti, M.,** A new high permeability grain oriented Fe-3% Si Steel, *IEEE Trans. Magn.*, MAG-14, 345, 1978.
3. **Hasegawa, R., O'Handley, R. C., and Mendelsohn, L. I.,** Advances in ferromagnetic metallic glasses, *AIP Conf. Proc.*, 34, 298, 1977.
4. **O'Handley, R. C. and Boudreaux, D. S.,** Magnetic properties of transition metal-metalloid glasses, *Phys. Status Solidi A*, 45, 607, 1978.
5. **Hasegawa, R. and Ray, R.,** Iron-born metallic glasses, *J. Appl. Phys.*, 49, 4174, 1978.
6. **Hasegawa, R. and Ray, R.,** Magnetization of glassy Co–B alloys, *J. Appl. Phys.*, 50, 1586, 1979.
7. **O'Handley, R. C., Chou, C.-P., and DeCristofaro, N.,** High induction low-loss metallic glasses, *J. Appl. Phys.*, 50, 3603, 1979.
8. **Hasegawa, R.,** Magnetic properties of glassy $(Fe\text{-}Ni)_{86}B_{14}$ alloys, *J. Phys. (Paris)*, 41, C8-701, 1980.
9. **Durand, J. and Yung, M.,** Electronic and magnetic properties of Fe-P-B alloys, in *Amorphous Magnetism II*, Levy, R. A. and Hasegawa, R., Eds., Plenum Press, New York, 1977, 725.
10. **Graham, C. D. and Egami, T.,** FeBC alloys for use in power transformers, *J. Magn. Magn. Mater.*, 19, 185, 1980.
11. **Mitera, M., Masumoto, T., and Kazama, N. S.,** Effect of silicon addition on the magnetic properties of Fe–B–C amorphous alloys, *J. Appl. Phys.*, 50, 7609, 1979.
12. **Hasegawa, R.,** Magnetic properties of glassy Fe Be–B alloys, *J. Appl. Phys.*, 52, 1847, 1981.
13. **Luborsky, F. E. and Walter, J. L.,** The Fe–B–Ge amorphous alloys, *IEEE Trans. Magn.*, MAG-17, 1204, 1981.
14. **Luborsky, F. E., Becker, J. J., Walter, J. L., and Liebermann, H. H.,** Formation and magnetic properties of Fe-B-Si amorphous alloys, *IEEE Trans. Magn.*, MAG-15, 1146, 1979.
15. **DeCristofaro, N.,** private communication, 1980.
16. **Davis, L. A., DeCristofaro, N., and Smith, C. H.,** Technology of metallic glasses, in *Proc. Conf. Metallic Glasses: Science and Technology*, Vol. 1, Hargitai, C., Bakonyi, I., and Kemény, T., Eds., Central Research Institute for Physics, Budapest, 1981, 1.

17. **Köster, U. and Herold, U.,** Crystallization of metallic glasses, in *Metallic Glasses,* Güntherodt, H.-J., Ed., Springer Verlag, Berlin, 1980, 229.
18. **DeCristofaro, N., Datta, A., Davis, L. A., and Hasegawa, R.,** An iron-base metallic glass alloy with improved 60 Hz properties, *Proc. 4th Int. Conf. Rapidly Quenched Metals,* Masumoto, T. and Suzuki, K., Eds., The Japan Institute of Metals, Sendai, 1982, 1031.
19. **Duwez, P.,** Rapid quenching techniques, in *Techniques of Metal Research,* Vol. 1, Bunshah, R. F., Ed., Interscience, New York, 1968, 347.
20. **Hasegawa, R.,** unpublished data, 1970.
21. **Paul, D. I., Marti, J., and Valadez, L.,** Ferromagnetic domain walls in amorphous iron alloy, *AIP Conf. Proc.,* 18, 1377, 1974.
22. **Hasegawa, R.,** Magnetization, anisotropy and coercivity of a glassy metallic alloy, *AIP Conf. Proc.,* 29, 216, 1976.
23. **Nesbitt, E. A., Hasegawa, R., O'Handley, R. C., and Sullivan, M. O.,** Field induced magnetic anisotropy of glassy iron-boron and iron-boron-silicon alloys near eutectic composition, *J. Appl. Phys.,* 51, 3328, 1980.
24. **Fujimori, H., Morita, H., Obi, Y., and Ohta, S.,** On the magnetically induced anisotropy in amorphous ferromagnetic alloys, in *Amorphous Magnetism II,* Levy, R. A. and Hasegawa, R., Eds., Plenum Press, New York, 1977, 393.
25. **Chikazumi, S.,** *Physics of Magnetism,* R. E. Krieger, Huntington, N.Y., 1978, chap. 17.
26. **Luborsky, F. E. and Walter, J. L.,** Magnetically induced anisotropy in amorphous alloys of Fe-Ni-P-B, *IEEE Trans. Magn.,* MAG-13, 1635, 1977.
27. **Takahashi, M. and Kim, C. O.,** Temperature dependence of the magnetic anisotropy in amorphous $Fe_{81}B_{19}$ alloy, *Jpn. J. Appl. Phys.,* 16, 2061, 1978.
28. **Luborsky, F. E.,** Perspective on application of amorphous alloys in magnetic devices, in *Amorphous Magnetism II,* Levy, R. A. and Hasegawa, R., Eds., Plenum Press, New York, 1977, 345.
29. **Hasegawa, R.,** Reorientation kinetics of the magnetization in a glassy ferromagnet, *J. Appl. Phys.,* 49, 5610, 1978.
30. **Callen, E.,** Magnetostriction, *J. Appl. Phys.,* 39, 519, 1968.
31. **O'Handley, R. C.,** Magnetostriction of transition metal-metalloid glasses: Temperature dependence, *Phys. Rev. B,* B18, 930, 1978.
32. **Jagielinski, T., Arai, K. I., Tsuya, N., Ohnuma, R., and Masumoto, T.,** Saturation magnetostriction and volume magnetostriction of Fe–Ni–Co amorphous ribbons, *IEEE Trans. Magn.,* MAG-13, 1553, 1977.
33. **Lee, E. W.,** Magnetostriction and magnetomechanical effects, *Rep. Prog. Phys.,* 18, 184, 1955.
34. **Fukamichi, K., Kikuchi, M., Arakawa, S., and Masumoto, T.,** Invar-type new ferromagnetic amorphous Fe–B alloys, *Solid State Commun.,* 23, 955, 1977.
35. **Ito, S., Aso, K., Makino, Y., and Uedaira, S.,** Magnetostriction and magnetization of iron-based amorphous alloys, *Appl. Phys. Lett.,* 37, 665, 1980.
36. **Friedberg, R. and Paul, D. I.,** New theory of coercive force of ferromagnetic materials, *Phys. Rev. Lett.,* 34, 1234, 1975.
37. **Marseglia, E. A.,** Kinetic theory of crystallization of amorphous materials, *J. Non-Cryst. Solids,* 41, 31, 1980.
38. **Avrami, M.,** Kinetics of phase change. I, *J. Chem. Phys.,* 7, 1103, 1939.
39. **Christian, J. W.,** *The Theory of Transformation in Metals and Alloys,* 2nd ed., Pergamon Press, New York, 1975, Part I.
40. **von Heimendahl, M. and Kuglstatter, G.,** The activation energies of crystallization in amorphous alloy Metglas 2826A, *J. Mater. Sci.,* 16, 2405, 1981.
41. **Ramanan, V. R. V. and Fish, G. E.,** Crystallization kinetics in Fe–B–Si metallic glasses, *J. Appl. Phys.,* 53, 2273, 1982.
42. **Chang, C.-F.,** private communication, 1982.
43. *Armco Oriented Electrical Steels,* Armco Steel Corporation, Middletown, O., 1974, 19.
44. **Suzuki, T., Nakayama, H., Yamamoto, T., and Taguchi, S.,** Transformer characteristics of Orientcore HiB, *IEEE Trans. Magn.,* MAG-8, 321, 1972; **Taguchi, T., Yamamoto, T., and Sakakura, A.,** New grain oriented silicon steel with high permeability "Orientcore HiB", *IEEE Trans. Magn.,* MAG-10, 123, 1974.
45. **Werner, F. E.,** Electrical steels: 1970-1990, in *Energy Efficient Electrical Steels,* The Metallurgical Society, American Institute of Mining, Metallurgical and Petroleum Engineers, Warrendale, Penn., 1981, 1.

Chapter 6

MAGNETIC MATERIAL PROPERTIES AND APPLICATIONS OF METALLIC GLASSES IN ELECTRONIC DEVICES

R. Boll, H. R. Hilzinger, and H. Warlimont

TABLE OF CONTENTS

I. INTRODUCTION

The particular suitability of metallic glasses for being employed as versatile magnetic materials is based on the absence of magnetocrystalline anisotropy and of structural defects such as grain boundaries and dislocations. The magnetization reversal in amorphous alloys is primarily affected by the counteraction of the magnetoelastic energy and of the induced magnetic anisotropy K_u. The magnetic and structural properties such as saturation induction, Curie temperature, magnetostriction, induced magnetic anisotropy and, thus, all characteristics of the hysteresis loop can be controlled and varied by alloying and suitable heat treatment. In this respect metallic glasses are tractable like crystalline soft magnetic materials. But the additional advantages of metallic glasses such as their high electrical resistivity, ''natural'' low thickness, high yield stress, and favorable surface properties give rise to novel property combinations which open up a new range of manufacturing procedures and applications that are not accessible to conventional crystalline soft magnetic materials.

In this chapter we discuss the magnetic properties as they pertain to applications in electronics, give information on manufacturing related properties and describe examples of applications which appear to be most likely to make use of metallic glasses in the near future. This presentation is based in part on related overviews published in recent years.[1-11]

II. MAGNETIC PROPERTIES

A. Magnetic Alloy System

The range of compositions which may be prepared in the glassy state by quenching from the melt has grown fast. For soft magnetic applications metal-metalloid based alloys are mainly being considered up to now. Recently glass formation has also been observed in the ferromagnetic metal-metal system $(Fe, Ni, Co)_{90}Zr_{10}$.[12]

The fundamental magnetic properties of amorphous alloys are the saturation induction B_s, the magnetostriction constant λ_s and the Curie temperature T_c. In a first approximation, the saturation induction of amorphous alloys is given by their metallic components.[13,14] As an example, Figure 1 shows the room temperature saturation induction, B_s, in the systems Fe–Co, Co–Ni, and Fe–Ni with 20 at.% B, respectively. In general, B_s is the highest for the Fe-rich alloys and decreases with increasing Ni and Co content. The maximum observed for crystalline Fe–Co alloys is only weakly developed and shifted to the Fe-rich side. Nonmagnetic additions such as Mo, Nb, Zr, Ti, or Cr decrease B_s by 0.03 to 0.1 T/at.%.[13,15] With the exception of the Fe-rich alloys,[16,17] B_s is reduced by increasing metalloid content in the order B, C, Si, Ge, P. The second fundamental property determined by the alloy composition is the magnetostriction constant, λ_s (see Figure 2), which is found to be isotropic in the amorphous state.[18] For the Fe rich alloys, λ_s is positive, whereas for the Co-rich alloys, λ_s has a negative sign. With increasing content of transition metals, such as Ni, Mo, Cr, or with increasing metalloid content, the absolute value of the room temperature magnetostriction decreases roughly as [18-20]$|\lambda_s| \sim B_s^2$. The disappearance of magnetostriction at about 65 at.% Ni (Fe–Ni) and 45 at.% Ni (Co–Ni) is thus only caused by the simultaneous disappearance of magnetization. A true zero magnetostriction has been found on the Co rich side of the Co–Fe-based[20-22] and the Co–Mn-based systems.[23]

The Curie temperatures of amorphous alloys are just as much affected by the metal as the metalloid content. This effect is highest for the Co alloys.[8] For the Fe alloys with low metalloid content, an anomalous decrease of T_c with increasing Fe content is observed.[24] This has far-reaching consequences for the development of amorphous materials with a high saturation induction B_s.

Recently it has been observed that these fundamental properties depend on the atomic

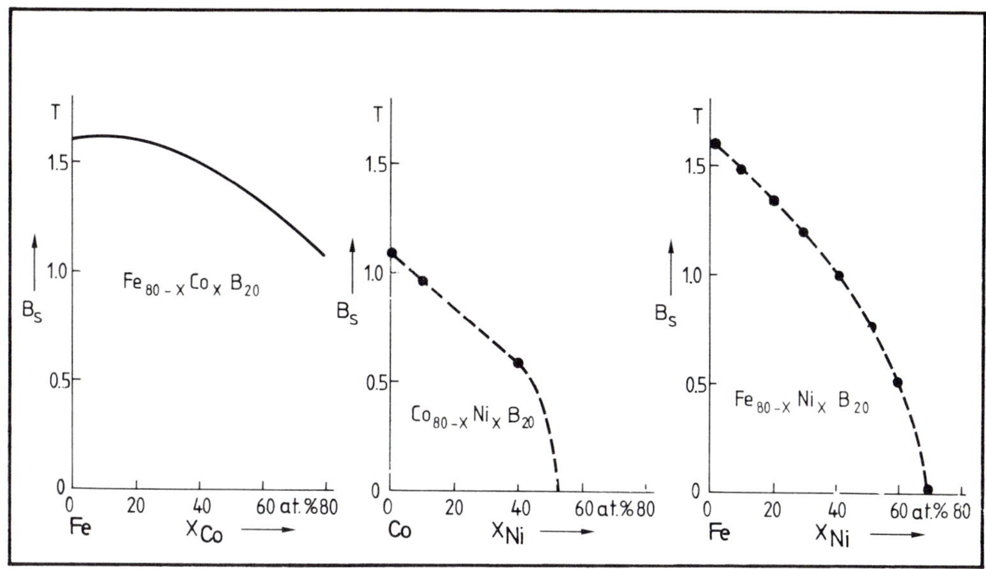

FIGURE 1. Saturation induction B_s of amorphous Fe–Co, Co–Ni, and Fe-Ni-based alloys with 20 at.% B, – R. C. O'Handley et al.[14], -●-H. R. Hilzinger et al.[7]

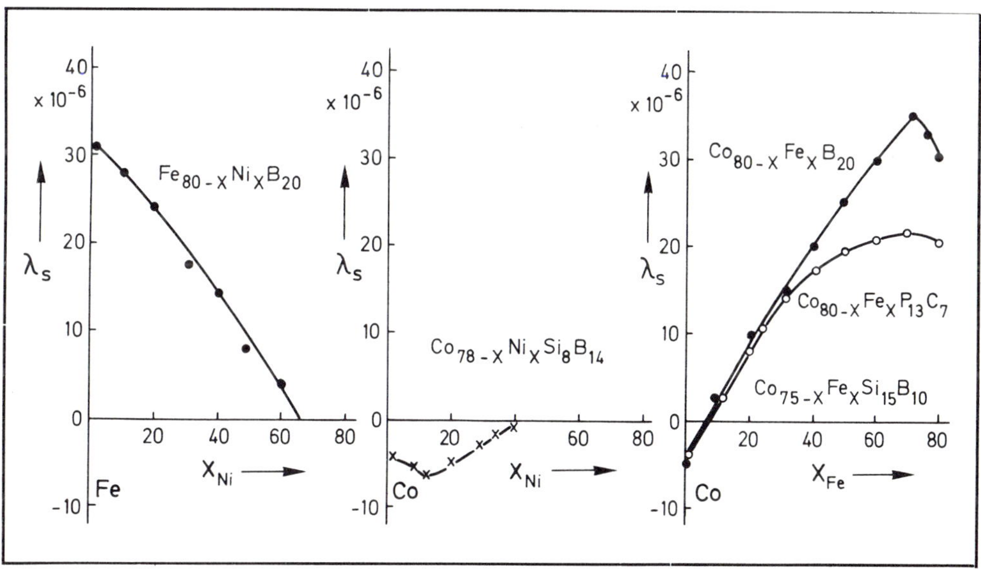

FIGURE 2. Saturation magnetostriction λ_s of amorphous Fe–Ni, Co–Ni, and Co–Fe-based alloys, ●R. C. O'Handley[19]; ○ H. Fujimori et al.,[22] x H. R. Hilzinger et al.[20]

short-range order in the amorphous state, which to some extent are affected by production parameters and subsequent heat treatments. The Curie temperature may increase after annealing by up to about 20 K, thus affecting the room temperature values of saturation induction[25] and magnetostriction.[26,27] Part of this variation is found to be reversible, depending on the annealing temperature.[28]

The development of amorphous materials with soft magnetic properties has been concentrated on three basic systems:

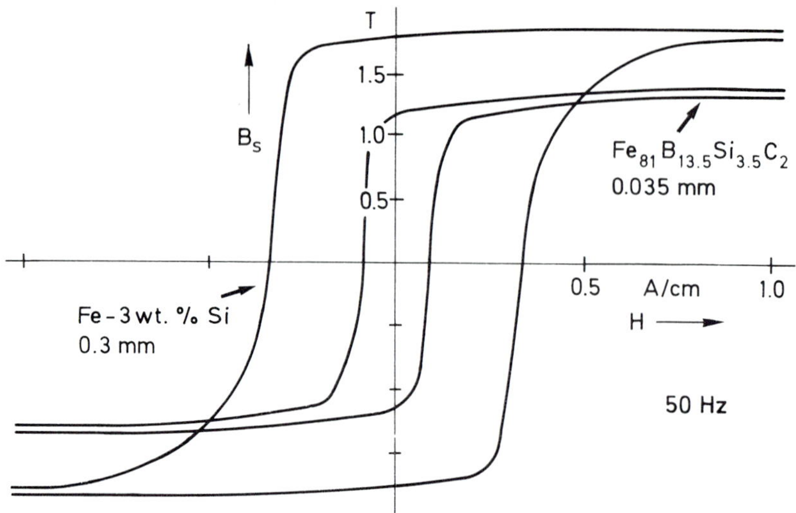

FIGURE 3. Hysteresis loops of strip-wound cores of amorphous $Fe_{81}B_{13.5}Si_{3.5}C_2$ and crystalline Fe-3 wt.% Si, grain oriented.

- Fe-rich alloys
- Fe–Ni-based alloys with about equal Fe and Ni content
- Co-rich alloys with low magnetostriction

which shall be considered in detail here.

The Fe-rich magnetic alloys, based on inexpensive raw materials, such as Fe, Si, C, and B, exhibit the highest saturation induction among the amorphous alloys (see Chapter 5). There have been numerous attempts to further improve B_s by an optimum choice of the metalloids,[16,25,29,30] by adding small amounts of Co or Ni[25] and by an annealing treatment.[25] Induction values of B_s = 1.7...1.75 T appear to be attainable now. Despite the high values of the magnetostriction constant ($\lambda_s \approx 30 \cdot 10^{-6}$), these amorphous alloys exhibit better soft magnetic properties than crystalline Fe-3 wt.% Si alloys. Amorphous alloys have coercive field strengths of about 20 to 50 mA/cm and substantially lower core losses.[2,3] A typical example is $Fe_{81}B_{13.5}Si_{3.5}C_2$ (see Figure 3). Recently, permeabilities of the Fe-rich amorphous alloys were reported[31] which exceed the corresponding data for crystalline Fe-50 wt.% Ni alloys. The second group of interesting and frequently investigated magnetic alloys are the amorphous Fe–Ni alloys with a saturation induction of 0.7 to 1.0 T. The addition of Ni improves the glass-forming ability and the ductility, thus facilitating the preparation process. Nickel also reduces the magnetostriction constant to about $10 \cdot 10^{-6}$. Very promising results have been reported for the metalloid combination (Si,B)[32,33] and for the addition of refractory metals such as Mo.[15,34] The latter additions increase the crystallization temperature T_{cr} by about 4 to 15 K/at.% and reduce the Curie temperature T_c thus permitting annealing treatments in the favorable temperature range between T_c and T_{cr}.

Among the amorphous materials, the low magnetostrictive Co base alloys exhibit the best soft magnetic properties. For improving the thermal stability, metalloid contents between 25 and 30 at.% are advantageous,[23,35] which on the other hand reduce the saturation induction of these high permeability alloys to B_s = 0.5 to 0.8 T. As an example, Figure 4 represents the permeability of strip-wound cores for the zero-magnetostrictive alloy $Co_{66}Fe_4(Mo,Si,B)_{30}$ in the as-quenched and annealed state. Even without any annealing treatment, initial permeabilities of 10,000 to 20,000 and maximum permeabilities of about 100,000 were obtained. On annealing, the maximum permeability (50 Hz) rises up to more than 400,000, whereas the coercive field strength decreases to less than 4 mA/cm.

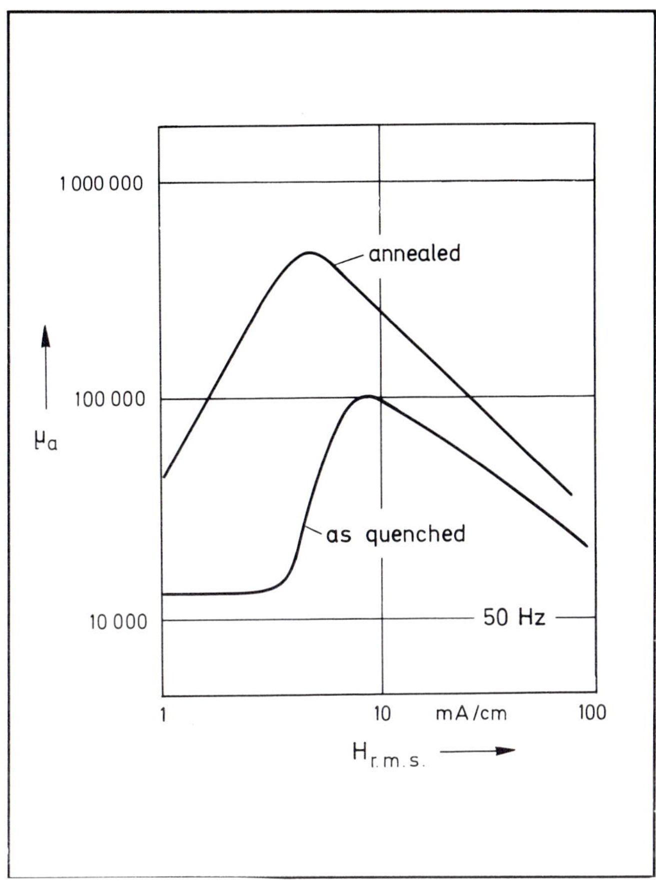

FIGURE 4. Amplitude permeability μ_a vs. applied field $H_{r.m.s.}$ for strip-wound cores of the low magnetostrictive alloy $Co_{66}Fe_4(Mo,Si,B)_{30}$.

B. Optimization by Heat Treatment

In the as-quenched state the magnetization curve is determined by mechanical stresses which are introduced during the preparation process or by winding the ribbons into toroids. These stresses may be relieved by subsequent annealing. Stress relief takes place over a wide temperature range. It sets in at about 150 to 250°C for most of the amorphous alloys and is completed near the glass transition temperature. The kinetics of stress relaxation is found to be strongly affected by the actual atomic configuration in the amorphous state,[36,37] as described for example by the excess free volume model.[38] That is why the rate of stress relaxation depends on a previous annealing treatment as well as on the preparation parameters.[39] In particular, the cooling rate in the lower temperature range and the contact time of the ribbon on the quenching wheel are of great importance.

In the amorphous metal-metalloid based alloys an uniaxial magnetic anisotropy, K_u, may be induced by annealing in an external magnetic field[40,41] or under a mechanical stress.[42,43] The equilibrium value of K_u at room temperature, reached after an infinitely long time, ranges from about 10 to 10^3 J/m^3, depending on composition and annealing temperature. Whereas annealing in a magnetic field is only effective below the Curie temperature, a mechanical stress induces an anisotropy even above T_c.

The induced anisotropy, K_u, permits predetermining the shape of the hysteresis loop by an appropriate annealing treatment, which, in general, consists of a stress relief anneal followed by cooling the material below T_c in a magnetic field parallel or perpendicular to

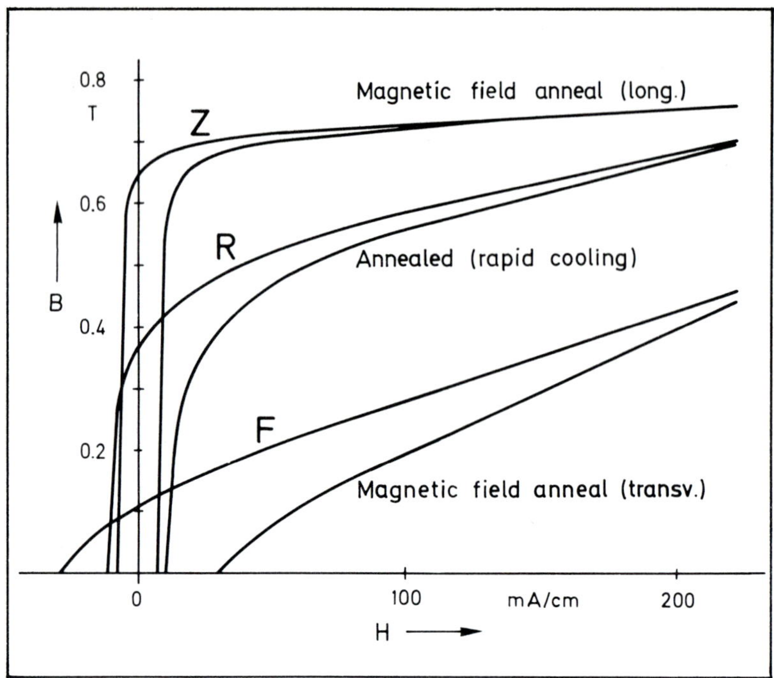

FIGURE 5. Hysteresis loops (d.c.) of strip-wound cores of $Fe_{39}Ni_{39}Mo_4Si_6B_{12}$ for different annealing treatments.

the ribbon axis.[41] The lower the cooling rate and the lower T_c, the higher is the absolute value of the induced anisotropy. Comparatively low values of K_u, which are required for high initial permeabilities, are obtained only if the stress relief anneal can be carried out at $T > T_c$, i.e. if $T_{cr} > T_c$. This condition has consequences for the optimum choice of the alloy composition, in particular for the Fe–Ni and Co-based alloys with high permeabilities.

Figure 5 shows the effect of different magnetic annealing treatments on the hysteresis loop of strip-wound cores of the alloy $Fe_{39}Ni_{39}Mo_4Si_6B_{12}$. After stress relief annealing and slow cooling below T_c (e.g., 100 K/hr), K_u is high. A nearly rectangular hysteresis loop is obtained with a small coercive field strength and an easy approach to saturation. Rapid cooling (e.g., 1000 K/hr) leads to a small value of K_u and to a domain structure with a greater number of domain walls, resulting in a round hysteresis loop with an increased initial permeability. Cooling in a transverse magnetic field yields transverse anisotropy and a flat hysteresis loop. The same type of magnetization curve is found after annealing in an oxidizing atmosphere, where a surface layer forms which generates mechanical stresses on subsequent cooling due to different coefficients of thermal expansion.[41] Since in both cases the magnetization takes place by domain rotation rather than by domain wall displacements, very low core losses without a contribution of anomalous eddy current losses are achieved. In addition the core losses of amorphous alloys are low due to the beneficial effect of a high electrical resistivity and small ribbon thickness.

Many applications require long term stability of a magnetic material in the temperature range near 100°C. Variations of the magnetic properties are mainly generated by changes of the anisotropy, K_u, which may occur in amorphous alloys even at comparatively low temperatures. The thermal stability may be enhanced by the addition of refractory metals such as Mo.[34,44] Figure 6 shows the variation of the virgin magnetization curves of two different alloys by an ageing treatment. The alloy $Fe_{40}Ni_{40}P_{14}B_6$ with a high content of P exhibits a drastic decrease of permeability by ageing for 20 to 50 days at a temperature of

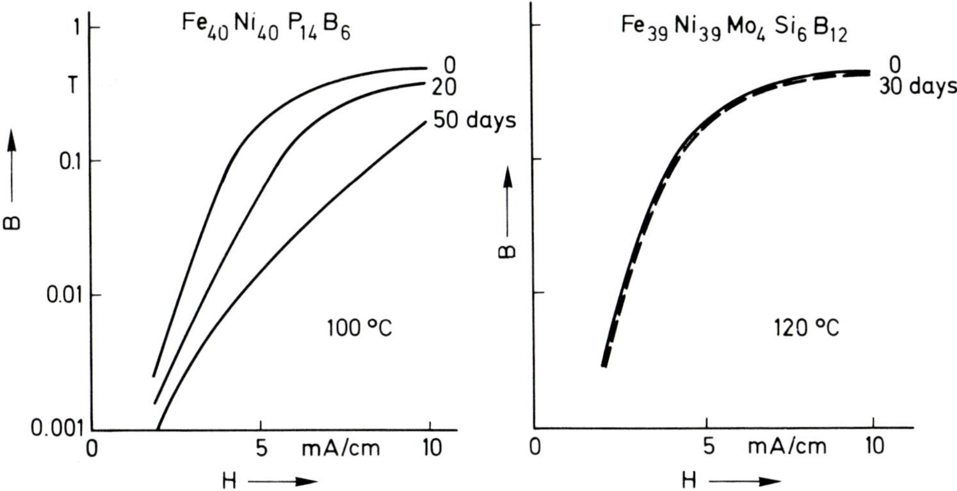

FIGURE 6. Ageing behavior of the virgin magnetization curve (d.c.) of Fe–Ni-based amorphous alloys.

100°C, whereas the alloy $Fe_{39}Ni_{39}Mo_4Si_6B_{12}$ without P and stabilized by Mo and Si remains stable even at 120°C.

III. TECHNOLOGICAL PROPERTIES

In the present context we shall treat some nonmagnetic technological properties which are of particular interest in the manufacturing, processing, and application of metallic glasses. Most prominent among these is the high elastic limit. Characteristic values for Fe–Ni and Co-based metallic glasses range from 2000 to 2500 N/mm^2. Typical values of Young's modulus are 150 kN/mm^2 for the same groups of alloys. This means that they can be subjected to four to five times the stresses and elastic strains that can be sustained by crystalline Fe–Ni-based soft magnetic materials without plastic deformation and have correspondingly high hardness. This exceptional position of the metallic glasses is illustrated in Figure 7, showing the combination of yield strength σ_Y and coercive field strength H_c for selected soft magnetic materials.

The static and dynamic plastic deformation are of similar importance. There is no work hardening in metallic glasses such that tensile straining beyond the elastic limit under increasing load leads to fracture at strains of only $\approx$1%. Fatigue life, however, is remarkably high. The bending fatigue limit (for N = 10^7 load cycles) has been found to range from 650 to 900 N/mm^2 for Fe–Ni-base and to be about 1200 N/mm^2 for Co-base metallic glasses. These favorable values permit applications where dynamic loading must be sustained.

Plastic deformation of metallic glass ribbons by rolling is possible. Since no work hardening occurs, intermediate softening treatments by annealing are not required. The only limitation in thickness reduction by rolling is given by the common relations between flow stress, strip thickness, and roll diameter. It should be noted that shear bands caused by the localized slip characteristic of metallic glasses can lead to the orange-peel-type surface roughness during rolling. Other types of forming operations involving tensile strain components have not been studied extensively as yet. Deep drawing at elevated temperatures has been shown to be possible to a limited extent[53] but general conclusions cannot as yet be drawn.

Bonding and compaction of metallic glasses are restricted due to their instability to crystallization and their high elastic limit. These two factors are interrelated because all bonding processes which involve metallic adhesion and diffusion (or wetting as during

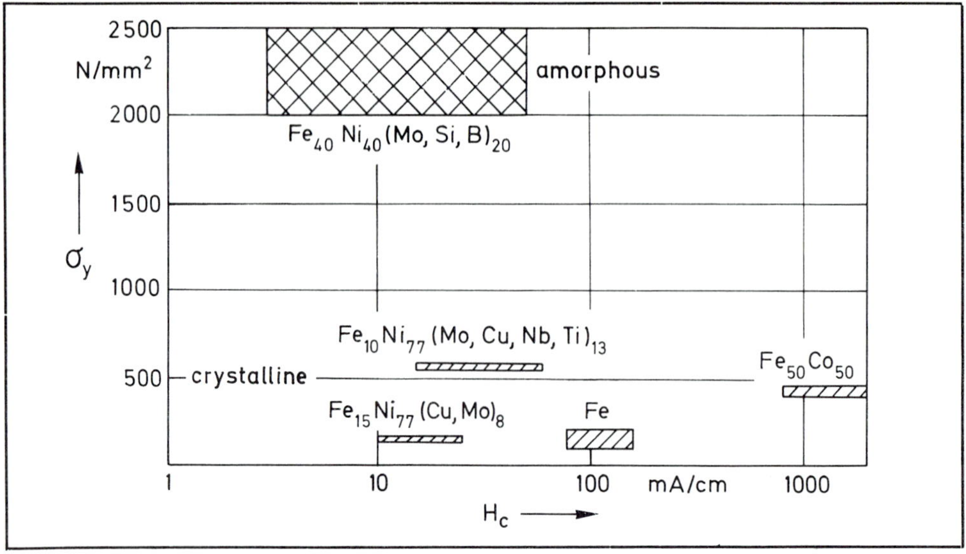

FIGURE 7. Mechanical and magnetic properties of amorphous and crystalline alloys.

explosive joining) require sufficient atomic mobility and a maximum area of atomic metallic contact, which are usually provided by heating and/or extensive plastic flow. Thus, bonding and compaction processes which require relatively high temperatures for long times, such as conventional fusion welding or sintering are not applicable. But processes involving short times at elevated temperatures and sufficient contact time and pressure have been applied successfully under controlled conditions. In the present context, the compilation given in Table 1 may suffice as a basis for further reference.

Corrosion and oxidation are treated in Chapter 8 of this book. It may be mentioned here that electrodeposition of bonding protective or contact layers of metals such as Au, Cu, Ni on metallic glasses poses no problem. The high and isotropic reactivity of a defect-free amorphous surface gives rise to uniform deposition.

IV. APPLICATIONS

A. 400 Hz Transformers

Apart from their potential use in 50/60 Hz power distribution transformers (see Chapter 5) the amorphous materials are of interest for small 400 Hz airborne and shipborne power transformers. The amorphous alloys yield a higher efficiency and a reduction in the dimensions and weight of the transformers as compared with those of commonly used grain-oriented silicon steel. Total power losses of $Fe_{81}B_{13.5}Si_{3.5}C_2$, 0.035 mm thick at 400 Hz are several times smaller than those of 0.1-mm-thick silicon steel.[55,56]

In order to achieve an optimum transformer design the maximum permissible temperature rise caused by iron and copper losses is of importance. The calculation of the maximum power output for a small transformer results in an optimum working induction for each core type, core material, and coil.[57] The power and apparent losses of the core material as a function of the peak induction are required as input data for the computation.

Figure 8 shows the dependence of the power output on the peak induction of a 400 Hz transformer designed with a toroidal strip-wound core. The ambient temperature assumed is 40°C, the temperature rise being 75°C. Using a core made of $Fe_{81}B_{13.5}Si_{3.5}C_2$ with 90 mm O.D., 50 mm I.D., and 25 mm height, a maximum power output of 780 VA is attained

Table 1
BONDING AND COMPACTION TECHNIQUES

Techniques	Conditions	Results	Ref.
Soldering	Limited composition and temperature ranges	Fair to good	
Spot welding	Close energy and time control	Good	54
Laser and electron beam glazing and joining	Limited compositional range, controlled pulse shape	Fair to good	51, 52
Warm compaction and cladding	Close temperature and time control	Fair	47
Explosive and other high rate compaction and joining techniques		Good	46, 48, 49, 50
Ultrasonic welding		Fair	45, 48
Adhesive bonding	Compatible adhesives	Good	

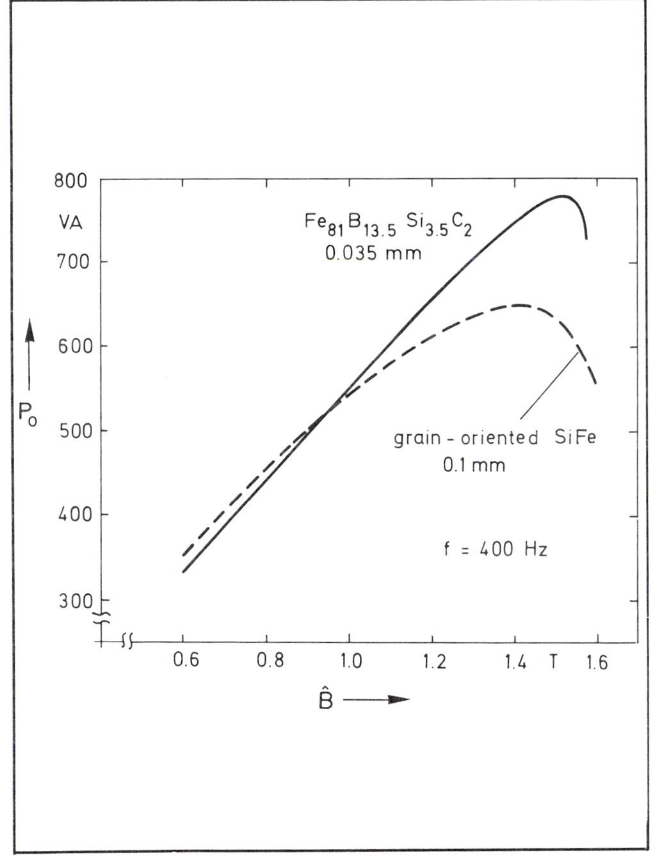

FIGURE 8. Output power P_o vs. peak of induction $\hat{B}$ of 400 Hz transformers with toroidal strip-wound cores (90 × 50 × 25 mm).

compared to 650 VA for a silicon steel core with the same dimensions. This means that the power output can be raised by 20% or else the core weight and volume can be reduced accordingly.

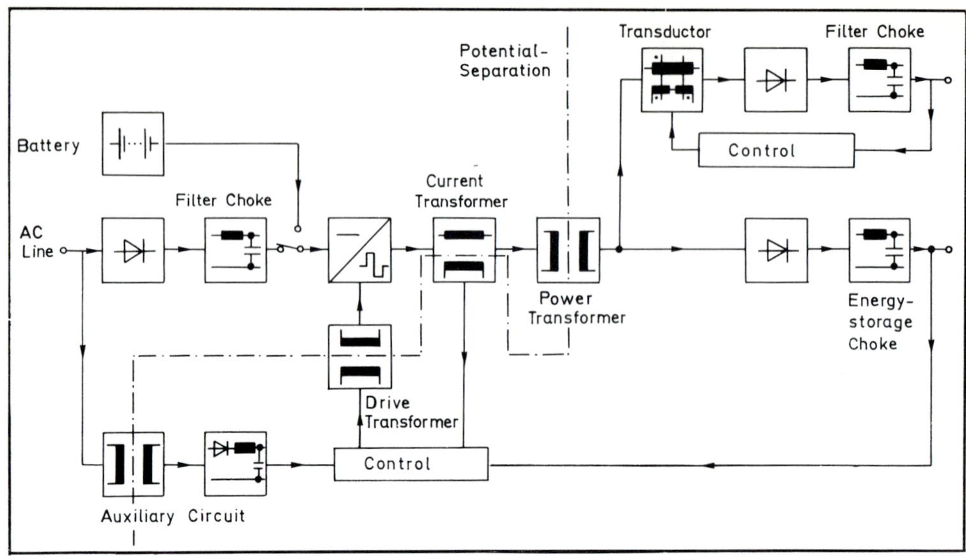

FIGURE 9. Inductive components in switched-mode power supplies.

B. Components for Switched-Mode Power Supplies

In power electronics the switched-mode power supplies are replacing the standard mains power supplies to an increasing extent. The operating frequency is shifted towards that of fast VMOS-transistors.[9,33,58] Therefore, greater attention is payed to the low power losses of amorphous materials in this frequency range.[3,58,59] Figure 9 gives a simplified survey of inductive components used in switched-mode power supplies: power transformers, saturable reactors and transductors, energy-storage chokes, filter chokes, driver transformers, etc. The selection and the design of the inductive components depend on the basic circuit and on the operational specifications.

1. Power Transformers (Nonsaturable Inverter Transformers) and Transductors

Low power losses and high saturation induction are the primary conditions for core materials operating in the frequency range above 20 kHz. Amorphous alloys are comparable with conventional thin crystalline Ni–Fe alloys but are superior to ferrites in the frequency range up to 100 kHz, see Figure 10. The low total losses of amorphous materials result from the low hysteresis losses, the high resistivity of the material and the low thickness of the strips.

Nonsaturable inverter transformers are designed as forward transformers or as push-pull power transformers. For forward transformers, a core material with low remanence but high saturation induction is required to yield a high induction range. The operation is comparable with that of a pulse transformer.

If a power supply has two or more outputs whose loads may vary over a wide range, these outputs can be stabilized independently by transductors. These transductors are reliable and robust components. In a simple circuit they achieve an efficiency of more than 90%. Although transductors are well known in 50/60 Hz techniques, their application in switched-mode power supplies introduces some new aspects.[58,60,61] Core materials for transductors should have a high loop squareness-ratio, lower coercivity, and low power losses.

2. Other Devices

A ferro-resonant inverter with an amorphous core was proposed.[62] The ferro-resonant circuit and the constant volt transformer have been used for various kinds of power supplies

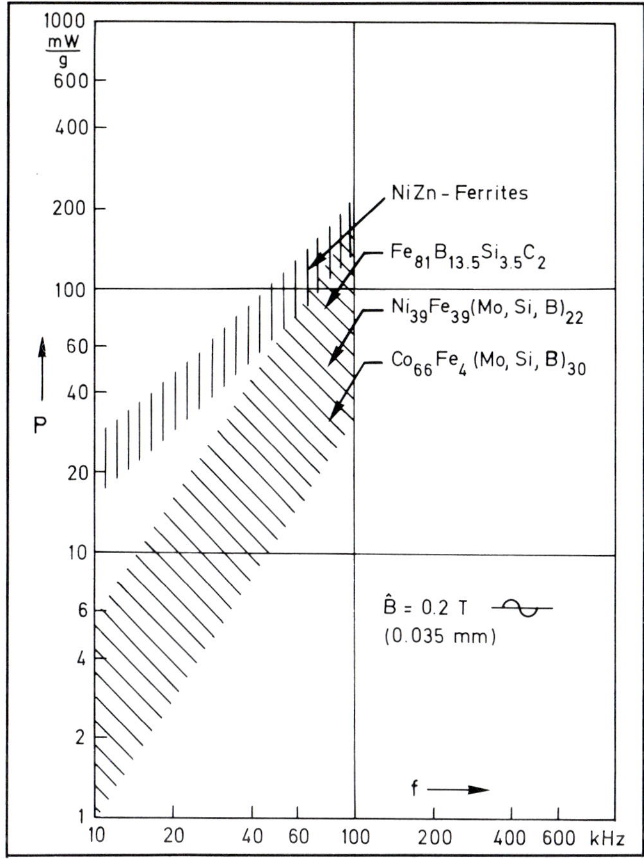

FIGURE 10. Total power losses (core loss density) vs. frequency at 25°C. Sinusoidal voltage drive. (References 55 and 58 and our results.)

at commercial frequency. The amorphous core is to be employed as a saturable core for a new design operating at higher frequencies. The saturable transformer is directly coupled to the switching transistors. The core material can be a Ni–Fe-based amorphous alloy with a rectangular hysteresis loop.

Another inductive component, an electronic current transformer using a Fe-based amorphous material core of reduced size has been described.[63] The new transformer core material is significantly superior to the common Fe–Si transformer core. This application appears to be justified by the higher permeability and the lower losses compared with crystalline Fe–Si. Crystalline Fe-50 wt.% Ni alloys could also be replaced by high saturation amorphous material of $Fe_{80}(B,C,Si)_{20}$ type for current transformers.

Two surveys on instrument and control devices have been given.[64,65] These papers deal with several types of flux control devices using toroidal cores of amorphous alloys and crystalline materials such as Fe-50 wt.% Ni and Supermalloy with rectangular loops. Amorphous materials for pulse transformers have been discussed.[66]

C. Magnetic Shielding

One of the first prototype applications proposed for amorphous metals was magnetic shielding. An efficient and easy to handle magnetic shielding wrap has been developed by weaving amorphous ribbons into mesh.[67,68,69]

1. *Flexible Shielding*

Metallic glasses permit the production of flexible and reusable shielding wrap due to their

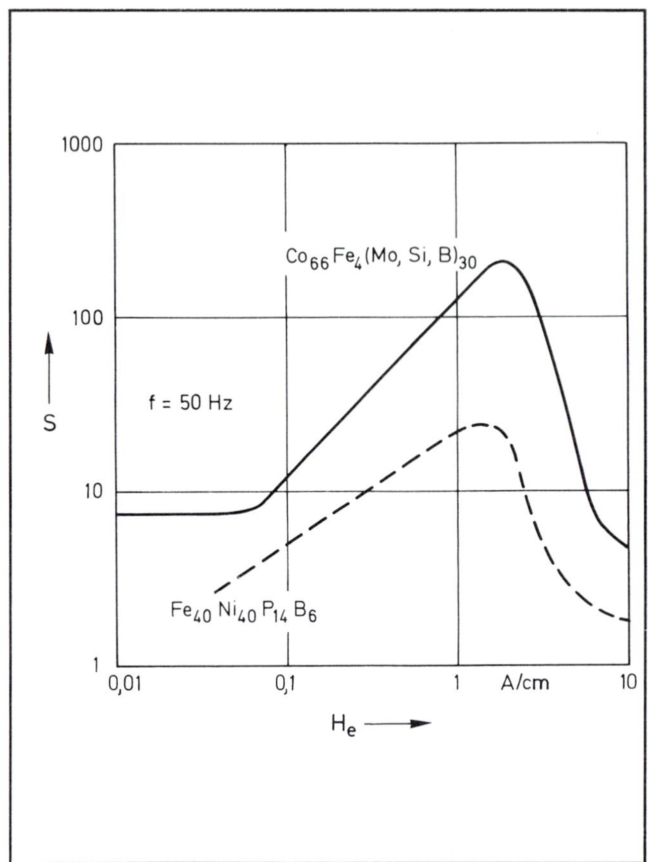

FIGURE 11. Shielding factor S of flexible shielding fabric vs. external field He (2-layered cylinder, ϕ 50 mm).

extraordinarily high yield strength. The first samples of amorphous shielding were woven from $Fe_{40}Ni_{40}P_{14}B_6$. But because of the low initial permeability and the large magnetostriction of this material, the efficiency of this shielding was poor, particularly in weak magnetic fields. The shielding efficiency can be improved considerably by using Co-based amorphous materials such as $Co_{66}Fe_4(Mo,Si,B)_{30}$ with zero magnetostriction and high initial permeability.

Shieldings made of this material are highly flexible and insensitive to mechanical deformation, like bending, due to their high elastic limit. This is a great advantage compared with crystalline materials such as thin Ni–Fe foils or strips which are damaged by rough handling.

For long cylinders and cable braids, the shielding factor, attenuation or shielding effectiveness S, is defined as $S = \mu_r \cdot (d/D)$, where D = diameter of the cylinder or cable braid, d = thickness of the shielding material, μ_r = relative permeability. In order to evaluate the magnetic shielding characteristics a Helmholtz-coil can be used according to ASTM standard A-698.

Shieldings for electron tubes or other components can be produced by winding one or more layers of amorphous shielding material over the tube diameter or over the component.

Figure 11 presents data obtained from cylinders with 50 mm I.D., for a two-layered shield of a Ni–Fe-based and of a Co-based amorphous alloy. The Co based alloy shows much better results in weak magnetic fields where the initial permeability is of importance but also in stronger fields where maximum permeability is relevant.

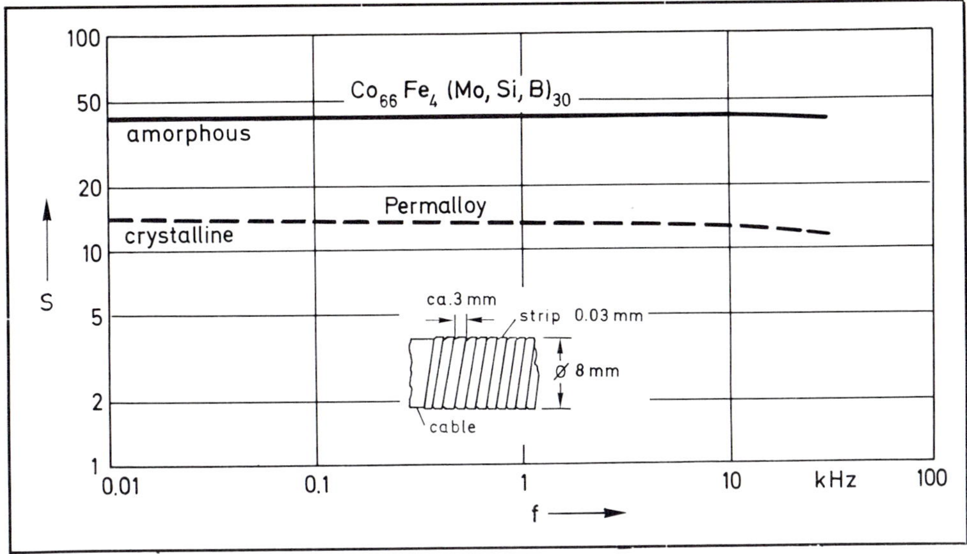

FIGURE 12. Shielding factor S vs. frequency f of thin cable braids.

Recently, in addition to the flat woven mesh, the use of amorphous strips and filaments as shielding envelopes has been suggested in place of conventional shielding materials.[70]

2. Cable Shielding

Another application of recent interest is the shielding of cables; this arises from a major EMI problem, the low-frequency interference in cable systems.[71]

There are various options available to eliminate the magnetic field pickup: thin walled convoluted tubing or braided sheaths of crystalline Ni–Fe. For highly flexible cables with small diameters braided sheaths are preferable.

Amorphous alloys with zero magnetostriction are materials uniquely suitable for this application because they can be braided or woven practically, without loss of material permeability. This is a large improvement compared with crystalline strips which are deformed plastically when wound in small diameters. Figure 12 shows the shielding factor vs. frequency for cables with 8 mm I.D. measured in a constant a.c. field of 1 A/cm. Here, we compare a cable braid made by contiguous helical windings of 3 mm × 0.03 mm strip of crystalline Permalloy and of Co-based amorphous material.[72-74]

D. Elastomagnetic Transducers

Elastomagnetic transducers and pickups made of conventional magnetic alloys have been in use for compression or tensile measurements for several years and the development appeared almost completed. The metallic glasses have given fresh impetus to many novel applications in this sensor area because of their excellent mechanical and magnetic properties.

1. Basic Systems

A survey of basic systems of elastomagnetic sensors is given in Figure 13. Three systems are considered here.

For force and displacement transducers, the simplest example is a narrow soft magnetic strip subjected to a tensile load in the longitudinal direction (Figure 13, a). The inductance change is used as an indication of force or displacement. A new magnetic core multivibrator applied to tensile-stress transducers was recently proposed.[75]

Both compressive and tensile stresses occur when a circular ring is deformed to an ellipse

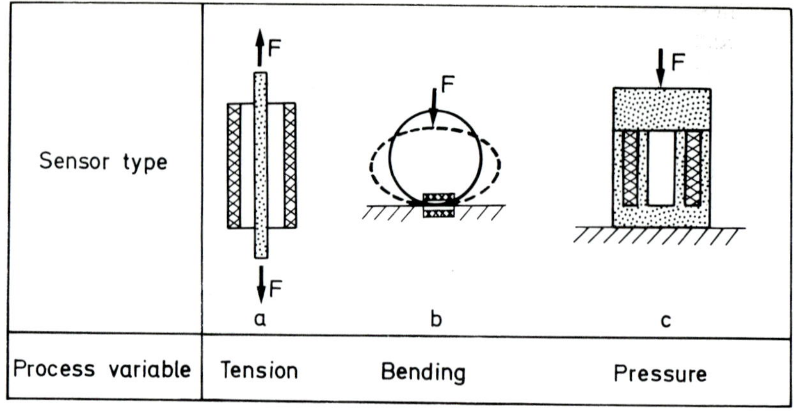

FIGURE 13. Basic principles of elastomagnetic sensors.

(Figure 13, b). Sensors working on this principle have been proposed for pressure and displacement measurements as toroidal core distortion transducers.[75,76,78] Different calculations of the relationship between distortions and change of magnetic properties have been given.[20,77]

A third basic system is the elastomagnetic force and pressure load cell (Figure 13, c). In principle, this transducer is an iron-closed choke; the inductance change is used as an indication of the load.[78,79]

2. Materials for Elastomagnetic Sensors

For this application a favorable combination of both property groups, magnetic and mechanical, is sought. For crystalline magnetic materials, magnetic and mechanical softness are mutually dependent so that, for example, high permeability and high tensile strength cannot be attained at the same time. Therefore, where crystalline materials are used for elastomagnetic sensors certain compromises must be made. Amorphous metals which allow the combination of good soft magnetic and very good mechanical properties, have opened up new possibilities in this field.

Since in amorphous metals no magnetocrystalline anisotropy exists and the uniaxial magnetic anisotropy can also be kept low, high magnetic isotropy is attainable. This is an important prerequisite for most transducers. Magnetostriction of amorphous alloys, as in crystalline metals, can essentially be adjusted by the alloy composition.[20,80,81] It can be made positive to a different degree with Ni–Fe-based and Fe-rich materials, negative or equal to zero with Co-based materials.

As an example, Figure 14 shows the modification of the magnetization curve of an amorphous alloy strip (composition $Co_{75}Si_{15}B_{10}$) under tensile load.[74]

E. Delay Lines

The high magnetostriction and the low magnetic anisotropy of Fe-rich metallic glasses make these materials attractive for applications in surface-acoustic wave (SAW) substrates. The use of metallic glasses in SAW delay lines is of particular interest because they yield a relatively large delay time variation with a small change in the bias field. An increasing bias field increases the ΔE effect of Fe-rich amorphous materials, this corresponds to a change in sound velocity and in delay time. Figure 15 shows a model of a magnetically variable delay line.[82]

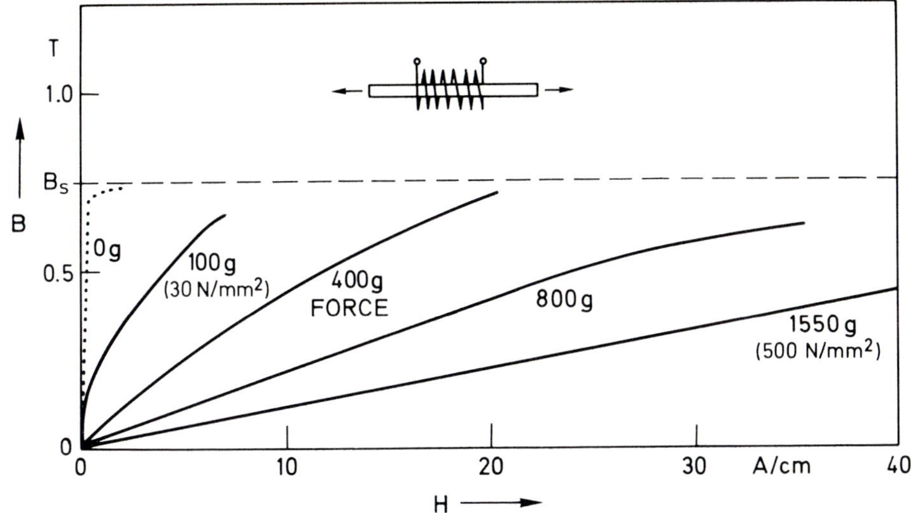

FIGURE 14. Variation of the magnetization curve by an applied tensile force (Co-based alloy).

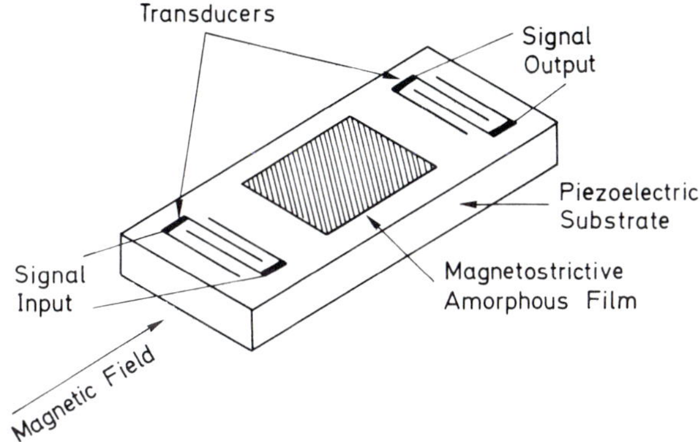

FIGURE 15. Magnetically variable SAW delay line. (From Webb, D. C., Forester, D. W., Ganguly, A. K., and Vittoria, C., *IEEE Trans. Magn.*, MAG-15, 77, 1979. With permission.)

The following materials have been investigated: Fe–B film of 0.8-μm thickness on ST quartz substrates,[82] $Fe_{0.7}Si_{0.3}$ films on YZ–$LiNbO_3$ substrate,[83] $Fe_{80}P_{13}C_7$ ribbons, and various other composition.[84,85]

The variation in delay time ranges from tenths of nanoseconds to tenths of microseconds for these films. The extent of the relative velocity change within the films is up to 0.05% and can be controlled by film composition and thermal treatment. For ribbons with dimensions of 2 mm × 30 μm made of $Fe_{80}P_{13}C_7$ or $Fe_{75}P_{15}C_{10}$ a maximum delay time variation of 10% was attained.

The applications suggested are various UHF and VHF systems and subsystems, adaptive array antenna systems, tunable oscillators, etc.

The propagation of the magnetization processes along thin strips of $Fe_{40}Ni_{40}P_{14}B_6$ was investigated by Eifrig et al.[86]

F. Magnetic Heads

Materials for magnetic heads must have high initial permeability, adequate saturation,

Table 2
DATA OF MAGNETIC HEAD MATERIALS

Material group	Composition	Permeability (μ_4)	Saturation induction (T)	Electrical resistivity ($\mu\Omega$·cm)	Hardness (HV)
Crystalline	77 Ni,Cu Mo Fe (Permalloy)	20,000	0.8	60	120
	77 Ni,Cu Mo Fe 5 (Nb,Ti)	40,000	0.5	90	240
	85 Fe, 9.6 Si, 5.4Al (Sendust)	25,000	1.1	88	450
	84 Fe, 16 Al	8,000	0.9	145	280
Amorphous	$Co_{66}Fe_4(Mo,Si,B)_{30}$	20,000	0.55	130	900
	$Co_{69}Mn_6(Si,B)_{25}$	10,000	0.8	130	900

high mechanical hardness and wear resistance. Zero magnetostriction is desirable. The unique combination of magnetic and mechanical properties of some amorphous alloys makes these materials attractive for magnetic heads. Long reliable operation of tape recorders, full recording with low distortion of high coercivity tapes and high frequency response were reported as the advantages of the new material group. The application of Co-based amorphous materials for magnetic heads has been discussed.[87]

Not only amorphous alloys but also thin crystalling Sendust ribbons produced by rapid quenching techniques have recently been investigated for magnetic heads.[88] Sendust is a well known material, the composition being approximately $Fe_{85}Si_{9.5}Al_{5.5}$. It also has excellent magnetic and mechanical properties, but it is generally very brittle in the as-cast state and cannot be rolled into a thin sheet. The rapid quenching technique has permitted the production of thin Sendust strips of 40 μm for the first time.

Table 2 shows material for magnetic heads. The table compares two Co-rich amorphous alloys, a Sendust ribbon, a Fe-16 wt.% Al alloy, and a typical Permalloy and an age-hardenable Permalloy.

G. Security Systems

Magnetized strips of amorphous materials are also employed to monitor the use of books in libraries and provide pilferage protection of expensive articles and products in stores. They make possible a silent system of security invisible to the naked eye.[89]

High permeability and great flexibility are the essential material characteristics for this application. To avoid altering of magnetic performance by mechanical strain, Co-based alloys with zero magnetostriction are advantageous. Because of the use of high-frequency detector systems, very thin strips with thickness of 10 to 20 μm are required.

V. OUTLOOK

Amorphous magnetic materials with optimized properties have been available in kilogram quantities, some even in tons, for several years now. In the meantime the development of many applications of the kinds discussed here has progressed. As with novel metallic materials in general, the introduction into industrial use is slow. Many technological and economic barriers have to be overcome first. But we are certain that the particular properties and unique property combinations of amorphous magnetic materials will lead to their entry into the field of electronics before too long.

In this context it should be noted that with some modifications crystalline strip materials may be produced by the same economic process as amorphous ribbons. Thus it is an open question whether amorphous magnetic materials will find broad application where substitution of existing materials is a major aspect. In fields where their novel properties are being exploited they will definitely find their place as special magnetic materials.

REFERENCES

1. **Kobe, S. and Ferchmin, A. R.,** Amorphous magnetism and magnetic materials: Bibliography 1950-1976, *J. Mater. Sci.*, 12, 1713, 1977.
2. **Luborsky, F. E.,** Perspective on application of amorphous alloys in magnetic devices, *Amorphous Magnetism II,* Levy, R. A. and Hasegawa, R., Eds., Plenum Press, New York, 1977, 345.
3. **Luborksy, F. E., Frischmann, P. G., and Johnson, L. A.,** The role of amorphous materials in the magnetics industry, *J. Magn. Magn. Mater.*, 8, 318, 1978.
4. **Graham, C. D. and Egami, T.,** Magnetic properties of amorphous materials, *Proc. 3rd Int. Conf. Rapidly Quenched Metals,* Vol. 2, Cantor, B., Ed., The Metals Society, London, 1978, 96.
5. **Luborsky, F. E., Becker, J. J., Frischmann, P. G., and Johnson, L. A.,** Potential of amorphous alloys for application in magnetic devices, *J. Appl. Phys.*, 49, 1769, 1978.
6. **Luborsky, F. E.,** Applications of amorphous alloys, *IEEE Trans. Magn.*, MAG-14 (5), 1008, 1978.
7. **Hilzinger, H. R., Mager, A., and Warlimont, H.,** Amorphe ferromagnetische Werkstoffe-Magnetische Grundlagen, Eigenschaften und Anwendungen, *J. Magn. Magn. Mater.*, 9, 191, 1978.
8. **Luborsky, F. E.,** Amorphous ferromagnets, *Ferromagnetic Materials,* Vol. 1, Wohlfarth, E. P., Ed., North Holland, Amsterdam, 1980, 451.
9. **Chen, D. Y.,** Amorphous magnetic alloys for high frequency power electronic applications, *Solid-State Powerconversion 7/8,* Intertex Communications, Oxnard, Calif., 1979, 44.
10. **Luborsky, F. E. and Johnson, L. A.,** Applications of magnetic amorphous alloys, *J. Phys. (Paris),* 41, C8-820, 1980.
11. **Hilzinger, H. R.,** Amorphe Magnetwerkstoffe, NTG-Fachberichte, 76, 283, 1980.
12. **Nose, M. and Masumoto, T.,** Characteristics of (Fe,Co and/or Ni)-Zr amorphous alloys, *Sci. Rep. Res. Inst. Tohoku Univ. Ser. A, Suppl.*, 222, 1980.
13. **Mizoguchi, T.,** Magnetism in amorphous alloys, *AIP Conf. Proc.*, 34, 286, 1976.
14. **O'Handley, R. C., Hasegawa, R., Ray, R., and Chou, C.-P.,** Ferromagnetic properties of some new metallic glasses, *Appl. Phys. Lett.*, 29(No. 6), 330, 1976.
15. **Hasegawa, R. and Chou, C.-P.,** Metallisches Glas, D-Offenlegungsschrift 27 56 920 v. 6.7.1978.
16. **Kazama, N. S., Mitera, M., and Masumoto, T.,** The size effect of metalloids on the magnetic properties of iron based amorphous alloys, *Proc. 3rd Int. Conf. Rapidly Quenched Metals,* Vol. 2, Cantor, B., Ed., The Metals Society, London, 1978, 164.
17. **Luborsky, F. E., Walter, J. L., and Wohlfarth, E. P.,** The saturation magnetization, Curie temperature and size effect of amorphous iron alloys, *J. Phys. F,* 10, 959, 1980.
18. **Ito, S., Aso, K., Makino, Y., and Uedaira, S.,** Magnetostriction and magnetization of iron-based amorphous alloys, *Appl. Phys. Lett.*, 37(7), 665, 1980.
19. **O'Handley, R. C.,** Magnetostriction of ferromagnetic metallic glasses, *Solid State Commun.*, 21, 1119, 1977.
20. **Hilzinger, H. R., Hillmann, H., and Mager, A.,** Magnetostriction measurements on Co-base amorphous alloys, *Phys. Status Solidi A,* 55, 763, 1979.
21. **Sherwood, R. C., Gyorgy, E. M., Chen, H. S., Ferris, S. D., Norman, G., and Leamy, H. J.,** Ferromagnetic behavior of metallic glasses, *AIP Conf. Proc.*, 24, 745, 1975.
22. **Fujimori, H., Kikuchi, M., Obi, Y., and Masumoto, T.,** New Co-Fe amorphous alloys as soft magnetic materials, *Sci. Rep. Res. Inst. Tohoku Univ. Ser. A,* 26, 36, 1976.
23. **Hilzinger, H. R. and Kunz, W.,** Magnetic properties of amorphous alloys with low magnetostriction, *J. Magn. Magn. Mater.*, 15-18, Part III, 1357, 1980.
24. **Luborsky, F. E., Liebermann, H. H., Becker, J. J., and Walter, J. L.,** The magnetic properties of Fe-B amorphous alloys, *Proc. 3rd Int. Conf. Rapidly Quenched Metals,* Vol. 2, Cantor, B., Ed., The Metals Society, London, 1978, 188.
25. **Hatta, S., Egami, T., and Graham, C. D., Jr.,** Amorphous alloys with improved room-temperature saturation induction, *IEEE Trans. Magn.*, MAG-14 (5), 1013, 1978.
26. **Narita, K., Yamasaki, J., and Fukunaga, H.,** Saturation magnetostriction and its annealing behavior of $Fe_{100-x}B_x$ and $Co_{100-x}B_x$ amorphous alloys, *J. Appl. Phys.*, 50, 7591, 1979.

27. **Kohmoto, O., Yamaguchi, N., Ohya, K., Fujishima, H., and Ojima, T.,** Effects of annealing on magnetostriction and permeability of zero magnetostrictive amorphous alloys, *IEEE Trans. Magn.*, MAG-14 (5), 949, 1978.
28. **Hilzinger, H. R.,** unpublished data, 1979.
29. **Luborsky, F. E., Walter, J. L., Becker, J. J., and Martin, D. L.,** The Fe-B-C Ternary Amorphous Alloys, Report 79CRD169, General Electric Technical Information Series, Schenectady, N.Y., 1979.
30. **Mitera, M., Masumoto, T., and Kazama, N. S.,** Effect of silicon addition on the magnetic properties of Fe–B–C amorphous alloys, *J. Appl. Phys.*, 50, 7609, 1979.
31. **Jaschinski, W., Koenig, U., and Wolf, W.,** Magnetische Eigenschaften von amorphen FeSiB-Legierungen, NTG-Fachberichte, 76, 307, 1980.
32. **Masumoto, T., Watanabe, K., Mitera, M., and Ohnuma, S.,** High magnetic permeability amorphous alloys of the Fe–Ni–Si–B system, *Amorphous Magnetism II*, Levy, R. A. and Hasegawa, R., Eds., Plenum Press, New York, 1977, 369.
33. **Kunz, W. and Graetzer, D.,** Amorphous alloys for switched-mode power supplies, *J. Magn. Magn. Mater.*, 19, 183, 1980.
34. **Hasegawa, R., Narasimhan, M. C., and DeCristofaro, N.,** A high permeability Fe–Ni base glassy alloy containing Mo, *J. Appl. Phys.*, 49, 1712, 1978.
35. **Ohnuma, S. and Masumoto, T.,** Amorphous magnetic alloys (Fe, Co, Ni)-(Si, B) with high permeability and its thermal stability, *Proc. Third Int. Conf. Rapidly Quenched Metals*, Vol. 2, Cantor, B., Ed., The Metals Society, London, 1978, 197.
36. **Luborsky, F. E. and Walter, J. L.,** Stress relaxation in amorphous alloys, *Mater. Sci. Eng.*, 35, 255, 1978.
37. **Chi, G. C., Chen, H. S., and Miller, C. E.,** The influence of quenching procedures on the kinetics of embrittlement in a $Fe_{40}Ni_{40}B_{20}$ metallic glass, *J. Appl. Phys.*, 49, 1715, 1978.
38. **Warlimont, H.,** unpublished data, 1980.
39. **Hilzinger, H. R. and Hock, S.,** Preparation of metallic glasses, *Proc. Int. Conf. on Metallic Glasses: Science and Technology*, Vol 1, Hargitai, C., Bakonyi, I., and Kemeny, T., Eds., Central Research Institute for Physics, Budapest, 1981, 71.
40. **Luborsky, F. E., Becker, J. J., and McCary, R. O.,** Magnetic annealing of amorphous alloys, *IEEE Trans. Magn.*, MAG-11, 1644, 1975.
41. **Pfeifer, F. and Behnke, W.,** Magnetische Eigenschaften von Ringbandkernen aus amorphen Legierungen auf Eisen-Nickel-Basis, *J. Magn. Magn. Mater.*, 6, 80, 1977.
42. **Egami, T., Flanders, P. J., and Graham, C. D., Jr.,** Amorphous alloys as soft magnetic materials, *AIP Conf. Proc.*, 24, 697, 1975.
43. **Nielsen, O. V. and Nielsen, H. J. V.,** Strain- and field induced magnetic anisotropy in metallic glasses with positive or negative λ_s, *Solid State Commun.*, 35, 281, 1980.
44. **Hasegawa, R.,** Reorientation kinetics of the magnetization in a glassy ferromagnet, *J. Appl. Phys.*, 49, 5610, 1978.
45. **Kreye, H., Hammerschmidt, M., and Reiners, G.,** Ultrasonic welding of metallic alloy glass to copper, *Scr. Metall.*, 12, 1059, 1978.
46. **Babul, T., Kozlowski, W., and Zieliński, P. G.,** Explosive welding of metallic glasses with copper, *Proc. Int. Conf. Metallic Glasses: Science and Technology*, Vol. 1, Hargitai, C., Bakonyi, I., and Kemeny, T., Eds., Central Research Institute for Physics, Budapest, 1981, 159.
47. **Liebermann, H. H.,** Warm-Consolidation and Cladding of Glassy Alloy Ribbons, Rep. 80CRD029, General Electric Technical Information Series, Schenectady, N.Y., 1980.
48. **Hammerschmidt, M., Kreye, H., and Meinck, B.,** Schweissverhalten metallischer Glaeser, *Metall (Berlin)*, 34, 23, 1980.
49. **Cline, C. F. and Hopper, W.,** Explosive fabrication of rapidly quenched materials, *Scr. Metall.*, 11, 1137, 1977.
50. **Morris, D. G.,** Compaction and mechanical properties of metallic glass, *Met. Sci. J.*, 14, 215, 1980.
51. **Kear, B. H., Breinan, E. M., and Greenwald, L. E.,** Laser-glazing - a new process for production and control of rapidly chilled metallurgical microstructures, *Met. Technol. (N.Y.)*, 6, 121, 1979.
52. **White, C. W. and Peercy, P. S.,** *Laser and Electron Beam Processing*, Academic Press, New York, 1980.
53. **Patterson, J., Greer, A. L., Leake, J. A., and Jones, D. R. H.,** Hot forming of a metallic glass, *Proc. 3rd Int. Conf. Rapidly Quenched Metals*, Vol. 2, Cantor, B., Ed., The Metals Society, London, 1978, 293.
54. **Kavesh, S. and Bretts, G. R.,** Welding of Glassy Metallic Materials, U.S. Patent 4,115,682, 1978.
55. **Nathasingh, D. and Smith, C. H.,** A new high-flux, low-loss magnetic material for high frequency applications, *Proc. POWERCON*, Vol. 7, Power Concept, Inc., Ventura, Calif., 1980, B2-1.
56. **Mas, J. A.,** Design and performance of power transformers with metallic glass cores, *IEEE Power Electronics Specialists Conf.*, IEEE, Piscataway, N.J., 1980, 68.
57. DIN 41 300 Part 1: Small transformers; characteristic data, 1979.

58. **Chen, D. Y.,** Magnetic properties of the amorphous magnetic metals between 50 kHz and 225 kHz, *Proc. 2nd Int. Powerconversion Conf., 4A.5-1,* Intertex Communications, Inc., Oxnard, Calif., 1980.
59. **Hasegawa, R. and O'Handley, R. C.,** Soft magnetic properties of metallic glasses - Recent developments, *J. Appl. Phys.,* 50, 1551, 1979.
60. **Graetzer, D.,** Transductor-regulators for switched-mode power supplies, *IEEE Trans. Magn.,* MAG-16, 922, 1980.
61. **Hiramatsu, R., Harada, K., and Ninomiya, T.,** Switch mode converter using high-frequency magnetic amplifier, *Powerconversion International,* Vol. 6(2), Intertex Communications, Oxnard, Calif., 1980, 75.
62. **Harada, K. and Murata, K.,** Ferro-resonant inverter with an amorphous core, *Proc. 2nd Int. Powerconversion Conf. 3A.7-1,* Intertex Communications, Inc., Oxnard, Calif., 1980.
63. **Milcovic, M., Luborsky, F. E., Chen, D., and Tompkins, R. E.,** Electronic transformers using amorphous materials, *IEEE Trans. Magn.,* MAG-13, 1224, 1977.
64. **Murakami, K.,** Applications of amorphous magnetic cores to instrument and control devices, *Sci. Rep. Res. Inst. Tohoku Univ. Ser. A,* Suppl., 221, 1978.
65. **Murakami, K., Watanabe, T., and Goto, A.,** A new bridge-connected magnetic circuit using amorphous magnetic core and its applications, *IEEE Trans. Magn.,* MAG-14, 966, 1978.
66. **Jouve, H.,** Les propriétés et les applications en électronique des verres metalliques, *L'onde Électrique,* 60, 49, 1979.
67. **Dismukes, J. P. and Sellers, G. J.,** Metallic glasses for improved magnetic shielding, *Proc. 3rd Int. Conf. Rapidly Quenched Metals,* Vol. 2, Cantor, B., Ed., The Metals Society, London, 1978, 205.
68. **Mendelsohn, L. I., Nesbitt, E. A., and Bretts, G. R.,** Glassy metal Fabric: A unique magnetic shield, *IEEE Trans. Magn.,* MAG-12, 924, 1976.
69. **Sellers, G. J.,** METGLAS alloys: an answer to low frequency magnetic shielding, *IEEE Int. Symp. Electromagnetic Compatibility,* IEEE, Piscataway, N.J., 1977, 129.
70. Electromagnetic Shielding Envelopes from Wound Glassy Metal Filaments, European Patent Application No. 0 007 701.
71. **Sellers, G. J., Dismukes, J. P., Shiau, T., Krstansky, J. J., Sharp, R. E., and Kincaid, J. W.,** Flexible braids for improved magnetic shielding of cables, *IEEE Int. Symp. Electromagnetic Compatibility,* IEEE, Piscataway, N.J., 1978, 151.
72. **Shiau, Y., Bridges, J., and Sellers, G. J.,** Wide-band high performance very flexible coaxial shield, *IEEE Int. Symp. Electromagnetic Compatibility,* IEEE, Piscataway, N.J., 1978, 155.
73. **Boll, R. and Borek, L.,** Elektromagnetische Schirmung, NTG-Fachberichte, *Band,* 76, 178, 1980.
74. **Boll, R.,** Metallische Glaeser aus industrieller Sicht, *Umsch. Wiss. Tech.,* 80, 403, 1980.
75. **Mohri, K. and Korekoda, S.,** An amorphous magnetic core multivibrator for tensile-stress transducers, *Techn. Engineering,* 7, 25, 1977.
76. **Mohri, K. and Sudoh, E.,** Sensitive force transducers using a single amorphous core multivibrator bridge, *IEEE Trans. Magn.,* MAG 15, 1806, 1979.
77. **Mohri, K. and Korekoda, S.,** New force transducers using amorphous ribbon cores, *IEEE Trans. Magn.,* MAG-14, 1071, 1978.
78. **Boll, R. and Borek, L.,** Magnetic sensors of new material, *Siemens Forsch. Entwicklungsber.,* 10, 83, 1981.
79. **Engl, W.,** Magnetoelastische Kraftmessdosen, *Siemens Z.,* 29, 219, 1955.
80. **O'Handley, R. C.,** Magnetostriction of transition-metal-metalloid glasses: Temperature dependence, *Phys. Rev. B,* 18, 930, 1978.
81. **O'Handley, R. C.,** Magnetostriction of metallic glasses, *Amorphous Magnetism II,* Levy, R. A. and Hasegawa, R., Eds., Plenum Press, New York, 1977, 379.
82. **Webb, D. C., Forester, D. W., Ganguly, A. K., and Vittoria, C.,** Applications of amorphous magnetic-layers in surface-acoustic-wave devices, *IEEE Trans. Magn.,* MAG-15, 77, 1979.
83. **Simpson, E. M. and Robbins, W. P.,** Magnetostrictive Fe–Si thin films for tunable SAW devices, *IEEE Trans. Magn.,* MAG-14, 919, 1980.
84. **Arai, K. I. and Tsuya, N.,** Magnetomechanical coupling and variable delay characteristics by means of giant ΔE effect in iron-rich amorphous ribbon, *J. Appl. Phys.,* 49, 1718, 1978.
85. **Arai, K. I., Tsuya, N., Yamada, M., and Masumoto, T.,** Giant ΔE effect and magnetomechanical coupling factor in amorphous $Fe_{80}P_{13}C_7$ ribbons, *IEEE Trans. Magn.,* MAG-12, 936, 1976.
86. **Eifrig, H., Grosse-Nobis, W., and Jansen, K.,** Propagation of the magnetizing processes along thin strips of a metallic glass, *J. Magn. Magn. Mater.,* 6, 73, 1977.
87. **Shiiki, K., Otomo, S., and Kudo, M.,** Magnetic properties, aging effects and application potential for magnetic heads of Co–Fe–Si–B amorphous alloys, *J. Appl. Phys.,* 52, 2483, 1981
88. **Tsuya, N., Arai, K. I., and Ohmori, K.,** Ribbon-form Sendust alloy, *IEEE Trans. Magn.,* MAG-15 (No. 4), 1979.
89. Technical staff, Development and Application of METGLAS Products, Allied Chemical Corporation, Morristown, N.J., 1980.

Chapter 7

RARE-EARTH-TRANSITION METAL-BASE ALLOYS

R. F. Soohoo

TABLE OF CONTENTS

I. INTRODUCTION

A single crystal can be thought of as composed of a regular repetition of an elementary building block known as the unit cell in all three dimensions. On the other hand, a polycrystal is a collection of single crystals with their crystalline axes oriented in different directions. If each of these single crystals constitute a grain, the boundaries between them will of course be synonymous with grain boundaries. Alternatively, a grain can encompass a number of single crystals.

In contrast, no finite building block or grain boundaries can be identified for an amorphous material; in other words, it lacks a regular crystalline structure. Nevertheless it is possible for some amorphous materials to exhibit ferromagnetic, antiferromagnetic or ferrimagnetic behavior. The possible existence of long range magnetic order in a disordered atomic lattice is not difficult to appreciate. As is well-known, the electron spins in a ferromagnet are coupled by exchange forces. Since these forces are extremely short range, usually only coupling between the nearest neighbors is significant. Although there is a distribution in the nearest neighbor distances in an amorphous material, exchange coupling between neighboring spins should be still effective providing the spacing between any two is not much different from that in the crystalline counterpart. This condition is usually satisfied as the density of amorphous materials has been found to be only a few percent smaller than that of corresponding crystalline samples. Since the exchange energy E_{ex} is a sensitive function of distance, a random distribution in spin spacings will lead to a random distribution in exchange energy. In other words, E_{ex} for an amorphous sample is given by:

$$E_{ex} = -2\sum_{i,j}{}' J_{ij}(r_{ij})\vec{S}_i \cdot \vec{S}_j \qquad (1)$$

where $J_{ij}(r_{ij})$ indicates that the magnitude of J_{ij} depends on the spacing r_{ij} between spins $\vec{S}_i$ and $\vec{S}_j$. Care must be taken to sum over a given nearest neighbor pair only once in Equation (1); also, the prime over the summation indicates that the cases $i = j$ should be excluded. Since the magnetization M(T) and the Curie temperature T_c are both related to E_{ex}, we can expect both M(T) and T_c of an amorphous sample to differ from its crystalline counterpart, as experimentally observed. Indeed, Gubanov in 1969, using molecular field theory, showed that translational invariance was not a requirement for ferromagnetic order.[1] In addition to random spin distribution, there could be more subtle effects such as charge transfer between atoms which can also affect M(T) and T_c in the amorphous state.

Amorphous ferromagnetic alloys can be divided into several groups; among them are: (1) transition metal (Fe,Ni,Co, etc.) -metalloid (B, C, Si, P, etc.) base alloys and (2) rare-earth (Gd, Ho, Tb, etc.) -transition metal-base alloys. Here, the emphasis is on the latter.

Ferromagnetic behavior was first observed in an electrodeposited amorphous Co–P film in 1950.[2] However, it was not until the mid-1960s that amorphous transition metal-metalloid alloys as a class including Fe–Pd–P, Ni–P, Co–P, Fe–B, Fe–Ni–B, and Fe–B–Si received much attention.[3] These materials, prepared by rapid quenching from the liquid state or by electrodeposition from solution, all require the presence of a glass former (B, C, Si, P, etc.) to produce the amorphous state.[4] In general, these metallic glasses are amorphous for only a limited compositional range, typically with about 75 to 85 at.% transition metal. In many cases, the magnetization is significantly reduced by exchange fluctuations and the materials generally possess low coercivity and no magnetic anisotropy.

RE-*TM* bulk alloys, prepared by d.c. sputtering,[5] and thin films prepared by both d.c. and r.f. sputtering,[6] appeared in 1972 to 73. These alloys, e.g., Gd–Co and Tb–Fe, have radically different properties compared to the metallic glasses mentioned above. First, they contain no glass former and can be prepared in amorphous form over a wide range of composition. Secondly, some of these materials possess a large low temperature coercivity and a giant uniaxial anisotropy perpendicular to the film surface rendering the samples useful for magnetic bubble applications.[6] Other applications for these alloys include thermomagnetic information storage,[7] as precursor materials for the preparation of permanent magnets,[8] and contact printing.[9]

In the following sections, the subjects of RE-*TM* amorphous alloy preparation, magnetic order, anisotropy, coercivity, resonance spectra, and homogeneity, as well as applications will be discussed.

II. PREPARATION

Amorphous RE-*TM* bulk alloys can be prepared by a high deposition rate d.c. triode sputtering system with krypton as the sputtering gas.[5,10] Both the target and the sample substrate are backed by copper for good heat dissipation; the target is made from compacted powders, mixed according to the compositions desired.

Thin foils (50-μm thick) can be made by first arc-melting pure constituent elements several times and then rapidly quenching the foil from the liquid by a hammer-anvil technique.[11]

Thin film samples can be prepared by either d.c. or r.f. sputtering.[6] As in the case of foils, the targets are also fabricated by arc-melting a mixture of the alloyed elements. Upon applying a negative bias voltage to the substrate (relative to the bias electrodes near the substrate) during deposition, film composition can be varied up to 10 at.% over the nominal composition expected from a given target. Films with a wide range of composition can be prepared. Furthermore, properties of these films are essentially independent of the nature of the substrate used. These substrates, initially near room temperature, include amorphous materials such as glass and fused quartz, polycrystalline materials such as copper and tungsten, single crystal substrates such as NaCl, Al_2O_3, Si, and mica, as well as flexible substrates such as heat-stable polymers. A variation of this sputtering technique uses multiple-targets.[12] This method has the advantage of controlling the film composition by merely varying the voltage applied to the individual targets.

RE-*TM* amorphous films can also be deposited by thermal evaporation,[13] although evaporated Gd–Co films exhibit no uniaxial perpendicular anisotropy. These films can be deposited by electron beam evaporation.[14]

Generally speaking, the amorphous phase is not a stable ground state of a solid. Upon heating it, it irreversibly changes into the most stable crystalline phase. On the other hand, we can vary the composition of the amorphous phase continuously without crystal structure restrictions. In some cases we can obtain combinations of elements unrealizable in the crystalline phase.

III. MAGNETIC ORDER

Neutron scattering, like X-ray scattering, provides a sensitive method of examining the atomic arrangement in a material. In addition, due to the interaction of the neutron magnetic moment with that of the magnetic electrons, it can be used for the determination of the magnetic spin distribution. Separation of nuclear and magnetic scattering contributions is accomplished either by comparing the elastic diffraction data taken above and below the Curie temperature or by comparing data sets with magnetic field applied parallel and perpendicular to the scattering vector. Furthermore, a neutron can be inelastically scattered by the magnetic structure with the creation or annihilation of a magnon or spin wave, thus enabling the determination of spin wave spectra.

Neutron diffraction and magnetization measurements have been made on sputtered amorphous specimen[15] of $Tb_{0.33}Fe_{0.66}$. The copper backing of the samples was removed by grinding

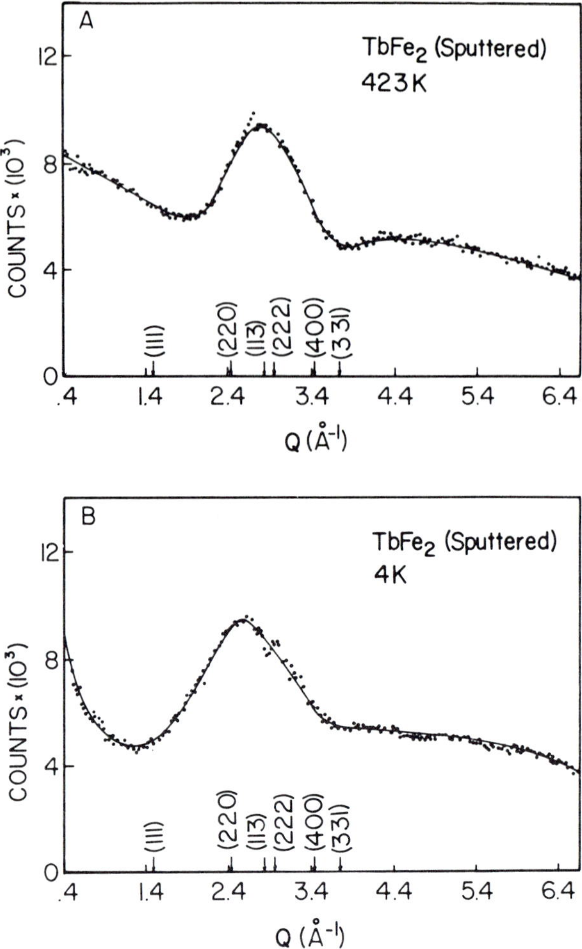

FIGURE 1. Total neutron diffraction intensity. (A) Above the magnetic ordering temperature. (B) At 4.2 K, as a function of scattering vector Q. The "blips" in the data near the peak are the result of remanent scattering from the aluminum sample container.

before neutron measurements were made to eliminate complication of the scattering pattern by the copper Bragg lines. Two 2.54-cm disks, 1 and 3-mm thick, were used; no differences were detected between these two samples. Their density was 8.3 g/cm^3 compared to the crystalline X-ray density of 9.1 g/cm^3 and 8.9 g/cm^3 measured on a cast and annealed bulk sample of crystalline $TbFe_2$. Magnetization data on these amorphous samples gave a T_c of 409 K compared with 697 K for the Laves phase ($MgCu_2$) crystalline $TbFe_2$. The room temperature resistivity of 220 μΩ cm (compared to 58 μΩ cm for crystalline $TbFe_2$) remains essentially unchanged at 77 K due to the large temperature-independent atomic disorder scattering.

Aside from the interaction between the magnetic moments of the impinging neutrons and those of the magnetic electrons, expression for the scattered neutron intensity should be similar in form to those for X-rays.[16] Scattering data taken as a function of scattering wave vector at 423 K (above T_c) and at 4 K (below T_c) are shown in Figure 1(A) and (B).[15] Positions of Bragg diffraction lines for the crystalline counterpart $TbFe_2$ are shown by the arrows.

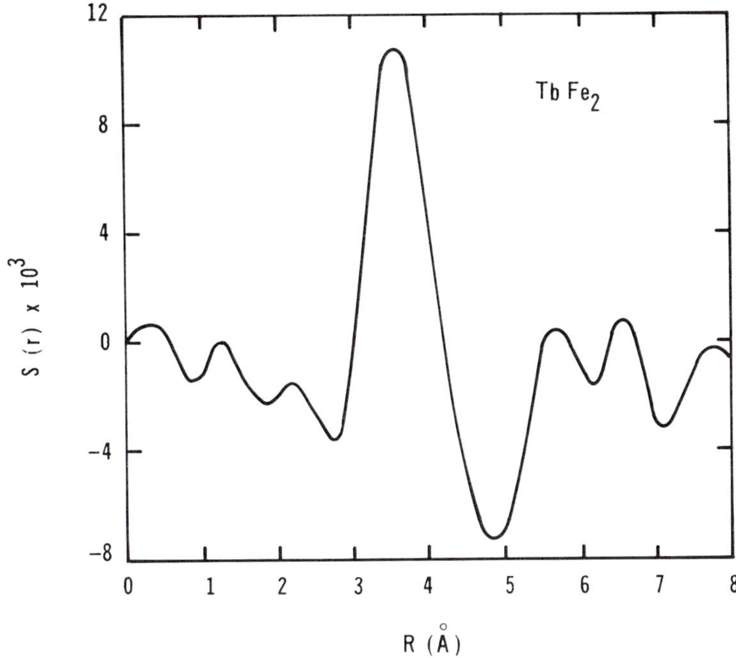

FIGURE 2. Magnetic spin correlation function obtained from the difference pattern of Figure 1 (A) and 1 (B).

The high temperature pattern is characteristic of atomic scattering from an amorphous material with the addition of a superimposed form factor from the diffuse paramagnetic scattering which alters the angular dependence. The broad maximum is observed to lie near the positions of the (113) and (222) Laves phase lines which are the strongest nuclear reflections in the crystalline compound. The amorphous pattern is, however, not the result of broadened Laves phase peaks since strong peak intensity would then occur both at higher and lower angles than we observe in Figure 1 assuming that the relative structure factors are unaltered. In addition, microcrystallinity on this scale would give rise to significant increased low angle scattering which is absent. An application of the Sherrer formula indicates a range of atomic correlations of 6 to 8 Å which is comparable to the $TbFe_2$ lattice parameter of 7.34 Å.[17] It is concluded, therefore, that the pattern results from truly amorphous atomic scattering.

Since T_c = 388 K, the material should be paramagnetic at 423 K. On the other hand, it should be ferromagnetic at 4 K. Thus, substracting the high temperature from the low temperature pattern numerically would give a pattern exhibiting the scattered intensity of purely magnetic origin. This procedure removes the nuclear (atomic) scattering contribution and also eliminates the effects of adsorption, nuclear incoherent scattering, and inelastic phonon scattering to the extent that they are temperature independent. The resulting intensity is again characteristic of an amorphous structure and displays directly the amorphous nature of the magnetic spin polarization. Taking the Fourier transform of this difference magnetic data yields the magnetic spin correlation function shown in Figure 2. It is seen that the resulting curve shows a strong positive spin correlation maximum near 3.6 Å, a large negative extremum near 4.8 Å, and other subsidiary peaks which may in part arise from statistical errors and truncation effects. Since this is a binary alloy, the distribution shown in Figure 3 involves a superposition of three pair-distribution functions in both space and spin coordinates. However, the RE-RE pair-distribution function would be expected to dominate the

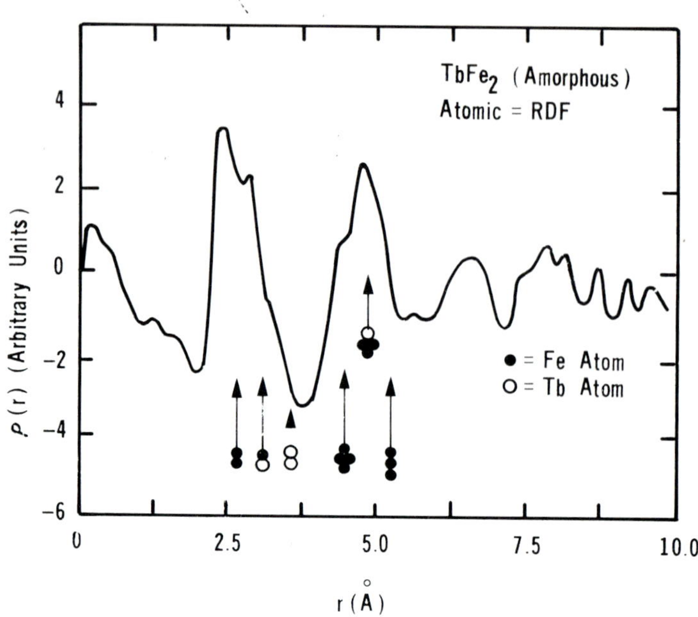

FIGURE 3. Multiple pair correlation function obtained from the data of Figure 1.

magnetic scattering due to the larger moment of the RE; it is noted that the positive peak occurs close to the nearest-neighbor Tb–Tb approach distance in the crystalline cell of 3.2 Å and the negative peak at one of the Tb–Fe distances. Dominance of the rare earth also partially accounts for the peak shift between Figures 1(A) and 1(B) since, in the high temperature (atomic scattering) case, all atomic correlations contribute essentially equally. To accurately determine the individual Fe and Tb spin distributions would require experiments with substituted nonmagnetic RE and *TM* atoms in a material retaining the exact atomic distributions.

The small angle scattering tail observed in Figures 1(B) and 2 and absent in the atomic scattering of Figure 1(A) is of magnetic origin and may reflect regions of inhomogeneity which are of longer range than the average near-neighbor spin correlations in the magnetic structure. These regions may be analogous to magnetic domains or regions of spin order transitions which give rise to the observed additional scattering. The strength of the low angle scattering is sensitive to magnetic fields and to temperature cycling.

The multiple atomic pair correlation function can be obtained by taking the Fourier transform of the nuclear scattering data. The latter in turn can be obtained by substracting the paramagnetic contribution from the high temperature data of Figure 1(A). Neutron scattering at large scattering angles (large Q) becomes a constant ($= Nb^{-2}$ for N atoms) thus removing the paramagnetic contribution. Assuming negligible nonmagnetic inelastic scattering, it follows that, at high Q, elastic, and conventional total scattering are equivalent aside from the Debye-Waller factor.[18] However, this is not a good approximation, because the inelastic cross-section has Q^2 dependence.

Thus from the composite of two experiments: (1) elastic scattering from Q = 0.4 to 6.4 A^{-1} and (2) total scattering from Q = 6.4 A^{-1} to 14.4 A^{-1}, the nuclear scattering intensity I(Q) at 433 K can be obtained. Taking the transform of this data, the multiple pair distribution function ρ(r) can be found as:

$$\rho(r) = \frac{2r}{\pi}\int_0^{Q_{max}} Q[I(Q) - C]\sin Qr dQ \qquad (2)$$

where C is the constant high Q asymptote of the scattering (equal to b^2/atom if there is no incoherent scattering). The result is shown in Figure 3. Due to the experimentally obtainable Q_{max} being less than infinity, peak broadening and oscillations at small r result in Figure 3.

It is seen that the first major peak is nearly resolved into two subpeaks and a shoulder corresponding to near-neighbor atom pairs of Fe–Fe, Fe–Tb, and Tb–Tb. The relative amplitudes of the three peaks are in accord with their total scattering amplitudes and relative number of atoms. The peak positions are strikingly close to the sums of the metallic radii for these atoms in contact as indicated by the arrows. Note that this is in direct contrast to the Laves phase crystalline $TbFe_2$ in which the Tb–Tb interatomic distance is significantly compressed from 3.6 Å possibly as a result of charge transfer to the Fe d-shell. Further confirmation that the Tb–Tb distance is indeed 3.6 Å is provided by the magnetic scattering result discussed above. Possible stacking arrangements of three and four atoms in contact are shown at positions of highest probability in the second peak in ρ(r). The weaker higher order peaks correspond to other possible planar and three-dimensional stacking arrangements. Atomic coordination numbers may be obtained from the first peak area by scaling the results to a vanadium standard scatterer. We thus conclude that the predominant features of the Laves phase unit cell, namely the tetrahedra of Fe atoms and the near planar arrangement of compressed rare earths are absent in the sputtered material. Rather these correlations are replaced by a predominance of pair interactions based solely on statistical factors representative of the dense random packing sequences of Polk[19] and others.

The nature of spin wave excitation in amorphous $TbFe_2$ was investigated using inelastic magnetic scattering.[20] Polarized beam time-of-flight data showed evidence for discrete spin wave excitation for $Q \leq 1.0\ A^{-1}$.

Neutron scattering measurements were extended to amorphous $HoFe_2$ and $GdFe_2$.[21] For $HoFe_2$, the scattered intensity for $T \geq T_c$ (195 K) is Lorentzian and can be analyzed for an inverse correlation length which remains finite at T_c. Below T_c a giant anomalous magnetic component appears due to magnetic inhomogeneity. Limited measurements indicate the absence of this component in s-state-ion ($GdFe_2$), suggesting an origin based on local crystalline field effects.

Hyperfine fields at the nuclei of amorphous materials can be measured using Mössbauer spectroscopy. Mössbauer spectra on $TbFe_2$, $DyFe_2$, $HoFe_2$ and $ErFe_2$ have been measured.[22] At room temperature, $TbFe_2$ exhibits Mössbauer pattern due to a broad distribution of hyperfine fields, similarly for $DyFe_2$ and $HoFe_2$ at 77 K. On the other hand, $DyFe_2$, $HoFe_2$, and $ErFe_2$ show a resolved doublet at room temperature due to well-defined quadrupolar interaction at iron nuclei. Low temperature measurements on $HoFe_2$ show broad pattern at 145 K and quadrupolar doublet at 230 K. These data indicate that iron moments experience exchange field of random orientation and magnitude and that crystalline field on iron ions is relatively uniform.

The Mössbauer spectrum of $DyFe_2$ was discussed in terms of a model in which Dy spins are more strongly coupled to their random-direction anisotropy axes than to neighboring Fe spins which also experience a weaker random-direction anisotropy.[23]

IV. MAGNETIZATION AND COERCIVITY

The magnetization of RE-*TM* amorphous alloys depends on the sign of the exchange integral J between RE-RE, RE-*TM*, and *TM-TM* pairs as well as on the spatial distribution of these pairs, since J is a sensitive function of the spacing between a given pair of spins. The overall temperature dependence of the magnetization can be treated in terms of molecular or mean field theory, providing T is sufficiently below T_c, whereby exchange is overcome by thermal fluctuations. The molecular field $H_m(i)$ acting on a spin S_i can be expressed in terms of the thermal average of the surrounding spins S_j as follows:

$$H_m(i) = <H_{eff}(i)> = -\left(\frac{1}{g\mu_B}\right)\sum_j 2J_{ij}<S_j> \quad (3)$$

For a crystalline sample, the thermal average $<S_j>$ should be unique. However, for the case of amorphous materials, J_{ij} has a spatial distribution so that both $<S_j>$ and $H_m(i)$ may vary from site to site. Thus, a second average over all sites i is required. Furthermore, in the case of a RE-*TM* binary alloys, there are two H_m's, one for each lattice.

Similarly, in the molecular field approximation, the Curie temperature is given by

$$T_C = \frac{2S(S+1)}{3k}\sum_{<i,j>} J_{ij} \quad (4)$$

We can therefore surmise from Equations 3 and 4 that M(T) and T_c will depend on $J_{RE\text{-}RE}$, $J_{RE\text{-}TM}$, and $J_{TM\text{-}TM}$ as well on the spin value S in the amorphous state. Two models, reduced charge transfer and local anisotropy, have been proposed to account for M(T) in RE-*TM* amorphous materials. Each model has generated a number of supporters citing experimental evidence to support their point of view. In what follows, a critical examination of these models and their validity in view of available experimental results is given. In this process we should keep in mind that according to aggregated data,[24] M(0) for amorphous materials can be larger or smaller than that for the crystalline counterpart and the same is true for T_c.

The change in T_c can be accounted for qualitatively by assuming that transfer of charge from rare earth ions to transition metal ions has an important influence on T_c and that less charge transfer takes place in the amorphous films.[25] In this connection, we note that in the Laves phase crystalline $TbFe_2$, the Tb–Tb interatomic distance is significantly compressed from 3.6 Å down to 3.2 Å possibly as a result of charge transfer to the Fe d-shell while in the amorphous state, no such compression is evident from the neutron scattering data.[15]

Charge transfer to the empty d-orbitals of Co or Fe atoms will affect both the magnitude of the moment and T_c. Since in the band model the majority spin states on Co atom are full, transfer of charge to Co should decrease M_{co} and T_c. Since Fe has unfilled majority as well as minority spin states, M_{Fe} and T_c could either increase or decrease. These predictions are consistent with observations.[24]

The lower density of the amorphous structure provides extra room for the valence electrons squeezed out of the rare earth atoms by the close packing. They need not go into the transition metal d-shell, but may also fill the holes in deficient coordination spheres. The decrease in charge transfer to the transition metal ions provides a mechanism for the increase of M_{co} and T_c in $Gd_{0.33}Co_{0.67}$ when it is made amorphous, and should also contribute to the decrease of T_c in $Tb_{0.33}Fe_{0.67}$.

The concept of local anisotropy energy in an amorphous material may at first appear to be a contradiction in terms; however, one must consider the Coulomb interaction of the highly nonspherical Tb 4f charge distribution located in an electric field of unknown and spatially varying symmetry produced by the neighboring ions. This plus the strong L-S coupling, provide a local anisotropy which, together with a spatially fluctuating exchange interaction, determines the equilibrium direction of a spin or group of spins satisfying the condition of minimum free energy. The bulk manifestation is then a material for which conceptually all directions are hard magnetic directions and energy is required to align spins parallel along any single magnetized direction. Such a local anisotropy model has been advanced.[26] It was shown that a fluctuating anisotropy can produce a reduced zero Kelvin magnetization as well as a reduced T_c. Reduction of the Curie temperature given by Equation 4 of almost 40% in amorphous $TbFe_2$ would require very large values of fluctuation in J_{ij} (r_{ij}), unreasonable fluctuations of the same order as J_{ij} (r_{ij}) itself, if this was the only operable mechanism. The inclusion of the randomized crystal field interaction then reduces the re-

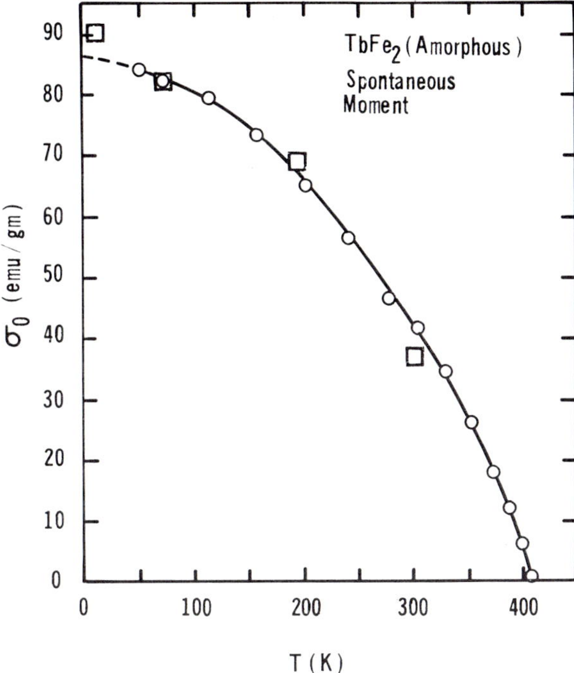

FIGURE 4. Spontaneous moment as a function of temperature, circles obtained from isothermal magnetization data, and squares obtained from elastic neutron diffraction.

quired exchange field disorder and also introduces a local anisotropy of second order in the spin variables which is forbidden in crystalline $TbFe_2$ by the cubic symmetry. However, since both M(0) and T_c can increase or decrease in going from the crystalline to the amorphous state, this theory cannot account for all cases of RE-*TM* alloys.

Magnetization measurements have been made on a specimen of sputtered $TbFe_2$ from 4 to 480 K in applied fields up to 18 kOe, and additional high field magnetization data have been taken at 4 and 77 K in fields of 100 kOe.[15] Below about 120 K an anomalous coercive-force-type behavior develops. H_c reaches almost 30 kOe at 4.2 K.[27] Significant remanent magnetization M_r appeared at 201 K and below and amounted to 74% of the saturation value M_s at 82 K. A small time-dependence of M_r was observed, reducing it by an additional 1% after two minutes. The sample was annealed above 250 K between each magnetization to remove M_r.

The spontaneous magnetization was obtained by extrapolation of the linear high field portion of the magnetization back to the demagnetizing field and is given in Figure 4. The dotted portion indicates the region in which a reliable extrapolation cannot be made due to the anomalous magnetization curves. The square symbols represent the magnetization obtained from the neutron elastic scattering magnetic intensity, by subtracting elastic scattering at 433 K from elastic scattering at temperature T normalized to the bulk magnetization at liquid N_2 temperature. The Curie temperature obtained from Figure 4 is 409 K as compared to the crystalline value of 710 K. The 0 K intercept of Figure 4 corresponds to a spontaneous moment of 4.2 μ_B per formula unit as compared to 4.7 μ_B for crystalline $TbFe_2$.[28] If the antiparallel arrangement of two iron moments and one terbium moment found in the Laves phase compounds is maintained here, then the minimum net obtainable magnetization would be either 5.6 μ_B or 4.6 μ_B assuming the full moment of 9 μ_B for Tb and respectively the 1.7 μ_B Fe moment of the Laves phase or the 2.2 μ_B of elemental iron. The low moment of the amorphous material implies either an alternate magnetic structure (e.g., one containing antiparallel or noncollinear Tb moments) or that a significant proportion of the magnetization is disordered.

The low temperature coercivity discussed above is anomalous. The rapid development of the large coercive field at very low temperatures does not suggest a single-ion anisotropy energy mechanism, for this would be dependent on a power law of the related magnetization. In contrast, the observed change in coercive force is greatest where the moment change is nearly zero. The low temperature behavior is suggestive of a field-induced type of magnetic phase transition having a thermal activation energy dependence.

Annealing studies of initially amorphous $TbFe_2$ shows a large increase in H_c at onset of recrystallization, from 100 Oe to 3.75 kOe after 525°C anneal.[29] Abnormally large H_c and time dependent magnetization have been observed in a series of bulk amorphous Tb_xFe_{1-x} samples at low temperatures.[30]

T_c of a series of amorphous RE(Gd,Tb,Dy,Ho,Er,Y)-Fe_2 alloys decrease smoothly from 500 K for $GdFe_2$ to 0 K for YFe_2, with decreasing DeGennes factor $G = (g-1)^2J(J+1)$ of the rare earth atom.[31] In contrast, T_c of the crystalline Laves phase compound (=535 K for YFe_2) show a much weaker dependence on G.

An extensive survey of R_xFe_{1-x} amorphous alloys (R = Gd,Tb,Dy,Ho,Tm,Yb,Lu) using the Mössbauer effect, magnetometer, and magnetooptic measurements indicates a decrease of Fe moment μ_{Fe} with increasing x according to the relation[32]

$$\mu_{Fe} = \{2.2 - \frac{x}{1-x}[3.2 - 0.7(g-1)J]\}\mu_B \tag{5}$$

where $(g-1)J$ is the effective rare earth spin.

The temperature dependence of M_s of vapor-deposited amorphous Gd–Fe films has been analyzed in terms of molecular field given by Equation 3.[33] This analysis yielded values of Fe spin and Fe–Fe and Fe–Gd exchange interaction energies which were then correlated with composition. It was found that temperature sensitivity of M_s of amorphous Gd–Fe is less than that of Gd–Co alloys. This fact may be important in device applications. Similarly, magnetization data for Ho–Co, Dy–Co, and Gd–Fe evaporated films have been fitted using mean field theory; exchange constants and atomic magnetic moments were determined.[34] The M_s-data confirms the importance of charge transfer. From X-ray photoemission studies of the electronic structure of Gd_xFe_{1-x} ($0 \leq x \leq 1$), it was concluded that charge transfer appears too small to account for magnitude of Fe moment.[35]

Magnetization data on evaporated $Gd_{18}Fe_{82-x}Ni_x$ with x ranging from 12 to 68 show that magnetic moment of Ni is strongly dependent on composition with a decrease in moment with increasing nickel content, to almost zero at 66% Ni.[36] The spins of Gd and Ni in sputtered Gd–Ni are found to be antiparallel.[37] The T_c reaches a pronounced minimum of 40 K at x = 0.2. In the case of Gd–Fe–Co sputtered films, it was found that increasing Fe by 12% lowers exchange between RE and TM by about 35%;[38] this trend is also true for evaporated Gd–Fe–Co films.[39]

We now turn to the discussion of Gd–Co alloys. It was observed that H_c of Gd–Co films ($4\pi M_s$ = 1000 gauss) evaporated in controlled oxygen atmosphere increases from 15 Oe at 2.5×10^{-8} torr to 150 Oe at 1.5×10^{-7} torr.[40] This is attributed to preferential oxidation of Gd. For Ni-substituted Gd–Co films, it was found that for a fixed Gd/TM ratio, T_{comp} (compensation temperature) and T_c decreases from 880 K to 750 K for 50% of Ni while $4\pi M_s$ remains unchanged.[41]

T_c of amorphous Nd_{64}–Co_{36}, Gd_{65}–Co_{35} and Er_{65}–Co_{35} foils were found to be 38 K, 170 K, and 12 K, respectively.[11] It appears that local random anisotropy is important in Nd–Co and Er–Co alloys.

The pressure dependence of T_c for electron beam evaporate RE-*TM* thin films has been measured.[14] Relatively large negative dT_c/dp values were found, similar to those for crystalline RE-*TM* intermetallic compounds. Magnetization data for RE(Gd,Dy,Ho,Er)-Co d.c. sputtered films show that T_c of four RE-Co_{35} alloys are very close to each other indicating absence of charge transfer.[42] At low temperatures, 15 to 20% reduction in RE contribution to M is probably due to local anisotropy fluctuations.

The magnetic moment per formula unit, Curie temperature as well as the compensation temperature of a number of RE-*TM* amorphous films (RE = Gd,Ho,Tb,Y; *TM* = Co,Fe,Ni,Mn) prepared across a broad compositional range have been measured by Lee and Heiman.[43] The samples were examined by polar Faraday rotation, polar Kerr rotation, and magnetometer techniques. Certain trends were observed to occur for all amorphous RE-*TM* alloys as follows:

1. T_c of Co and Ni compounds are higher for amorphous than crystalline alloys.
2. T_c of Fe alloys are lower for amorphous than crystalline alloys.
3. Amorphous Mn compounds do not appear to be magnetically ordered.
4. The 4.2 K magnetic moments of the amorphous alloys (with the notable exception of Y–Fe) are more *TM* dominated than in crystalline alloys.
5. T_{comp} of all amorphous alloys are lower than T_{comp} of the crystalline materials.
6. Just as in the case of crystalline materials, amorphous Gd alloys have higher T_c than Tb alloys, which in turn have higher T_c than Ho alloys.

With the exception of Y-*TM* films (YFe_5 films exhibited microcrystalline bcc Fe), Lee and Heiman[24] concluded that their data can best be explained in terms of a picture for amorphous RE-*TM* films in which:

1. The *TM* magnetic moments at 4.2 K are larger than in crystalline compounds.
2. The RE magnetic moments at 4.2 K are near their crystalline values.
3. Co–Co and Ni–Ni exchange is enhanced, but Fe–Fe and Mn–Mn exchange are reduced.
4. RE-RE exchange and/or RE-*TM* exchange is reduced.

Thus, this picture rules out the local anisotropy model[26] and to some extent supports the concept of reduced charge transfer.[25,44] However, it appears that the effects resulting from altered atomic dimensions and coordination also cannot be neglected.

Although the anisotropy and orbital moment of the RE does not play a significant role in the altered exchange in amorphous RE-*TM* alloys, it strongly affects H_c. Above room temperature all amorphous alloys exhibit H_c on the order of a few oersteds. As T is reduced to T_{room} or slightly below, Tb–*TM* alloys developed H_c greater than 20 Oe. At very low T, Ho–Fe compounds also develop large H_c.

In order to tailor the magnetic characteristics for magnetic bubble applications, Mo is frequently added to Gd–Co. In the $Gd_{1-x-y}Co_xMo_y$ alloy, the Co moment was found by Hasegawa et al. to be given by[45]

$$\mu_{Co} = \left[1.65 - \frac{5.0\,y}{x + y}\right] \mu_B \tag{6}$$

so that μ_{Co} can be greatly reduced by the addition of Mo. For $(Gd_{1-x}Co_x)_{1-y}Mo_y$, they obtained the following values for the exchange integrals: $J_{Co\text{-}Co} \simeq (105 \pm 15)$ K, $J_{Co\text{-}Gd} \simeq (-17 \pm 1)$ K, and $J_{Gd\text{-}Gd} \sim 1.4$ K. Thus, the Co–Co interaction dominates and the Gd–Gd interaction is very small. The direct observations of the temperature dependence of the sublattice magnetization has been achieved using the polar Kerr effect.[46]

Isothermal annealing, at 150 to 350°C, of Gd–Co–Mo thin films shows that all fundamental magnetic properties are retained after anneal.[47] Similarly, it was found that annealing at various temperatures enhances M_s and exchange constant A in all GdCo, GdFe, and GdCoMo films studied.[48]

Finally, it has been found that M_s and T_{comp} changes only slightly by implantation damage.[49]

V. ANISOTROPY

In the Fourier transform of observed scattered X-ray or neutron intensity by amorphous materials, isotropic atomic arrangement was assumed *a priori*. Nevertheless, uniaxial anisotropy with its easy axis perpendicular to the film surface does exist in many RE-*TM* samples. This anisotropy must be due to slight deviation from the isotropic random distribution of atoms; such anisotropic distribution can be due to nonspherically symmetric preparation conditions. In principle, it is possible to discern any anisotropic arrangement of atoms, if it exists, by comparing the scattered intensity in reflection and transmission. This was attempted in the case of Gd–Co film, but the result was not conclusive.[50]

In general, anisotropy in magnetic materials can arise from several sources: (1) magnetic crystalline, (2) stress, (3) shape (4) dipolar interaction, and (5) short-range ordering.

Since the orbital motion is influenced by the electrostatic field set up by the central and neighboring atoms of a crystal lattice, the electron spin is coupled to the lattice through spin-orbit coupling. This gives rise to an anisotropic exchange interaction and single ion anisotropy. The former is of the pseudo-dipolar type and is responsible for magnetocrystalline anisotropy in ordered materials. Since long range order is absent in amorphous materials, this type of interaction cannot account for the observed anisotropy. As previously discussed, single ion anisotropy gives rise to randomly oriented local anisotropy axis; this also cannot account for the existence of a uniaxial anisotropy.

Planar stress at the substrate-film interface due to difference in coefficient of expansion can give rise to a perpendicular anisotropy providing the magnetostriction is of the proper sign. If stress is a significant source of anisotropy, then removal of the film from the substrate should change the anisotropy observed. However, it was found that the stripe width of Cd–Co film remained unchanged when its NaCl substrate was removed. Since the stripe width w at constant thickness is proportional to the characteristic length ℓ where[51]

$$\ell = \frac{\sqrt{AK_u}}{\pi M_s^2} \qquad (7)$$

constancy of ℓ implies that K_u is not related to the interface stress. Equation 7 can be simply derived as follows. First, define the characteristic length ℓ as equal to $\sigma_w/4\pi M_s^2$ where σ_w is the wall energy density. For a planar Bloch wall $\sigma_w = 4\sqrt{AK_u}$;[52] substituting this expression for a σ_w gives Equation 7. The proportionality between σ_w and ℓ can be shown by utilizing the energy expression for straight stripe domains derived by Malek and Kambersky.[53] On the other hand, Tsunashima et al. concluded that 10 to 30% of the perpendicular anisotropy is due to stress for Gd–Co films deposited onto three different kinds of substrates.[54] In the case of Gd–Fe and Tb–Fe films where magnetostriction may be large, there is agreement that stress does contribute to anisotropy.[54-57]

Anisotropic microstructures, accompanying spatial variation in magnetization, can also produce effective magnetic anisotropies (internal shape effect). However, perpendicular easy-axis anisotropies produced in films by this mechanism can have H_k no larger than $4\pi M_s$. This means that if this is the only source of anisotropy, M_s in the remanent state will lie in the plane even if the easy axis is perpendicular to the plan as $K_u < 2\pi M_s^2$ (recall that $N_x + N_y + N_z = 4\pi$). Such samples will clearly not be useful for bubble memory applications. It is well known that no perpendicular anisotropy appears in vapor deposited GdCo films or in sputtered films without negative bias. Argon inclusion is thought to result in shaped void formation leading to K_u in GdCo.[58,59] Similarly, K_u in GdCo is assumed to arise from anisotropically shaped Co-rich regions in a transition stage between the homogeneously amorphous and crystalline states.[60,61] Such phase separation also occurs in GdFe.[55] Compositional fluctuations in the form of rods and thin sheets of Gd-rich material nearly perpendicular to the surface were found in GdCo and GdCoMo films.[62]

It was also found that oxygen contamination causes Gd in GdCo to preferentially oxidize and form magnetically inactive inclusions of presumably amorphous Gd_2O_3 of anisotropic shape due to evaporation process itself.[40] Similar effects are found in GdFe where O_2 contamination is assumed to form nonmagnetic Gd and Fe oxides arranged in sponge-like manner.[63] Also, it was found that formation of nonmagnetic inclusions can occur with H_2 and N_2 contamination.[64]

Angle of incidence effect has been observed in a number of RE-*TM* films. The incident atomic beam direction appears to define the anisotropy axis in evaporated films while the electric field direction at the substrate surface is the dominant influence in bias sputtering of RE(Gd,Ho,Tb)-*TM*(Fe,Co) films.[65] It was also found that the uniaxial hard axis in GdCo can be tilted toward the direction of incidence.[66] Similarly, it was found that column orientation depends upon vapor flux orientation.[67] An isotropic microstructure is composed of columnar regions surrounded by a low density network produced by shadowing of incident vapor by the growing film.[68]

Ion implantation can be used to drastically change K_u.[69] A relatively low level of radiation damage seems to both destroy and induce anisotropy when the sample is in a strong external field.[70] The implantation result is thought to be most consistent with the short-range atomic ordering model for anisotropy.[49]

Similarly, films prepared under various conditions show that K_u in GdCo is induced by resputtering.[56] Experiments also show that negative bias is required to produce K_u in GdCo but not for GdFe.[71] Change of position and shape of X-ray pattern for GdCo when M becomes in plane is interpreted to support the short-range ordering model.

Annealing at various temperatures was found to reduce K_u in GdCo, GdFe, and GdCoMo films with the rate of reduction increasing with Mo content.[48] Isothermal annealing from 150 to 350°C of GdCoMo showed all functional magnetic properties were retained even after annealing at 350°C. This result of course has implications in bubble applications.[47]

In studying a series of RE(Gd,Dy,Ho,Er)-Co films, it appeared that K_u was independent of RE content.[42] There is no apparent change in K_u with Fe/Co ration in sputtered Gd–Co–Fe films.[38]

We now turn to the fourth source of anisotropy, i.e., dipolar interaction, listed at the beginning of this section. Simple expressions were derived for K_u due to dipolar interaction in amorphous films with slight structural anisotropies on either atomic or microstructural scales.[72] Relative importance of spin-orbit and dipolar interaction in GdCo remains uncertain. Although K_u data can be fitted by the dipolar equation $\Sigma C_{ij}M_iM_j$ where C_{ij} is the coupling coefficient and M_i,M_j sublattice magnetization, it may be that pair ordering merely approximates a more complex form of structural ordering.[34] Similarly, temperature dependence of K_u was fitted by the same dipolar equation.[73]

From the discussion above, we see that enormous amount of data have been accumulated on the origin of anisotropy. Unfortunately, these data are not always consistent with each other. It is quite possible that no single source can explain all the results observed. It is quite likely that the magnetic anisotropy in these alloys is *closely related to the preparation conditions*.

We are now left with atomic short range ordering as the remaining anisotropy mechanism in sputtered amorphous Gd–Co films. The simplest kind of short-range order in a binary alloy is single atomic pairs, e.g., Co–Co pairs in amorphous Gd–Co alloys.[74] Studies of hcp cobalt suggests that the easy direction associated with a Co–Co pair will be perpendicular to the pair axis and the anisotropy will have a magnitude of the order of 10^{-15} to 10^{-16} ergs/pair. The number of pairs required to account for the observed anisotropy in amorphous Gd–Co (10^5 erg/cm^3, for example) should thus be in the order of 10^{20} to 10^{21} pairs/cm^3, a small percentage of the total number of Co atoms/cm^3 in the material.

But, how does this preferential alignment come about in the deposition process? To understand this, it is necessary to first understand the process of thin film growth. In vapor deposition, the adatom arrives at the film surface with considerable kinetic energy due to the high temperature of the melt. However, since the mass of the adatom is small compared to that of the sample as a whole, it quickly condenses and attains the temperature of the substrate. If the substrate temperature is high enough, the adatom will move by surface diffusion to an equilibrium sticking site, usually the growth step of a crystal nucleus. On the other hand, if the substrate temperature is sufficiently low and the deposition rate is sufficiently high, the adatom will have insufficient time to move away before being crowded in by incoming impinging atoms. Thus the adatom essentially sticks where it lands and an amorphous rather than a crystalline film is produced.

Let us now consider a somewhat intermediate case. For example, assume that the adatom can undergo a jump to any of its nearest neighbor (nn) surface sites. The jump probability for each site would be expected to be proportional to the energy difference between the initial site and the nn surface site. In these circumstances, the adatom would be expected to preferentially occupy lower energy sticking sites, but not equilibrium crystalline sites. The energy differences between the initial site and the nn surface site can arise from chemical, geometric, or magnetic considerations. For example the bond energy of a Co–Co bond is undoubtedly considerably different from that of a Gd–Co bond. Geometric considerations may also be important because Gd and Co differ in atomic size by about 30%. The magnetic energy difference is apparently important in the bulk amorphous solid as evidenced by the presence of magnetic annealing effects and may be even more important in the ordering of adatoms at the surface.

A special case of thin film vapor deposition is bias sputtering. In bias sputtering, the growing film is made a secondary target and is bombarded with low energy (e.g., 50 eV) ions which causes some fraction of the surface atoms to sputter away. The probability that a surface atom will be represented should be proportional to its site energy. Since atoms in high energy sites are preferentially removed, bias sputtering can greatly enhance the probability of incorporated atoms residing in lower energy sticking sites. In the Gd-Co case, resputtering should greatly enhance the probability of incorporating Gd bonded to Co.

The model for the growth-induced anisotropy in sputtered amorphous Gd–Co alloys emerges as follows. The uniaxial perpendicular anisotropy arises from Co–Co pairs in the plane of the film in excess of a random distribution. More complex short range order clusters may be more important, but Co–Co pairs are the simplest cluster which will explain the data. The excess of pairs in the plane is caused by the higher resputtering probability of an adatom that is bonded to a like surface atom, i.e., Co–Co or Gd–Gd bonds are less stable than Gd–Co bonds. The statistical probability of a Gd–Co pair oriented perpendicular to the plane of the film is high, and therefore, the concentration of Co–Co pairs oriented perpendicular is low. The excess Co–Co pairs in the plane results from the decreased concentration of the pair perpendicular to the plane.

The observations that support this model are the following. The anisotropy of a film is relatively independent of temperature in the vicinity of room temperature indicating that it is associated with the Co sublattice. The anisotropy induced during growth increases with increased resputtering as the model would predict. At high deposition rates or low substrate temperatures, conditions which would be expected to reduce site selection, the growth-induced anisotropy is reduced. The observation that the anisotropy increases with thickness then reaches a limiting value suggests that a process akin to magnetic annealing is taking place during growth. The initial layers develop a small perpendicular component by the process described above. The component of magnetization perpendicular to the plane of the film provides an exchange field which decreases the stability of vertical Co–Co pairs. The excess Co–Co pairs in subsequent layers contribute an increased exchange field which further destabilizes vertical Co–Co pairs. A limiting value of anisotropy is reached because once $K_u > 2\pi M_s^2$ the full $4\pi M_s$ magnetization is oriented in the vertical direction so that further increases in K_u no longer increase the probability of in-plane Co–Co pair formation. One would expect K_u to be a function of M_s and to pass through a minimum at the compensation composition; this is found to be the case.

VI. FMR AND SAMPLE INHOMOGENEITY

It turns out that FMR is a very good way to determine the inhomogeneity in a sample. Inhomogeneities give rise to peak broadening and/or extra peaks in the ferromagnetic resonance spectra. Dimensional resonance for spin waves with their $\vec{k}$-vector perpendicular to the film can lead to a determination of exchange constant A. In the following discussion, we shall first begin with a theoretical discussion of FMR in amorphous films.

First consider the case where the static field $H_{\|}$ is applied parallel to the film plane.[75] Specifically, let the film lie parallel to the x-y plane and the (anisotropy) easy axis parallel to the z-direction. If $\vec{H}_{\|}$ and the microwave magnetic field $\vec{h}$ are applied parallel to the x- and y- direction, respectively, the free energy of the system is given by:

$$E = K_u \sin^2\theta - H_{\|} M_s \sin\theta\cos\phi - hM_s \sin\theta\sin\phi + 2\pi M_s^2 \cos^2\theta \tag{8}$$

where K_u is the uniaxial anisotropy constant, M_s the saturation magnetization, and θ and ϕ are the angles $\vec{M}_s$ makes with the z- and x axis, respectively. The first term in Equation 8 represents the energy density of a magnetic system with a z-directed easy axis and a hard plane perpendicular to it. The second and third terms represent Zeeman energy densities whereas the last term represents the demagnetizing energy density.

The torque $\vec{\tau}$ active on $\vec{M}$ may be derived from the energy E by the relations

$$\vec{\tau} = \vec{r} \times \vec{F} = -\vec{r} \times \vec{\nabla} E \tag{9}$$

where $\vec{F} = -\vec{\nabla}E$ is the generalized force and $\vec{r}$ is the radius vector in the direction of $\vec{M}$.

On solving the equation of motion of the magnetization $\vec{M}$:

$$\frac{d\vec{M}}{dt} = \gamma\vec{\tau} \tag{10}$$

where γ is the gyromagnetic ratio and with the help of Equations 8 and 9, we obtain the resonance relation for the uniform mode in the $H_{\|}$ configuration as

$$\omega = |\gamma| \, [H_{\|}(H_{\|} - H_k + 4\pi M_s)]^{1/2} \tag{11}$$

where ω is the resonance frequency and $H_k = 2K_u/M_s$ is the anisotropy field. In the derivation of Equation 11, we have assumed that $H_{\|}$ is sufficient to saturate the film so that $\theta \rightarrow \pi/2$ and $\phi \rightarrow 0$.

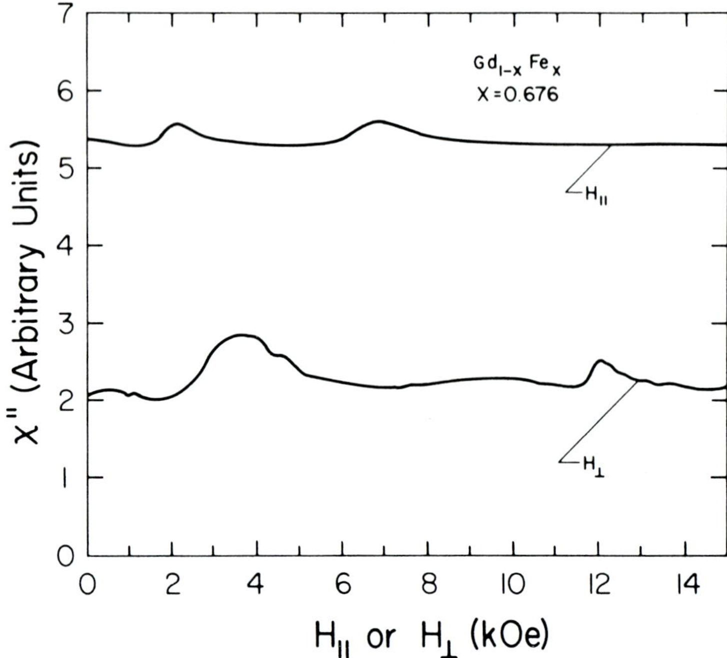

FIGURE 5. Absorption spectra of an amorphous $Gd_{0.324}Fe_{0.676}$ bubble film approximately 5000 Å thick with static field applied parallel and perpendicular to the film plane.

Consider next the case where the static field $\vec{H}_{\perp}$ is applied perpendicular to the film plane or along the z-axis. The resonance relation can be found exactly as for the parallel orientation except that the second term in Equation 8 should be replaced by $-H_{\perp}M_s \cos\theta$. The result is

$$\omega = |\gamma|(H_{\perp} + H_k - 4\pi M_s) \tag{12}$$

On combining the resonance relations Equations 11 and 12, we obtain

$$H'_k = \frac{-H_{\parallel} - 2H_{\perp} + (4H_{\parallel}H_{\perp} + 5H_{\parallel}^2)^{1/2}}{2} \tag{13}$$

and

$$|\gamma|_{\perp} = \frac{\omega}{H_{\perp} + H'_k} \tag{14}$$

where $H'_k = H_k - 4\pi M_s$. Equations 13 and 14 will be used to find H'_k and $|\gamma|$ from the measured resonance spectra.

Ferromagnetic resonance spectra of a series of evaporated $Gd_{1-x}Fe_x$ amorphous bubble films with x ranging from 67.6 to 81.6% have been measured.[75] Typical absorption and differential absorption spectra for these films are shown in Figures 5 and 6. From these spectra the g-factor (g_{eff}) was deduced and compared with g_{eff} calculated from Wangsness formula.[76]

$$g_{eff} = \frac{g_{Gd}(S - R)}{S - \alpha R} \tag{15}$$

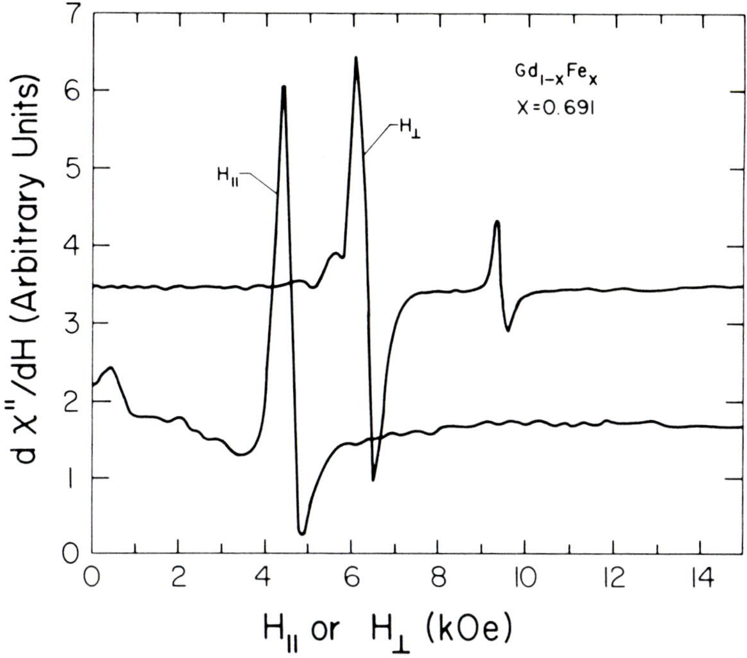

FIGURE 6. Differentiated absorption spectra of an amorphous $Gd_{0.309}Fe_{0.691}$ bubble film approximately 5000 Å thick with static field applied parallel and perpendicular to the film plane.

where $S = (1-x)/x$, $R = M_{Fe}/M_{Gd}$ and $\alpha = g_{Gd}/g_{Fe}$. R can in turn be determined from the formula $M_s = (1-x)M_{Gd} - xM_{Fe}$. As shown in Figure 7, the agreement between the theoretical and experimental values of g_{eff} is very good. Whereas FMR can provide information on the anisotropy distribution in the film, vibrating sample magnetometer yields at best an average value for H_k. In our experiments, several resonance peaks are found for a given film, as shown in Figures 5 and 6, indicating the existence of a number of regions having different values of $H'_k = H_k - 4\pi M_s$. Interface effects on the substrate side and oxidation at the air-film interface may result in layers of different chemical composition. Sometimes there was also some fine structure, even in the major peak, indicating inhomogeneity even in the "bulk". In this connection, Auger electron spectroscopy depth profiles showed that Gd–Fe exposed to air possessed an oxidized surface layer 500 to 1000 Å thick.[77] On the other hand, four line FMR spectra in $TbFE_2$, $GdFe_2$, and YFe_2 are explained in terms of a spin cluster model.[78]

A large number of FMR experiments have been performed on GdCo and GdCoMo films. For example, FMR showed the existence of regions differing slightly in M_s and therefore Gd and Co content.[79] From etching experiments, it was shown that the inhomogeneity increased with increasing bias and was spread throughout the volume of the film. From the parallel and perpendicular resonance spectra, K_u was determined as a function of composition and bias.[80] $K_u(T)$ and g_{eff} for GdCoMo films were measured and g-value compared to Wangsness formula.[73,81] It has also been found that the exchange constant A for GdCo and GdCoMo films is only 10 to 20% as large as the crystalline case.[82] A was determined from perpendicular spectra exhibiting spin wave peaks due to dimensional resonance of perpendicularly directed spin waves by the formula

$$A = \frac{(H_{\perp i} - H_{\perp j})M_s}{2(n_j^2 - n_i^2)} \left(\frac{d}{\pi}\right)^2 \tag{16}$$

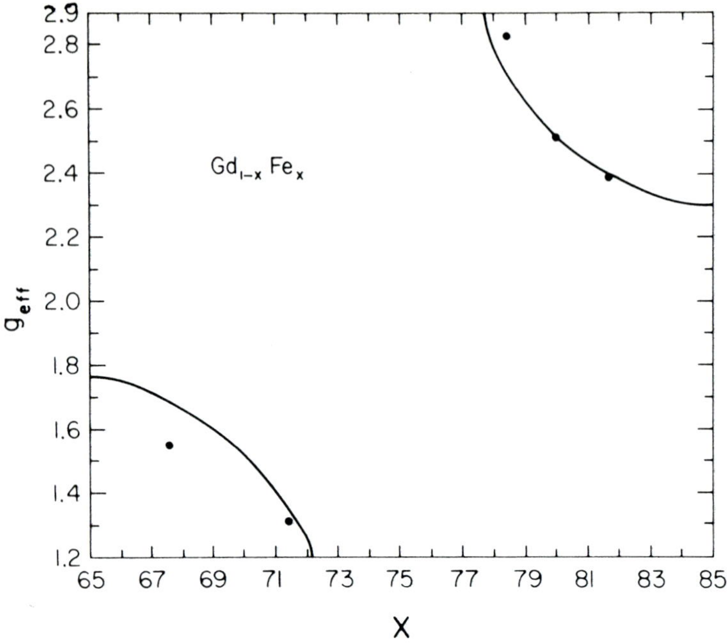

FIGURE 7. g_{eff} as a function of Fe content in at.%. The solid curve was calculated using the Wangsness formula; the dots represent experimental points.

where i and j denote spin wave peaks and $k_{i,j} = n_{i,j}\pi/d$ and d is the thickness of the film. Damping frequency λ for $GdCo_3$, Gd-, Ho-, Tb- $DyFe_2$ has also been measured.

FMR spectra of implanted GdCo films showed that $H'_k = H_k - 4\pi M_s$ decreases with intensity of ion implantation and ultimately changed sign.[69]

Uniformity in composition, thickness, and magnetic properties has been achieved in sputtered GdCo and film composition has been controlled to within 1%. However, significant variation of magnetic properties from film to film was observed.[83] Spatial uniformity of $H_{collapse}$ (the field at which the bubble diameter goes to zero), H_k, d, stripewidth w for sputtered GdCoMo films from 10 to 12 cm arc-melted targets has also been reported.[84]

A method has been devised to relate the shape of the transverse or hard-plane magnetization curve to the dispersion of uniaxial anisotropy (in magnitude and direction) and M_s in a bubble film.[85] In this method, a field parallel to the easy axis sufficient to saturate the film as well as a varying field transverse to the easy axis (or in the film plane) are applied.

From SWR, it was found that $\lambda = 1.36 \times 10^{-8}$ sec and A equal 1/4 of crystalline value for γ-Co alloys.[86] Although the authors do not agree on the explanation, there is agreement on the existence of a surface layer in γ-Co and γ-Fe alloys.[87-89]

Lastly, we should mention that a simple semiclassical theory of spin wave propagation in an amorphous Heisenberg ferromagnet has been developed.[90] It was found that both extended and localized spin waves may exist in these systems giving rise to significant broadening of resonance lines. An effective medium approximation for ferromagnetic spin waves in amorphous systems and alloys have also been presented.[91]

VII. APPLICATIONS

The most important application for RE-*TM* amorphous films is as bubble memory material. Briefly stated, a magnetic bubble is in fact cylindrical in shape; it represents a region of reversed magnetization in a magnetic material. If $K_u > 2\pi M_s^2$ in a given film possessing a uniaxial anisotropy with its axis perpendicular to the film, $\vec{M}_s$ in the remanent state will be aligned in either direction perpendicular to the film with equal probability. Under these conditions, the domains, known as stripe domains, are serpentine in shape as shown in Figure 8(a); such patterns can be obtained by shining polarized light through the film and adjusting the analyzer for extinction for domains oriented in the same direction. Such pictures are known as Faraday micrographs since it utilizes the fact that the Faraday rotation is dependent on the angle between the magnetization and the wave vector $\vec{k}$ of the incident optical radiation. If a bias field $\vec{H}_B$ is applied in the direction shown in Figure 8(b), the domains with their $\vec{M}_s$ oriented upward will expand while those with $\vec{M}$ oriented downward will contract. At some appropriate values of H_B, the downward-oriented domains will shrink into cylindrical domains called bubbles as shown in Figure 8(b). Whether a magnetized region should expand or contract is determined by the condition of minimum energy according to the expression $E = -\vec{M}_s \cdot \vec{H}_B$. Thus, if $\vec{M}_s$ and $\vec{H}_B$ are aligned in the same direction, the region expands while if $\vec{M}_s$ and $\vec{H}_B$ are opposite directed, the region contracts, consistent with the picture depicted in Figure 8.

In order to construct a digital bubble memory, it is necessary to restrict the bubbles to certain storage locations; when this is done, absence and presence of a bubble at a given location can be considered as storing a "0" and a "1", respectively. In other words, a bubble memory must have a bubble generator (in contrast to the random bubble nucleation of Figure 8(b)), a propagating structure, and a bubble detector. Bubble propagation is again based on the energy minimization principle in that bubbles move in the direction of decreasing energy. We shall return to this subject after a brief discussion of bubble materials.

There are two kinds of bubble materials, crystalline and amorphous. The former composes of a garnet film grown by the liquid phase epitaxy (LPE) method onto a $Gd_3Ga_5O_{12}$ (GGG) nonmagnetic substrate. Garnets have the general formula $M_{3\text{-}y\text{-}x}N_yP_zGa_xFe_{5-x}O_{12}$ in which M, N, and P are rare-earth ions, and Ga can also be replaced by Al. By properly choosing the constituents of the garnet compound, it is possible to tailor the important physical parameters: $M_s(T)$, $K_u(T)$, and α.

The main advantage of using amorphous instead of crystalline films lies in lower potential cost. Its deposition method is compatible with integrated circuit (IC) processing technology and its small bubble diameter can significantly increase storage density. Unlike crystalline garnets which must be grown by LPE onto an essentially defect-free single crystal GGG substrate, amorphous RE-*TM* film on the other hand, can be deposited onto a glass substrate by vacuum deposition or sputtering. Indeed, if Si substrate is used instead, the film deposition process is analogous to deposition of Al connections for an IC circuit. Thus, the electronics required to drive the bubble memory can be on the same chip and this type of integration clearly represents a technological advantage. The smaller bubble diameter d (typically tenths of microns instead of microns as in a garnet material) gives rise to higher storage density ($\propto d^{-2}$). However, submicron diameter bubbles are difficult to observe because they are too small to be resolved by optical means; this is a disadvantage during research and development stages since their shape and motion cannot be seen visually. Other problems connected with RE-*TM* amorphous materials such as Gd–Co are (1) how to produce a sample that is chemically homogeneous throughout, and (2) how to produce samples with coercivity comparable to that of crystalline garnet films.

$Gd_{1-x}Co_x$ or $Gd_{1-x}Fe_x$ films are good examples of amorphous bubble films. The resulting magnetization arises from the antiferromagnetic coupling between the spins of the Gd and Co atoms in the amorphous alloy. The net magnetization is given by

$$M = (1 - x)M_{Gd} - xM_{Co} \tag{17}$$

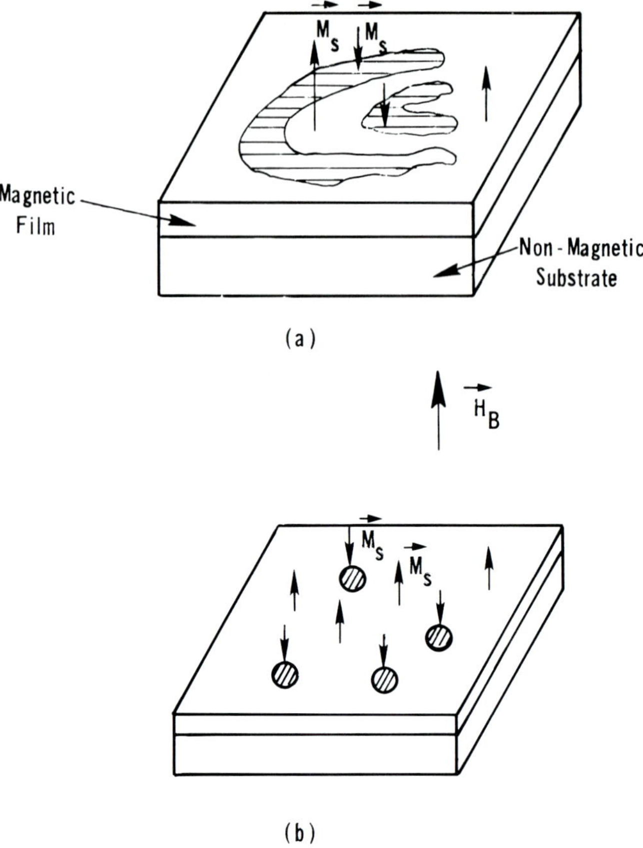

FIGURE 8. Bubble domains.

Thus, at some value of x, M goes to zero; the corresponding composition is known as the compensation composition. For example, $Gd_{0.195}Co_{0.805}$ represents the compensation composition for Gd–Co films. Since M_{Gd} and M_{Co} are both functions of temperature, M also goes to zero at some temperature (the compensation temperature T_{comp}). By varying x, it is seen that M(T) can be varied over a very wide range without developing segregated phases.

Figure 9 shows the organization of bubble mass memory with one major loop and a multitude of minor loops. It is seen that a generator, a detector, and propagation structures are required. In addition, a transfer gate to transfer bubbles between major and minor loops is also needed. Keeping this structure in mind, we shall discuss the material parameters which influence the operation of a bubble memory.

For film uniformly magnetized in an easy direction (perpendicular to the film), the free energy is given by

$$E = K_u \sin^2\theta + 2\pi M_s^2 \cos^2\theta \qquad (18)$$

where θ is the angle between the easy axis and M_s. Noting that $\cos^2\theta = 1 - \sin^2\theta$, Equation 18 can be rewritten as

$$E = 2\pi M_s^2 + (K_u - 2\pi M_s^2) \sin^2\theta \qquad (19)$$

Thus we see that minimum energy corresponds $\theta = 0,\pi$ providing $K_u > 2\pi M_s^2$. In other

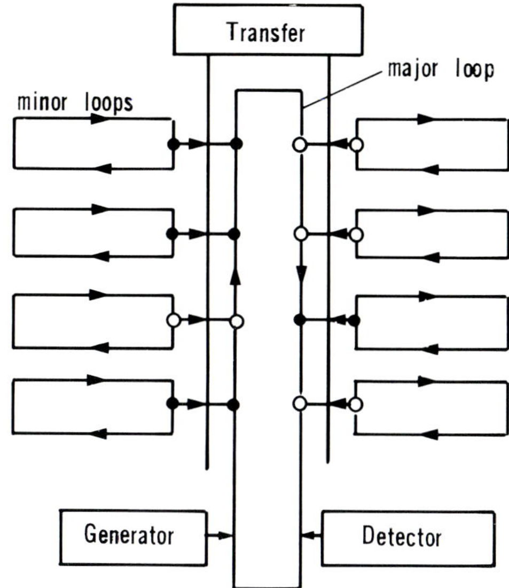

FIGURE 9. Organization of bubble mass memory with one major loop and a multitude of minor loops.

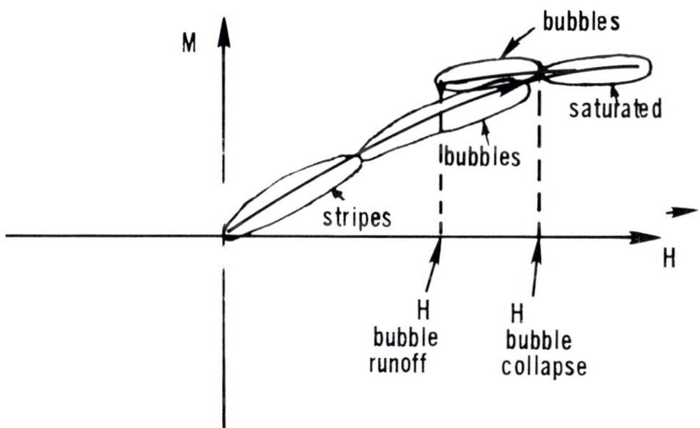

FIGURE 10. Hysteresis loop of bubble materials.

words, this inequality must be satisfied if directions perpendicular to the film were to the stable direction of the magnetization, as in the case of bubbles and their surroundings. However, as mentioned earlier, the film will not be uniformly magnetized in the remanent state since the free energy of a stripe domain state is lower. These domains are of course separated by domain walls which also have energy. In the case of parallel stripe domains magnetized alternately in the + and − directions of the film normal, the wall energy density σ_w is equal to $4\sqrt{AK_u}$ while the wall width is $\pi\sqrt{A/K_u}$ found by setting $\partial E/\partial\theta = 0$ with E of Equation 19 augmented by a wall energy density term.

When a bias field is applied perpendicular to the film, the stripe domains magnetized opposite to the bias field gradually shrinks and changes into cylindrical domains. As shown in Figure 10, if the field is increased beyond that required for bubble formation, the diameter of the bubble will decrease with increasing field until the bubble collapse field is reached at which point the bubble disappears and the medium is magnetized for $H > H_{b.c.}$ Now, if the field is decreased, bubbles will reappear and when the field is decreased to a low enough value, called the bubble run off field, the bubble becomes elliptical. Thus, we see that a bubble memory operating margin can be defined as the field interval between $H_{b.r.}$ and $H_{b.c.}$. Of course, $H_{b.r.}$ and $H_{b.c.}$ in turn depends on the material parameters.

Consider now a single bubble domain of diameter d existing in a thin film of height h. The free energy E_T can be written as the sum of wall energy E_w, Zeeman interaction energy E_H, and demagnetizing energy E_D due to magnetic poles on the surface of the film. Thus,

$$E_T = \sigma_w \pi dh + \frac{1}{2} M_s H \pi d^2 h + E_D \tag{20}$$

where E_T is in fact equal to E_T (with bubble) $-E_T$ (without bubble). Introducing the characteristic length $\ell = \sigma_w/4\pi M_s^2$ as in Equation 7, Equation 20 becomes

$$E_T = 4\pi^2 M_s^2 dh\ell + \frac{1}{2} M_s H \pi d^2 h + E_D \tag{21}$$

At equilibrium, $\partial E_T/\partial d = 0$ giving

$$4\pi^2 M_s^2 h\ell + \pi M_s H dh + \frac{\partial E_D}{\partial D} = 0 \tag{22}$$

Before proceeding further, we note from Equation 21 that both E_W and E_H decreases with decreasing d. Although E_D is not explicitly given here, we can readily see that it increases with decreasing d, keeping in mind that $E_D = (E_D)_{W.B.} - (E_D)_{W.O.B.}$ Since $(E_d)_{W.O.B.} = 2\pi M_s^2$, the maximum value attainable for the demagnetizing energy, E_D will also be negative. Thus, as d decreases, E_D will become less negative or increases since $E_D \rightarrow 0$ as $d \rightarrow 0$. When we divide Equation 22 by $4\pi^2 M_s^2 h^2$ we obtain the normalized force equation of Thiele[92] (note that since the terms in Equation 22 are that of $\partial E_T/\partial d$, they must have the unit of force):

$$\frac{\ell}{h} + \frac{d}{h}\frac{H}{4\pi M_s} - F\left(\frac{d}{h}\right) = 0 \tag{23}$$

in which

$$F\left(\frac{d}{h}\right) = -\frac{1}{4\pi^2 M_s^2 h^2}\frac{\partial E_d}{\partial d} \tag{24}$$

Since $\partial E_d/\partial d < 0$ as discussed above, F(d/h) is expected to be positive. Thus, the 1st, 2nd, and 3rd terms of Equation 23 are positive, positive, and negative, respectively.

The diameter d of the bubble can be determined by solving Equation 23 graphically. To facilitate this, rewrite Equation 23 as:

$$\frac{\ell}{h} + \frac{d}{h}\frac{H}{4\pi M_s} = F\left(\frac{d}{h}\right) \tag{25}$$

If we plot both sides of the equation as a function of d/h, the stable bubble diameter can be determined by their crossover. The left-hand side would simply give a straight line of slope $H/4\pi M_s$ terminating at $d/h = 0$ with a value equal to ℓ/h, as shown in Figure 11. If F(d/h) is also plotted on the same graph as shown, intersection of the straight line and the F(d/h) curve gives two bubble diameters, the smaller one being unstable while the large one being stable, as can be readily justified by referring to Equation 25. When $H = 0$, negative, or larger than the value at which the straight line is tangent to the F(d/h) curve (bubble

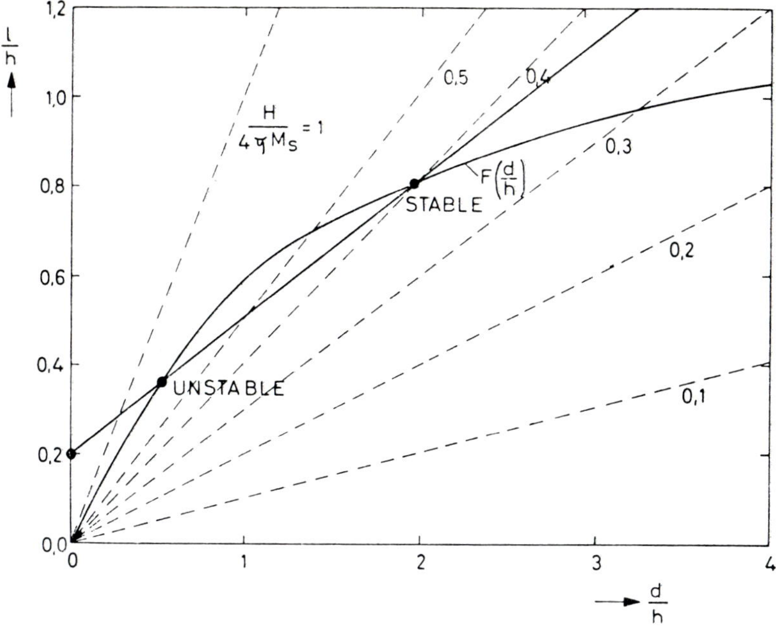

FIGURE 11. Graphical determination of the bubble diameter. Solid line indicates solution for $\ell = 0.2$ h and a bias field $H = 1.2\pi M_s$.

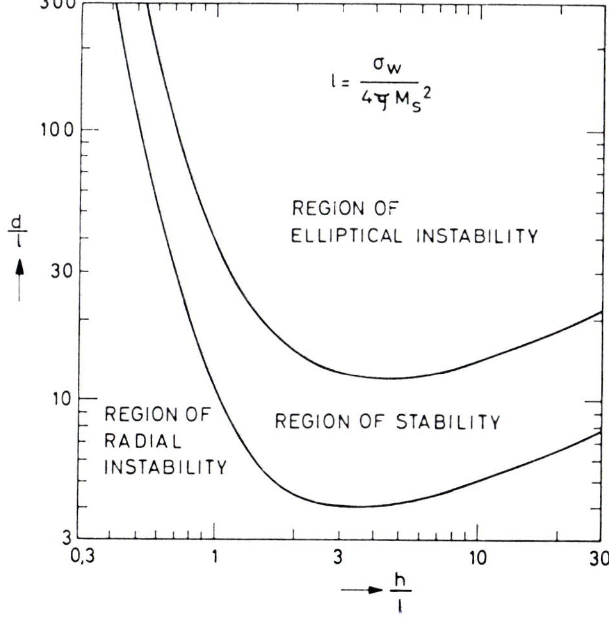

FIGURE 12. Bubble diameter d as a function of layer thickness h both expressed in units of the characteristic length ℓ for the three regions of stability as calculated by Thiele.

collapses), there are no stable solutions. It is also possible to calculate the largest bubble diameter at which a bubble will transform into an elliptical shape (bubble run out).

The results of these calculations are summarized in Figure 12. The upper curve shows

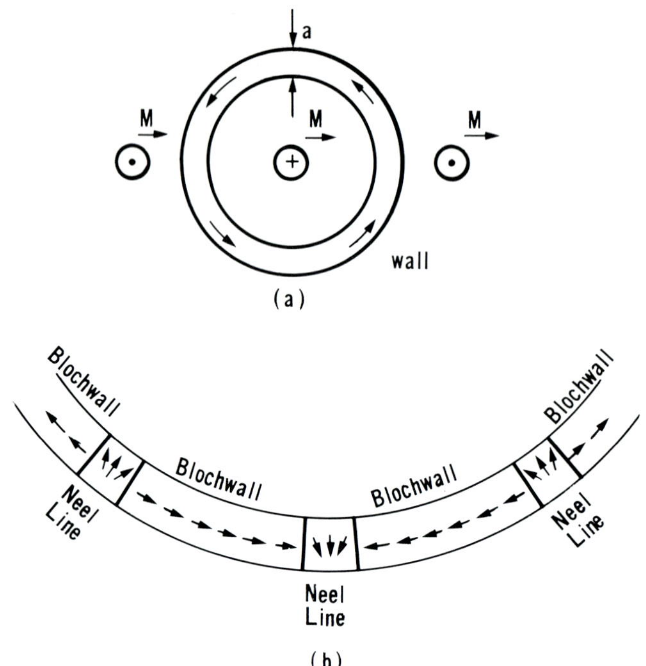

FIGURE 13. (a) Bubble Bloch walls. (b) Schematic representation of a bubble wall consisting of Bloch walls with different polarities separated by Néel lines.

the normalized bubble diameter at which a circular bubble transforms into an elliptical one as a function of the normalized film thickness. At the lower curve the bubble collapses. From this plot, minimum d and corresponding h, etc. can all be determined.

Another important parameter is the wall velocity which depends on the applied field gradient and on the rotating speed of the spins in the wall. In general, spin rotation is described by the magnetodynamics equation:

$$\frac{d\vec{M}}{dt} = \gamma MXH - \frac{\gamma\alpha}{M}\frac{d\vec{M}}{dt} \quad (26)$$

where γ is the gyromagnetic ratio and α is the damping constant. From this equation we find the wall velocity as:

$$v = \frac{\gamma}{\alpha}\frac{a}{\pi} H \quad (27)$$

Since $a = \pi(A/K_u)^{1/2}$, Equation 27 becomes

$$v = \frac{\gamma}{\alpha}\left(\frac{A}{K_u}\right)^{1/2} H \quad (28)$$

Thus, bubble velocity can be increased by lowering K_u providing $K_u > 2\pi M_x^2$ as previously discussed. In the wall of one kind of bubble, M_s rotates from one normal to the other while in the middle of the wall M_s lies in the plane of the film as shown in Figure 13(a). In the case of other bubbles, the wall may consist of Bloch walls with different polarities separated by Néel lines as shown in Figure 13(b). These bubbles can be distinguished by observing

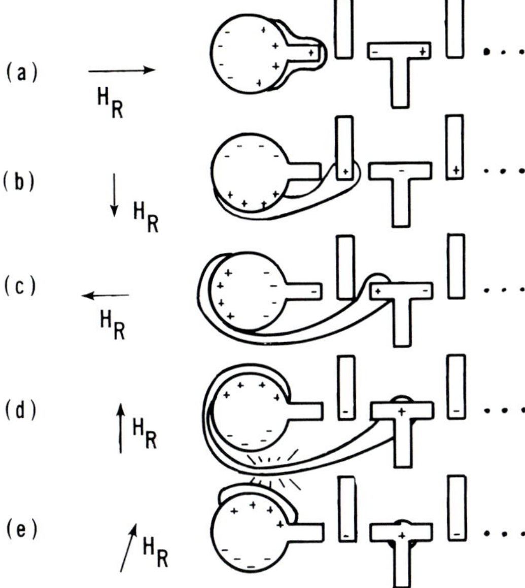

FIGURE 14. Domain generation.

their angle of propagation relative to the direction of the field gradient. Indeed, the "bubble lattice file" uses bubbles with zero and two Bloch lines to represent "0" and "1". Such lattice has the advantage of higher storage density as all positions in a two-dimensional hexagonal close-packed lattice sites are filed. No propagating structure that limits the bubble spacing is required in this case.

We are now in a position to discuss the bubble storage system further. Refer now to Figure 14. Bubbles are generated by the unusual shape of the Permalloy® overlay. As the in-plane field rotates, the bubble tends to follow the north pole as shown and eventually breaks free to enter the propagation structure. Once in the structure, it will again follow the north pole until it reaches the detector. The detector is usually a magnetoresistor also made of Permalloy®. The presence of the bubble is signified by a change in the magnetoresistance of the sensor and the change of current through it (with a constant voltage source) can be detected.

How a bubble follows the positive pole can be understood by referring to Figure 15. We see that the bubble tends to get close to the north pole and away from the south pole created by the Permalloy® overlay because the energy is minimum there. Since H_B is in fact reduced due to the fringing field of the Permalloy® $E = -\vec{M}_s \cdot (\vec{H}_B)_{eff} = M_s(H_B)_{eff}$ is also reduced near the north pole. The converse of course happens at the south pole. Also, it should be noted that Permalloy® tends to magnetize easily along the long dimension due to low demagnetization. Conversely, it does not magnetize easily in a short dimension unless field values much larger than the rotation field is applied.

Let us now turn to the subject of wall motion coercivity in amorphous materials.[74] Having eliminated grain boundaries in an amorphous material, what other discontinuities or fluctuations are there which can contribute to coercivity?

Consider a magnetic bubble domain of edge length D and height h (the thickness of the film) moving through an amorphous solid. If the bubble moves distance dx the change in volume is 2Dhdx. The work done is

$$\Delta E_i = 2MH(2Dhdx) \quad (29)$$

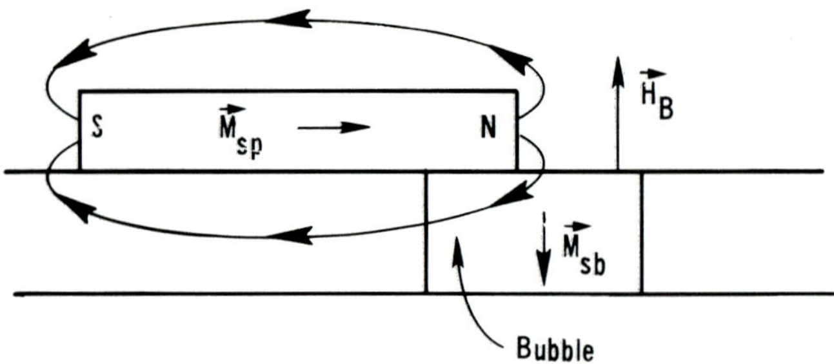

FIGURE 15. Effect on local field due to Permalloy® overlay.

where M and H are the magnetization and external field, respectively. If the domain wall is in a local potential well, it can be shown that the energy to move it is $\Delta E = (\partial E/\partial x)dx$. We now equate ΔE_i and ΔE to obtain in the H field at which the bubble moves: this field is H_c.

$$4MH_cDhdx = \frac{\partial E}{\partial x}\,dx \tag{30}$$

The energy of the wall is[52]

$$E_x = 4\sigma_w hD = 16\,(AK_u)^{1/2}\,hD \tag{31}$$

Taking the appropriate partial derivatives we can express $\partial E/\partial x$ in terms of $\partial K_u/\partial x$, $\partial A/\partial x$, etc. and arrive at the following expression for H_c:

$$H_c = \frac{\sigma_w}{M}\left[\frac{1}{2}\left(\frac{1}{K_u}\frac{\partial K_u}{\partial x} + \frac{1}{A}\frac{\partial A}{\partial x}\right) + \frac{1}{D}\frac{\partial D}{\partial x} + \frac{1}{h}\frac{\partial h}{\partial x}\right] \tag{32}$$

Assume a sawtooth shape potential well of half width δ which is equal to the domain wall width. The latter assumption is valid since if δ is very much greater than the width of the potential well, the wall will not be greatly perturbed by it. We rewrite Equation 32 to obtain

$$H_c = \frac{\sigma_w}{M\delta}\left[\frac{1}{2}\left(\frac{\Delta K_u}{\delta K_u} + \frac{\Delta A}{\delta A}\right) + \frac{\Delta D}{\delta D} + \frac{\Delta h}{\delta h}\right]$$

$$= \frac{\sigma_w}{M\delta}\left[\frac{1}{2}\left(\frac{\Delta K_u}{K_u} + \ldots.\right)\right] \tag{33}$$

where

$$\frac{\sigma_w}{\delta} = \frac{4\sqrt{AK_u}}{\sqrt{\frac{A}{K_u}}} = 4K_u$$

substituting $4k_u$ for σ_w/δ in Equation 33 leads to

$$H_c = \frac{2K_u}{M}\sum_i \frac{\Delta_{\chi i}}{\chi_i} \quad \text{where } \chi i = K_u, A, D, h \tag{34}$$

The fluctuation terms $\Delta M/M$ and $\Delta A/A$ would be expected to be small in an amorphous material that is not significantly phase separated. The mean square fluctuations in anisotropy can be estimated to be[70]

$$\frac{\Delta K_u}{K_u} = 2\left(\frac{1-c}{Nc}\right)^{1/2} \tag{35}$$

where N is the number of cobalt atoms in a volume corresponding to the wall-width region and c is the fraction of cobalt atoms giving rise to the observed uniaxial anisotropy energy. Assuming 10^{20} pairs/cm^3 and typical values of K_u and M for a 2-μm bubble in amorphous Gd–Co film, we calculate an H_c of the order of 0.1 Oe.

It has been found that thermal and magnetic annealing at about 180°C and with a field of 3 kOe, or an ion-radiation level of 10^{12} ions/cm^3 are sufficient to lower the wall-motion coercivity H_c significantly.[70] In accordance with Equation 34, H_c is reduced by decreasing K_u as evidenced by a decrease in stripe width. However, part of the decrease in H_c cannot be accounted for by a decrease in K_u alone. Indeed annealing and radiation experiments seem to indicate that fluctuations in the exchange constant, $\Delta A/A$, dominate over other fluctuations such as $\Delta K_u/K_u$ in determining H_c. In any event, H_c of amorphous bubble films is about an oersted or more, considerably higher than H_c for single crystal garnets. This factor, more than others, has rendered amorphous bubble applications unfeasible to date.

In Equation 30, the coercive force is determined by simply equating the change in the magnetostatic field energy to the maximum height of the energy barrier. Thus, the effects of the width of the domain wall relative to the width of the energy barrier are neglected. Friedberg and Paul have enlarged the theory of the coercive force of ferromagnetic materials within the framework of micromagnetics to include the wave nature of the ferromagnetic domain walls and have formulated a new theoretical approach to determine H_c.[93] However, since the emphasis of their paper is on grain boundaries in crystalline materials and not on amorphous materials, the reader is referred to their paper for details.

We now turn to a survey of actual amorphous bubble devices. An 8,000-bit GdCoMo storage chip employing electron-beam-fabricated T-bars, Y-bars, and chevrons of 1 μ linewidth has been constructed; the bubble diameter and film thickness are both about 2 μ's.[94] Operation of all function in an 8,000-bit bubble domain device with 10^7 bits/in.^{2} at 1 MHz, complete with replicate-transfer switches and chevron expander detector, was later reported.[95] Film composition can be easily changed using a multiple-target sputtering system.[12] Such fine tuning for application purposes has been achieved for Gd–Co–X (Au, Cu, Mo).[96]

The dynamic properties of GdCoCu film has been studied giving H_c = 2 to 5 Oe, v = 1,000 to 2,000 cm/sec at 25 Oe of drive field; the mobility (μ) is equal to 100 to 300 cm/sec-Oe.[97] Similarly, a laser system was used to study propagation and expansion of magnetic bubble domains in a 2-μm stripwidth GdCoMo film device operating up to 1 MHz.[98] Results obtained are: μ = 2 m/sec-Oe at low drive, μ_{max} = 20 m/sec-Oe and v = 500 m/sec at high drive, and error-free propagation for more than 10^9 cycles at 1 MHz.

Hall, Kerr, and magnetoresistance effects in RE-*TM* amorphous alloys have been studied by a number of investigators.[36,46,99-105] It was found that $Gd_{0.2}Fe_{0.8}$ is suitable as a Hall sensor.[105]

GdCo films can be used for thermomagnetic devices.[106,107] Spots are thermomagnetically written and erased by using a He–Ne laser in conjunction with H_{ext}. Annealing studies of $TbFe_2$ give an energy product of 10^6 ergs/cm^3 indicating considerable promise for recrystallized RFe_2 materials in permanent magnetic applications.[29] Velocity of sound below T_c in $DyFe_2$ has been measured.[108] Finally, RE-*TM* films can also be used in contact printing.[9]

REFERENCES

1. **Gubanov, A. I.,** *Fiz. Tverd. Tela,* 2, 502, 1960.
2. **Brenner, A., Couch, D. F., and Williams, E. K.,** *J. Res. Natl. Bur. Stand.,* 44, 109, 1950.
3. **Hooper, H. O. and de Graaf, A. M., Eds.,** *Amorphous Magnetism,* Plenum Press, New York, 1973; **Levy, R. A. and Hasegawa, R., Eds.,** Plenum Press, New York, 1976.
4. Exceptions are found in amorphous alloys prepared by evaporation onto a cooled substrate. However, the alloys so prepared often recrystallize when heated to room temperature. See **Fujime, S.,** *Jpn. J. Appl. Phys.,* 6, 305, 1967, and **Mader, S. and Norwich, A. S.,** *Appl. Phys. Lett.,* 7, 57, 1965.
5. **Stewart, R. W.,** Battelle Northwest Laboratories Research Report, Seattle, 1972, unpublished.
6. **Chaudhari, P., Cuomo, J. J., and Gambino, R. J.,** Amorphous metallic films for bubble domain applications, *IBM J. Res. Dev.,* 17, 66, 1973.
7. **Chaudhari, P., Cuomo, J. J., and Gambino, R. J.,** Amorphous metallic films for magneto-optic applications, *Appl. Phys. Lett.,* 22, 337, 1973.
8. **Garrett, H. J. and Frederick, W. G. D.,** Magnetic properties of sputter deposited rare-earth-cobalt permanent magnets. I, *AIP Conf. Proc.,* 10, 582, 1973.
9. **Imamura, N., Mimura, Y., and Kobayashi, T.,** *J. Appl. Phys.,* 48, 2634, 1977.
10. **Rhyne, J. J., Pickart, S. J., and Alperin, H. A.,** Amorphous spin polarization in a Tb-Fe compound, in *Amorphous Magnetism,* Hooper, H. O. and de Graaf, A. M., Eds., Plenum Press, New York, 1973, 373.
11. **Gerber, J. A., Miller, D. J., and Sellmyer, D. J.,** Magnetic ordering in amorphous Nd-Co, Gd-Co, and Er-Co alloys, *J. Appl. Phys.,* 49, 1699, 1978.
12. **Burilla, C. T., Bekebrede, W. R., and Smith, A. B.,** The multiple-target method for sputtering amorphous films for bubble domain devices, *AIP Conf. Proc.,* 34, 340, 1977.
13. **Dirks, A. G. and Leamy, H. J.,** Microstructure and magnetism in amorphous rare-earth-transtion metal thin films, *J. Appl. Phys.,* 49, 1735, 1978.
14. **Biesterbos, J. W., Brouha, M., and Dirks, A. G.,** Pressure dependence of magnetic properties of amorphous RE-TM thin films, *AIP Conf. Proc.,* 29, 184, 1976.
15. **Rhyne, J. J., Pickart, S. J., and Alperin, H. A.,** Amorphous spin polarization in a Tb-Fe compound, in *Amorphous Magnetism,* Hooper, H. O. and de Graaf, A. M., Eds., Plenum Press, New York, 1973, 373; **Rhyne, J. J., Pickart, S. J., and Alperin, H. A.,** Magnetism in amorphous terbium-iron, *AIP Conf. Proc.,* 18(Part 1), 563, 1974.
16. **Boudreaux, D. S.,** Structure of metallic glass alloys, Chapter 1 of this volume.
17. **Warren, B. E.,** *J. Appl. Phys.,* 12, 375, 1941.
18. **Leadbetter, A. J., Wright, A. C., and Apling, A. J.,** Neutron scattering in amorphous solids, in *Amorphous Materials,* Douglass, R. W. and Ellis, B., Eds., Wiley-Interscience, New York, 1972, 423.
19. **Polk, D. E.,** *Scr. Metall.,* 4, 117, 1970; **Polk, D. E.,** *Acta Metall.,* 20, 485, 1972.
20. **Rhyne, J. J., Price, D. L., and Mook, H. A.,** Inelastic magnetic scattering from amorphous $TbFe_2$, *AIP Conf. Proc.,* 24, 121, 1975.
21. **Pickart, S. J., Rhyne, J. J., and Alperin, H. A.,** "Critical neutron scattering in amorphous $HoFe_2$ and $GdFe_2$, *AIP Conf. Proc.,* 29, 117, 1976.
22. **Sarkar, D. and Segnan, R.,** Mössbauer effect studies in amorphous $TbFe_2$, $DyFe_2$, $HoFe_2$ and $ErFe_2$, *AIP Conf. Proc.,* 18(Part 1), 636, 1973.
23. **Forester, D. W., Abbundi, R., Segnan, R., and Sweger, D.,** Magnetic hyperfine structure in amorphous $DyFe_2$, *AIP Conf. Proc.,* 24, 115, 1975.
24. **Lee, K. and Heiman, K.,** Magnetism in rare earth-transition metal amorphous alloy films, *AIP Conf. Proc.,* 24, 108, 1975.
25. **Tao, L. J., Kirkpatrick, S., Gambino, R. J., and Cuomo, J. J.,** Charge transfer and magnetic properties of amorphous $Gd_{0.33}Co_{0.67}$, *Solid State Commun.,* 13, 1491, 1973.
26. **Harris, R., Plischke, M., and Zuckerman, M. J.,** New model for amorphous magnetism, *Phys. Rev. Lett.,* 31, 160, 1973.
27. **Clark, A. E.,** *Appl. Phys. Lett.,* 23, 642, 1973.
28. **Buschow, K. H. and VanStapele, R. P.,** *J. Appl. Phys.,* 41, 4066, 1970.
29. **Savage, H. T., Clark, A. E., Pickart, S. J., Rhyne, J. J., and Alperin, H. A.,** Effects of annealing on the coercivity of amorphous $TbFe_2$, *IEEE Trans. Magn.,* MAG-10, 807, 1974.
30. **Alperin, H. A., Cullen, J. R., and Clark, A. E.,** Magnetic properties of bulk amorphous Tb_xFe_{1-x}, *AIP Conf. Proc.,* 29, 186, 1976.
31. **Rhyne, J. J., Jr.,** Curie temperature of amorphous RFe_2 alloys (R=Gd, Tb, Dy, Ho, Er, and Y), *AIP Conf. Proc.,* 29, 182, 1976.
32. **Heiman, N., Lee, K., and Potter, R. I.,** Exchange coupling in amorphous rare earth-iron alloys, *AIP Conf. Proc.,* 29, 130, 1975.

33. **Gangulee, A. and Taylor, R. C.,** Mean field analysis of the magnetic properties of vapor deposited amorphous Fe–Gd thin films, *J. Appl. Phys.*, 49, 1762, 1978.
34. **Roberts, G. E., Wilson, W. L., Jr., and Bourne, H. C., Jr.,** Magnetic properties of Ho–Co, Dy–Co, and GdFe amorphous films prepared by dual-source evaporation, *IEEE Trans. Magn.*, MAG-13, 1535, 1977.
35. **Güntherodt, G. and Sherchik, N. J.,** XPS studies of co-sputtered Gd–Fe alloys, *AIP Conf. Proc.*, 29, 174, 1976.
36. **McGuire, T. R., Taylor, R. C., and Gambino, R. J.,** Magnetic and galvano-magnetic properties of amorphous thin film Gd–Fe–Ni alloys, *AIP Conf. Proc.*, 34, 346, 1977.
37. **McGuire, T. R. and Gambino, R. J.,** Magnetic and transport properties of Gd–Ni alloys, *IEEE Trans. Magn.*, MAG-14, 838, 1978.
38. **Chen, C. T. and Wilson, W. L., Jr.,** Magnetic properties of bias-sputtered Gd–Co–Fe, *J. Appl. Phys.*, 49, 1756, 1978.
39. **Taylor, R. C. and Gangulee, A.,** Magnetic anisotropy in evaporated amorphous films of the ternary system $Gd_x(Fe_{1-y}Co)_{1-x}$, *J. Appl. Phys.*, 48, 358, 1977.
40. **Brunsch, A. and Schneider, J.,** Evaporated amorphous GdCo-film with perpendicular anisotropy by controlled oxygen contamination, *IEEE Trans. Magn.*, MAG-13, 1606, 1977.
41. **Meyer, R., Jouve, H., and Rebouillat, J. P.,** Effect of Ni-substitution in Gd–Co amorphous thin films, *IEEE Trans. Magn.*, MAG-11, 1335, 1975.
42. **Jouve, H., Rebouillat, J. P., and Meyer, R.,** Magnetic properties of rare earth (Gd,Dy,Ho,Er)-cobalt amorphous films, *AIP Conf. Proc.*, 29, 97, 1976.
43. **Lee, K. and Keiman, N.,** Magnetism in rare earth-transition metal amorphous alloy films, *AIP Conf. Proc.*, 24, 108, 1975.
44. **Heiman, N. and Lee, K.,** Magnetic properties of Ho–Co and Ho–Fe amorphous films, *Phys. Rev. Lett.*, 33, 778, 1974.
45. **Hasegawa, R., Argyle, B. E., and Tao, L. J.,** Temperature dependence of magnetization in amorphous Gd-Co-Mo films, *AIP Conf. Proc.*, 24, 110, 1975.
46. **Argyle, B. E., Gambino, R. J., and Ahn, K.,** Polar Kerr rotation and sublattice magnetication in GdCoMo bubble films, *AIP Conf. Proc.*, 24, 564, 1975.
47. **Kobliska, R. J. and Gangulee, A.,** Annealing behavior of amorphous Gd-Co-Mo thin films, *AIP Conf. Proc.*, 24, 567, 1975.
48. **Katayama, T., Hasegawa, H., Kawanishi, K., and Tsushima, T.,** Annealing effects of magnet properties of amorphous GdCo, GdFe, and GdCoMo films, *J. Appl. Phys.*, 49, 1759, 1978.
49. **Mizoguchi, T., Gambino, R. J., Hammer, W. N., and Cuomo, J. J.,** Effects of ion implantation damage on the magnetic properties of amorphous GdCoMo films, *IEEE Trans. Magn.*, MAG-13, 1618, 1977.
50. **Cargill, G. S., III,** Ferromagnetism in amorphous solids, *AIP Conf. Proc.*, 24, 138, 1975.
51. **Fowlis, D. C. and Coperland, J. A.,** Rapid method for determining the magnetization and intrinsic length of magnetic bubble domain materials, *AIP Conf. Proc.*, 5, 240, 1972.
52. **Chikazumi, S.,** *Physics of Magnetism*, John Wiley & Sons, New York, 1964, 189.
53. **Malek, Z. and Kambersky, V.,** *Czech. J. Phys.*, 8, 1956, 416.
54. **Tsunashima, S., Takagi, H., Kamegaki, K., Fujii, T., and Uchiyama, S.,** Magnetoelastic contribution to perpendicular anisotropy in amorphous Gd-Co and Gd-Fe films, *IEEE Trans. Magn.*, MAG-14, 844, 1979.
55. **Gill, H. S., Chen, M., and Judy, J. H.,** Electron microscopy of in-situ ion beam sputtered amorphous Gd-Fe alloy thin films, *J. Appl. Phys.*, 49, 1741, 1978.
56. **Katayama, T., Hirano, M., Koizumi, Y., and Kawanishi, K.,** Different origin of the perpendicular anisotropy in amorphous Gd-Fe and Gd-Co films, *IEEE Trans. Magn.*, MAG-13, 1603, 1977.
57. **Takagi, H., Tsunashima, S., and Uchiyama, S.,** Stress induced anisotropy in amorphous Gd-Fe and Tb-Fe sputtered films, *J. Appl. Phys.*, 50, 1642, 1979.
58. **Esho, S. and Fujiwara, S.,** Growth induced anisotropy in sputtered GdCo films, *AIP Conf. Proc.*, 34, 331, 1977.
59. **Burilla, C. T., Bekebrede, W. R., Kestigian, M., and Smith, A. B.,** Influence of inert gas incorporation on uniaxial anisotropy of sputtered GdCo thin films, *J. Appl. Phys.*, 49, 1750, 1978.
60. **Herd, S. R.,** Phase separation as source of perpendicular anisotropy in amorphous GdCo, *J. Appl. Phys.*, 49, 1744, 1978.
61. **Herd, S. R.,** On nature of perpendicular anisotropy in sputtered GdCo thin films, *J. Appl. Phys.*, 50, 1645, 1979.
62. **Graaczyk, J. F.,** Structure of argon sputtered Gd-Co and Gd-Co-Mo amorphous thin films, *J. Appl. Phys.*, 49, 1738, 1978.
63. **Schneider, J. and Brunsch, A.,** O_2 contaminated amorphous FeGd films, *J. Appl. Phys.*, 49, 1747, 1978.
64. **Brunsch, A. and Schneider, J.,** H_2 and N_2 contaminated amorphous CoGd and FeGd films, *IEEE Trans. Magn.*, MAG-14, 731, 1978.

65. **Heiman, N., Onton, A., Kyser, D. F., Lee, K., and Guarnieri, C. R.,** Uniaxial anisotropy in rare-earth (Gd,Ho,Tb)-transition metal (Fe,Co) amorphous films, *AIP Conf. Proc.*, 24, 573, 1975.
66. **Onton, A. and Lee, K.,** Magnetic anisotropy in thermally evaporated Gd-Co, *AIP Conf. Proc.*, 34, 328, 1977.
67. **Dirks, A. G. and Leamy, H. J.,** Microstructure and magnetism in amorphous rare-earth-transition metal thin films, *J. Appl. Phys.*, 49, 1735, 1978.
68. **Dirks, A. G. and Leamy, H. J.,** Magnetic and structural properties of some amorphous rare-earth-transition metal thin films, *IEEE Trans. Magn.*, MAG-14, 835, 1978.
69. **Venturni, E. C., Richards, P. M., Borders, J. A., and Eer Nisse, E. P.,** Effect of ion implantation on amorphous GdCo, *AIP Conf. Proc.*, 29, 119, 1975.
70. **Hasegawa, R., Gambino, R. J., Cuomo, J. J., and Ziegler, J. F.,** *J. Appl. Phys.*, 45, 4036, 1974.
71. **Covault, M. L., Doctor, S. R., and Comstock, C. S.,** Effect of substrate bias on anisotropy in low voltage multitarget sputtered Gd-Co and Gd-Fe thin films, *AIP Conf. Proc.*, 29, 337, 1976.
72. **Cargill, G. S., III and Mizoguchi, T.,** Dipolar mechanisms for magnetic anisotropy in amorphous ferrimagnetic alloys, *J. Appl. Phys.*, 49, 1753, 1978.
73. **Chaudhari, P. and Cronemeyer, D. C.,** The temperature dependence of the uniaxial anisotropy of $Gd_{1-x-y}Co_xMo_y$ amorphous alloy films on glass substrates, *AIP Conf. Proc.*, 29, 113, 1976.
74. **Gambino, R. J., Chaudhari, P., and Cuomo, J. J.,** Amorphous magnetic materials, *AIP Conf. Proc.*, 18(Part 1), 578, 1974.
75. **Soohoo, R. F. and Moorish, A. H.,** FMR measurement of anisotropy dispersion in amorphous GdFe films, *J. Appl. Phys.*, 50, 1639, 1979.
76. **Wangsness, R. K.,** *Am. J. Phys.*, 24, 60, 1956.
77. **Gill, H. S. and Judy, J. H.,** Atomic diffusion in amorphous Gd-Fe thin films, *J. Appl. Phys.*, 50, 1648, 1979.
78. **Bhagat, S. M. and Paul, D. K.,** Magnetic resonance in $TbFe_2$, $GdFe_2$ and YFe_2, *AIP Conf. Proc.*, 29, 176, 1976.
79. **Prasad, S., Krishman, R., Suran, G., and Sztern, J.,** Inhomogeneous and FMR spectra in some d.c. sputtered amorphous Gd-Co films, *J. Appl. Phys.*, 50, 1623, 1979.
80. **Cronemeyer, D. C.,** Perpendicular anisotropy in $Gd_{1-x}Co_x$ amorphous films prepared by R.F. sputtering, *AIP Conf. Proc.*, 18(Part 1), 85, 1973.
81. **Lubitz, P., Schelleng, J., Vittoria, C., and Lee, K.,** FMR in some amorphous RE-3d transition metal, *AIP Conf. Proc.*, 29, 178, 1976.
82. **Chen, L. Y. and Soohoo, R. F.,** unpublished.
83. **Bourne, H. C., Jr., Goldfarb, R. B., Wilson, W. L., Jr., and Zwingman, R.,** Effect of d.c. bias on the fabrication of amorphous GdCo r.f. sputtered films, *IEEE Trans. Magn.*, MAG-11, 1332, 1975.
84. **Kobliska, R. J., Ruf, R., and Cuomo, J. J.,** Uniformity of H_{coe}, H_c, H_k, h, stripewidth w_s for Gd-Co-Mo, *AIP Conf. Proc.*, 24, 570, 1975.
85. **Soohoo, R. F. and Lee, K.,** Transverse magnetization curve in bubble films with parameter dispersions, *AIP Conf. Proc.*, 24, 590, 1975.
86. **Bhagat, S. M., Lloyd, J. N., and Heiman, N.,** Spin resonance in amorphous alloy films, *J. Appl. Phys.*, 49, 1683, 1978.
87. **Krishnan, R., Suran, G., and Sztern, J.,** Easy axis and easy plane magnetization in d.c. sputtered amorphous YCo_3 films, *J. Appl. Phys.*, 49, 1765, 1978.
88. **Suran, G. and Prasad, S.,** Evidence for an impurity like relaxation mechanism in oxygen contaminated amorphous YCo_3 films, *J. Appl. Phys.*, 50, 1617, 1979.
89. **Lloyd, J. N., Bhagat, S. M., Kundson, A. R., and Gossetl, C. R.,** Evidence for a surface ferromagnetic layer in amorphous YFe_2, *J. Appl. Phys.*, 50, 1614, 1979.
90. **Henderson, R. G. and de Graaf, A. M.,** Landau-Lifshitz theory of spin waves in an amorphous ferromagnet, *AIP Conf. Proc.*, 18(Part 1), 510, 1973.
91. **Roth, L. M.,** Effective approximation for ferromagnetic spin waves in amorphous systems and alloys, *AIP Conf. Proc.*, 29, 150, 1976.
92. **Thiele, A. A.,** Theory of the static stability of cylindrical domains in uniaxial platelets, *J. Appl. Phys.*, 41, 1139, 1970.
93. **Friedberg, R. and Paul, D. I.,** New theory of coercive force of ferromagnetic materials, *Phys. Rev. Lett.*, 34, 1234, 1975.
94. **Kryder, M. H., Ahn, K. Y., Almasi, G. S., Keife, G. E., and Powers, J. V.,** Bubble to T-I bar coupling in amorphous film small bubble devices, *IEEE Trans. Magn.*, MAG-10, 825, 1974.
95. **Kryder, M. H., Ahn, K. Y., and Powers, J. V.,** Amorphous film magnetic bubble domain devices, *IEEE Trans. Magn.*, MAG-11, 1145, 1975.
96. **Chaudhari, P., Cuomo, J. J., Gambino, R. J., Kirkpatrick, S., and Tao, L. J.,** Ternary amorphous alloys for bubble domain applications, *AIP Conf. Proc.*, 24, 526, 1975.

97. **Porter, R. I., Minkiewicz, V. J., Lee, K., and Albert, P. A.,** Dynamic properties of magnetic bubbles in GdCoCu films, *AIP Conf. Proc.*, 29, 76, 1976.
98. **Kryder, M. H., Tao, L. J., and Wilts, C. H.,** Dynamics of amorphous film bubble devices, *IEEE Trans. Magn.*, MAG-13, 1626, 1977.
99. **Okamoto, K., Shirakawa, T., Matsushita, S., and Sakurai, Y.,** Hall effect in Gd-Co sputtered films, *IEEE Trans. Magn.*, MAG-10, 799, 1974; *AIP Conf. Proc.*, 24, 113, 1975.
100. **Ogawa, A., Katayama, T., Hirano, M., and Tsushima, T.,** Reversal of Hall effect and Kerr roation in ferrimagnetic rare-earth-cobalt system, *AIP Conf. Proc.*, 24, 575, 1975.
101. **Minkiewicz, V. J., Albert, P. A., Potter, R. I., and Guarnieri, C. R.,** Magnetic properties of sputter deposited GdCoCu amorphous bubble films, *AIP Conf. Proc.*, 29, 107, 1976.
102. **Cutes, O. S., Holmen, J. O., Kooyer, R. J., and Audland, O. S.,** Inverted and biased loops in amorphous Gd-Co films, *IEEE Trans. Magn.*, MAG-13, 1615, 1977.
103. **Amatsu, M., Honda, S., and Kusuda, T.,** Anomalous hysteresis loops and domain observation in Gd-Fe co-evaporated films, *IEEE Trans. Magn.*, MAG-13, 1612, 1977.
104. **Shirakawa, T., Nakajima, Y., Okamoto, K., Matsushita, S., and Sakurai, Y.,** The Kerr and the Hall effect in amorphous-magnetic films, *AIP Conf. Proc.*, 34, 349, 1977.
105. **McGuire, T. R., Gambino, R. J., and Taylor, R. C.,** Galvanomagnetic effects in amorphous alloys, *IEEE Trans. Magn.*, MAG-13, 1598, 1977.
106. **Shirakawa, T., Okamoto, K., Onishi, K., Matsushita, S., and Sakurai, Y.,** Properties of sputtered Gd-Co films for magnetic bubbles, *IEEE Trans. Magn.*, MAG-10, 795, 1974.
107. **Matsushita, S., Sunago, K., and Sakurai, Y.,** Thermomagnetic writing in Gd-Co sputtered films, *IEEE Trans. Magn.*, MAG-11, 1109, 1975.
108. **Hathaway, K. and Melamud, M.,** Ultrasonic studies below T_c in amorphous rare-earth-iron alloys, *J. Appl. Phys.*, 50, 1636, 1979.

Chapter 8

CORROSION PROPERTIES OF AMORPHOUS ALLOYS

Koji Hashimoto and Tsuyoshi Masumoto

TABLE OF CONTENTS

I. INTRODUCTION

From the corrosion point of view, amorphous alloys are new materials. Because of the amorphous structure, they are free of defects associated with the crystalline state such as grain boundaries, dislocations, and stacking faults. Furthermore, rapid quenching of molten alloys to prepare amorphous alloys prevents solid state diffusion during quenching. Thus they are also free of defects formed by solid state diffusion such as second phases, precipitates, and segregates. The amorphous alloys are, therefore, regarded as ideally chemically homogeneous alloys. Since ordinary crystalline metals often suffer localized corrosion owing to the presence of chemically heterogeneous sites on the metal surface, the chemically homogeneous amorphous alloys should be resistant against localized corrosion attack.

The amorphous alloys prepared by rapid quenching from the liquid state consist at least of two components and often contain about 20 at.% metalloids. These complicated compositions, particularly a large amount of metalloid contained in the alloys greatly affects their corrosion behavior.

II. ELECTROCHEMISTRY OF CORROSION

The knowledge of electrochemistry necessary for an understanding of metallic corrosion will be described before characterization of corrosion behavior of amorphous alloys.

Corrosion of metals is an electrochemical reaction of metals with environments, and hence the corrosion behavior of various metals can be compared by polarization measurements. Typical examples are shown in Figures 1 and 2. A dissolution reaction of metal is expressed as

$$Me \rightarrow Me^{n+} + ne \qquad (1)$$

If a stable surface film is not formed, the dissolution rate measured in current density i_A generally increases logarithmically with increasing polarization potential E;

$$i_A = i_A^0 \exp \alpha(E - E_m) \qquad (2)$$

where E_m is the equilibrium potential of Reaction 1. When a potential rise results in the formation of a new corrosion product, which is a stable solid in the corrosive media and able to cover the metal surface, the dissolution rate of the metal substantially decreases. This corresponds to passivation and the corrosion product film covering the metal surface is called a passive film. If the solution contains aggressive ions such as halide ions, then further rise in potential leads to pitting corrosion for crystalline corrosion-resistant metals such as stainless steels; a major area of metal surface is covered by a passive film, but its localized breakdown by halide ions results in the formation of pits, where preferential corrosion attack proceeds deeply into the material. The nucleation and growth of pits are easily detected by a sharp increase in the current density by anodic polarization. Besides pitting corrosion, further increase in potential sometimes causes transpassive dissolution as a result of the formation of soluble cations which usually have higher valence than the cations of the same metal forming the passive film.

On the other hand, hydrogen ions and/or oxygen dissolved in the solution undergo cathodic reactions on the metal surface such as

$$2H^+ + 2e \rightarrow H_2 \qquad (3)$$

$$O_2 + 2H_2O + 4e \rightarrow 4OH^- \qquad (4)$$

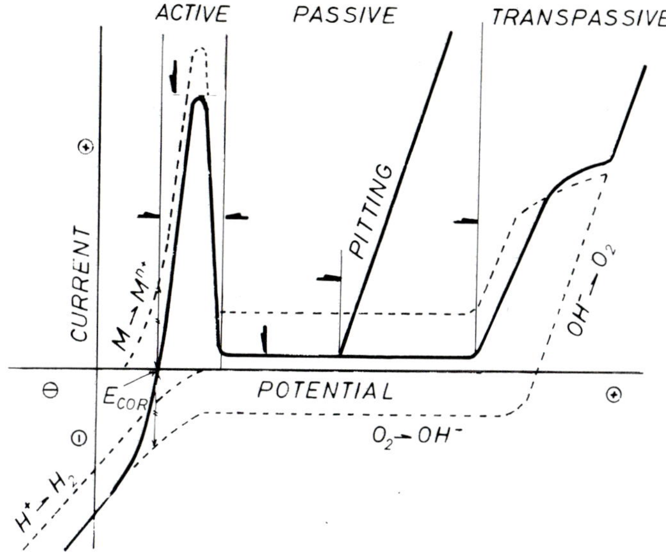

FIGURE 1. Schematic representation of polarization curves. Solid curves: polarization curves that are able to be measured. Dotted curves: true anodic and cathodic polarization curves. E_{cor}: an open circuit corrosion potential of the specimen immersed in the solution. Arrows indicate characteristic potentials and currents.

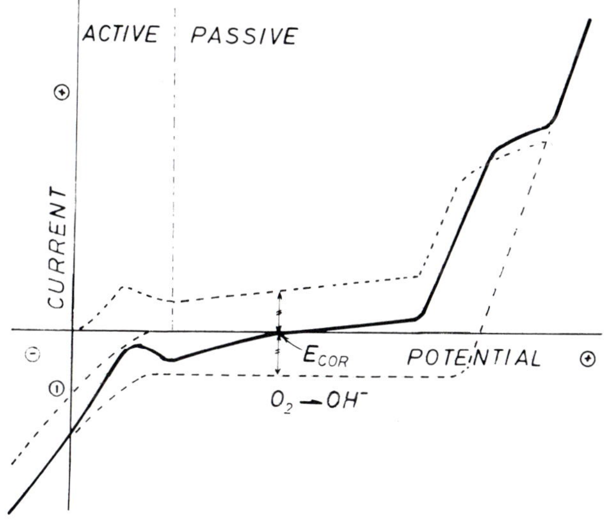

FIGURE 2. Schematic representation of polarization curve when the specimen passivates spontaneously.

The rate of these cathodic reactions rises logarithmically with lowering potential E;

$$i_C = i_C^0 \exp\left[-\beta(E - E_C^0)\right] \tag{5}$$

where E^0_C corresponds to the equilibrium potential of Reaction 3 or 4. When a specimen is immersed in an aqueous solution, both metal dissolution and cathodic reactions occur simultaneously on the metal surface with the same rates to maintain the electroneutrality of both the specimen and solution, and accordingly,

$$i_A = i_C \tag{6}$$

at the open circuit corrosion potential. If the active dissolution current is very low as shown by dotted curve in Figure 2, a positive current will be balanced with negative current in the passive region of the alloy. In other words, passivation occurs spontaneously by immersion in the solution due to dissolved oxygen.

The corrosion behavior of various metals can be compared from characteristic potentials and currents indicated by arrows in Figure 1. Shift of these values towards direction indicated by arrows implies an improvement of the corrosion resistance. In particular, spontaneous passivation as shown in Figure 2 is most desirable to possess the high corrosion resistance in the environment to which the specimen is exposed.

The corrosion behavior of amorphous alloys will be interpreted by using results of polarization measurements along with other data.

III. EXTREMELY CORROSION-RESISTANT ALLOYS

Figure 3[1,2] shows changes in the corrosion rates of amorphous iron-metalloid alloys measured in 1 *N* HCl solution at 30°C with increasing additive metallic elements. The corrosion rates of crystalline binary Fe–Cr and Fe–Ni alloys, various stainless steels and high nickel alloys are also shown for comparison. The corrosion rates at $x = 0$ correspond to those of crystalline iron and amorphous iron-metalloid alloys without addition of second metallic elements. As can be seen from the comparison of these corrosion rates at $x = 0$, amorphous iron-metalloid alloys without adding a second metallic element generally show higher corrosion rates than crystalline iron. However, their corrosion resistance is improved by addition of almost all metallic elements[3] such as titanium, zirconium, vanadium, niobium, tantalum, chromium, molybdenum, tungsten, cobalt, nickel, copper, rhodium, palladium, and platinum. In particular, the addition of chromium is quite effective. Chromium is more active than iron, and hence the addition of chromium to crystalline iron gives rise to an increase in corrosion rate in the aggressive hydrochloric acids. On the contrary, the addition of chromium to amorphous iron-metalloid alloys significantly decreases the corrosion rate[4]: amorphous Fe–Cr–13P–7C alloys containing 8 at.% or more chromium do not show detectable weight loss even in 2 *N* HCl at ambient temperature, due to spontaneous passivation.[5] As is well known, except for tantalum, any commercial corrosion-resistant metals such as stainless steels, high nickel alloys, etc. are not able to passivate spontaneously even in 1 *N* HCl at room temperature. Furthermore, crystalline stainless steels suffer pitting corrosion, hence, passivation by anodic polarization is quite difficult in 1 *N* HCl. For instance, a ferritic stainless steel must contain 28 wt.% or more chromium to avoid pitting by anodic polarization up to transpassive region of chromium in 1 *N* HCl.[6] Such behavior of crystalline alloys is in marked contrast with the fact that the amorphous Fe–8Cr–13P–7C alloy is able to passivate spontaneously even in 2 *N* HCl.

The combined addition of chromium and molybdenum is further effective in improving the corrosion resistance; some amorphous Fe–Cr–Mo–metalloid alloys passivate spontaneously even in hot concentrated hydrochloric acids such as 12 *N* HCl at 60°C.[5] Figure 4 shows the effect of the combined addition of chromium and molybdenum on the anodic polarization curves measured in 6 *N* HCl.[7] The Fe–25Cr–20B alloy is in the active state when immersed in 6 *N* HCl, but with increasing molybdenum content, the active current density decreases with a consequent increase in the corrosion potential. When the molybdenum content attains 5 at.%, the current density in the active state becomes unobservable as a result of spontaneous passivation.

The addition of chromium is also effective in increasing the corrosion resistance of amorphous cobalt- and nickel-base alloys. Figure 5 shows the variation of corrosion rates

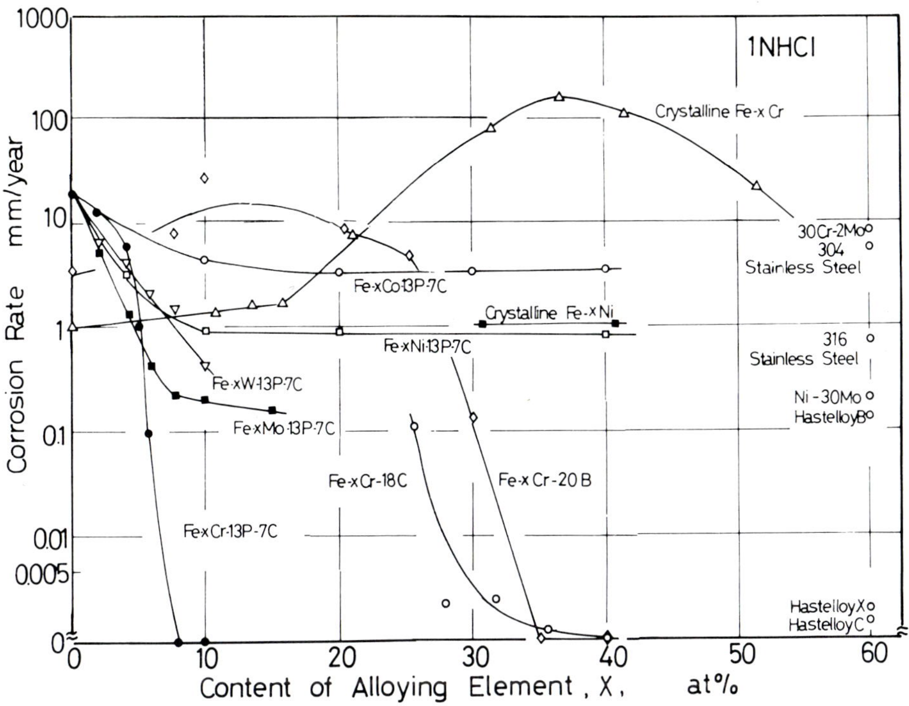

FIGURE 3. Changes in average corrosion rates of amorphous Fe–M–metalloid alloys in 1 *N* HCl at 30°C as a function of alloying element M. Corrosion rates of various crystalline alloys are also shown for comparison.[1,2]

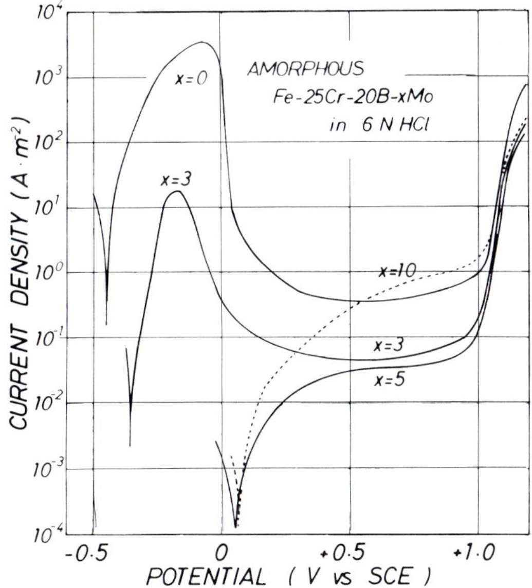

FIGURE 4. Potentiodynamic polarization curves of amorphous Fe–25Cr–20B–xMo alloys in 6 *N* HCl. The current density increase at negative potentials on each polarization curve corresponds to the cathodic current density. (From Asami, K., Naka, M., Hashimoto, K., and Masumoto, T., *J. Electrochem. Soc.*, 127, 2130, 1980, Reprinted by permission of the publisher, The Electrochemical Society, Inc.)

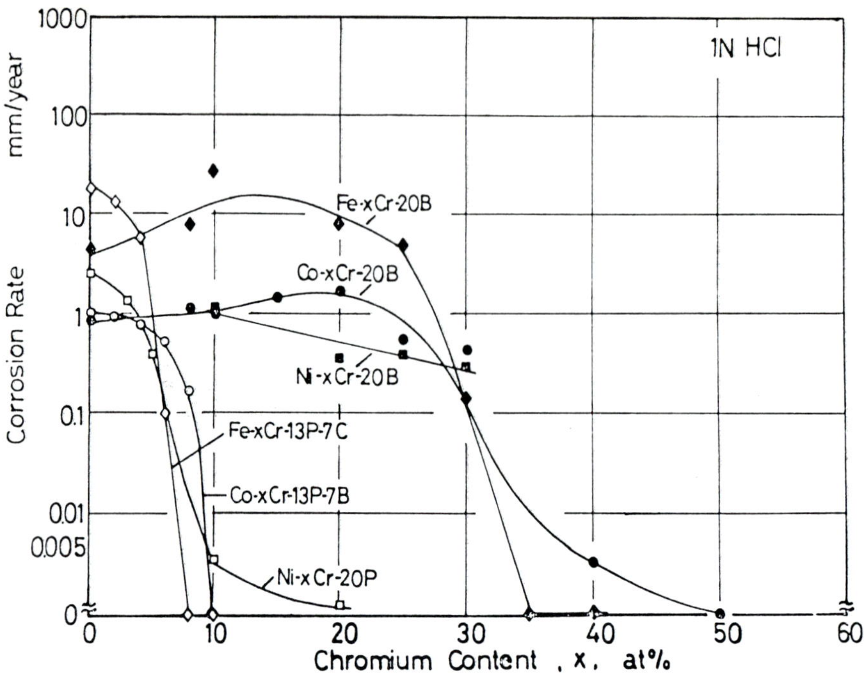

FIGURE 5. Change in average corrosion rates of amorphous iron-, cobalt- and nickel-base alloys in 1 *N* HCl at 30°C as a function of chromium content of alloys. (From Naka, M., Hashimoto, K., and Masumoto, T., *J. Non-Cryst. Solids*, 34, 257, 1979. With permission.)

of amorphous iron-, cobalt- and nickel-base alloys in 1 *N* HCl at 30°C as a function of the chromium content.[8] Both the cobalt- and nickel-base alloys as well as the iron-base alloys possess a high corrosion resistance by the addition of a definite amount of chromium. The effects of metalloids and main metallic components of alloys on their corrosion behavior are observed in Figure 5 and will be described later.

An example of the corrosion rate of amorphous metal-metal alloy is shown in Figure 6.[9] The formation of the amorphous structure is considerably effective in improving the corrosion resistance of metal-metal alloys. However, a decisive factor for improving the corrosion resistance of metal-metal alloys is the content of an element having a higher corrosion resistance than the other components; such as the zirconium content in zirconium-copper alloys.

The corrosion behavior of amorphous iron-silicon alloys prepared by sputtering appears to be similar to that of amorphous metal-metal alloys. As shown in Figure 7, the corrosion rate of crystalline Fe–30Si alloy is 1/50 as high as that of crystalline Fe–20Si alloy due to passivation by silicon.[10] The formation of the amorphous structure leads to a decrease in the corrosion rate of Fe–30Si alloy to 1/10 of the crystalline alloy. In this manner, an increase in the silicon content of the alloy is more effective than the formation of the amorphous structure in improving the corrosion resistance of iron-silicon alloys.

IV. PASSIVE FILM

Except a few noble metals, almost all metallic species have naturally been found in oxides or salts, indicating that the oxidized state is more stable than the metallic state in natural environments. In other words, corrosion is unavoidable for the majority of metals in natural environments. However, when the metal surface is covered by corrosion products, they are

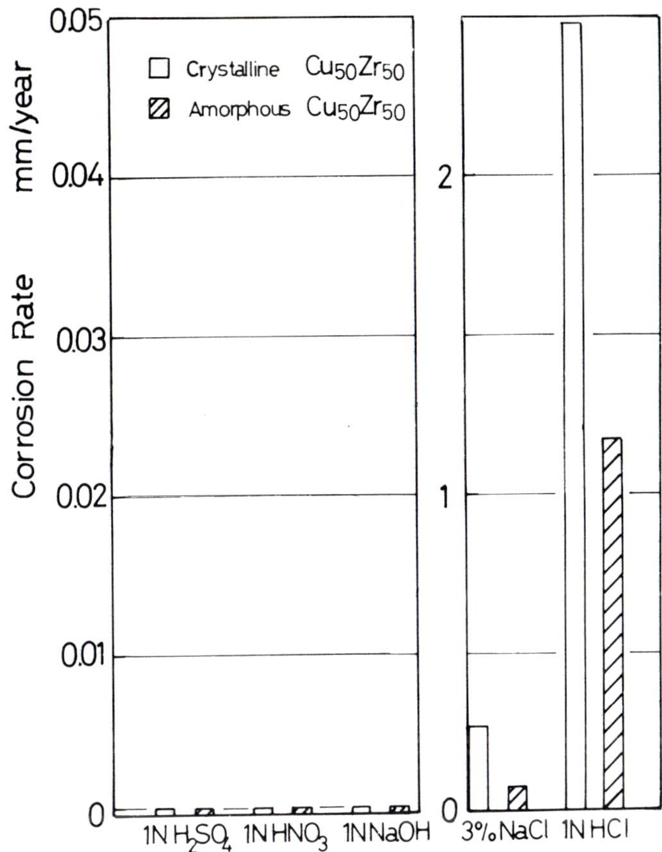

FIGURE 6. Average corrosion rates of amorphous and crystalline Cu–50Zr alloys measured in various solutions at 30°C. (From Naka, M., Hashimoto, K., and Masumoto, T., *J. Non-Cryst. Solids*, 30, 29, 1978. With permission.)

able to act as a barrier against further corrosion attack. The protective quality of the corrosion product films is determined by their chemical stability and uniformity.

As can be seen in Figure 3, the corrosion resistance of amorphous iron-metalloid alloys without adding a second metallic element is generally lower than the crystalline iron. Nevertheless, the corrosion resistance of amorphous metal-metalloid alloys increases by the addition of a second metallic element. This is due mainly to the formation of a protective corrosion product film in which cations of the second metallic element are concentrated. The metallic elements which play such a role in amorphous iron-metalloid alloys are titanium, zirconium, vanadium, niobium, tantalum, chromium, molybdenum, tungsten, etc.[11] Among them, titanium, zirconium, and chromium form their own passive films which have a high protective quality, and hence the alloys containing these metallic elements possess a quite high corrosion resistance.

One of differences between passive films and ordinary corrosion product films is that the beneficial cations in improving the protective quality of the films are highly concentrated in the passive films.

For example, the corrosion resistance of stainless steels, in which chromium is added to iron, is provided by the formation of a passive film consisting mainly of hydrated chromium oxyhydroxide $CrO_x(OH)_{3-2x} \cdot nH_2O$, which is highly stable and protective.[12] The concentration of hydrated chromium oxyhydroxide in the film and the values of x and n depend upon the alloy composition and the condition of the film formation. It must be noted that the higher

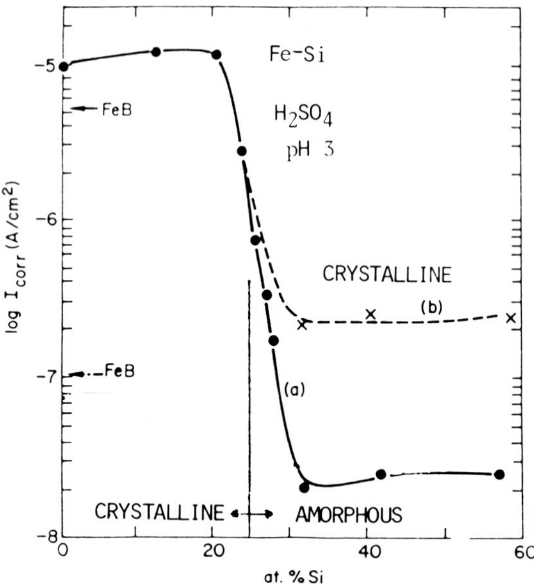

FIGURE 7. Corrosion current of Fe–Si alloys measured in diluted sulfuric acid (pH 3) as a function of silicon content. (a) alloys with Si $\gtrsim$ 25% were amorphous; (b) all alloys were crystalline. (From Brusic, V., MacInnes, R. D., and Aboaf, J., Passivation properties of amorphous Fe–Si thin films, in *Passivity of Metals,* Frankenthal, R. P. and Kruger, J., Eds., The Electrochemical Society, Princeton, N.J., 1978, 170. This figure was originally presented at the Fourth International Symposium on Passivity held in October of 1977 in Airlie, Virginia. Reprinted by permission of the publisher, The Electrochemical Society, Inc.)

Table 1
CHROMIUM CONTENTS OF PASSIVE FILMS FORMED ON AMORPHOUS ALLOYS AND STAINLESS STEELS IN 1 *N* HCl

Alloys	Cr^{III}/total metallic ions in films	Passivation method	Ref.
Amorphous alloys			
Fe–10Cr–13P–7C	0.97	Spontaneous	13
Fe–3Cr–2Mo–13P–7C	0.57	Polarization	14
Co–10Cr–20P	0.95	Spontaneous	15
Ni–9Cr–15P–5B	0.74	Spontaneous	16
Ferritic stainless steels			
Fe–30Cr	0.75	Polarization	17
Fe–19Cr	0.58	Polarization	18

the concentration of hydrated chromium oxyhydroxide in the film, the better is the protective quality of the film. Table 1 summarizes the cationic fraction of chromium in passive films formed on chromium-bearing amorphous alloys and ferritic stainless steels in 1 *N* HCl. These amorphous alloys except Fe–3Cr–2Mo–13P–7C alloy passivated spontaneously by immersion in 1 *N* HCl. In contrast, these ferritic stainless steels were in the active state in 1 *N* HCl, and hence their passivation was carried out by anodic polarization. In Table 1 are shown the highest concentrations of chromium among those found in the passive films formed on the ferritic stainless steels at various potentials. As can be seen from Table 1, amorphous metal-metalloid alloys are able to remarkably accumulate protective hydrated chromium oxyhydroxide in the passive film, and accordingly the passive films formed on them have a quite high protective quality. These results clearly reveal that the effect of beneficial alloying element is predominant for the amorphous alloys in comparison with the crystalline alloys. This is attributable to the rapid formation of a protective film and the chemical homogeneity of the amorphous alloys, which will be described in the following sections.

V. RAPID FILM-FORMATION

When a protective surface film covering metals is mechanically or chemically broken, rapid repassivation prevents serious corrosion attack. The high corrosion resistance will be provided by rapid passivation and high concentration of beneficial elements in the passive film. Both these two factors are affected by the activity (active dissolution rate) of the alloy, since the passive film is formed as a result of corrosion. In a mild corrosive environment such as neutral solutions without aggressive anions, even an air-formed film may be able to act as a passive film. However, in aggressive environments an accumulation of beneficial species in the surface film is necessary for passivation. For the accumulation of beneficial species, useless components of the alloy must dissolve into the solution. Therefore, active dissolution preceding passive film formation is generally required in aggressive solutions. Figure 8 exhibits a good example showing a correlation between active dissolution and passivation.[19] Polarization of amorphous Pd–16Ti–19P alloy at a high potential in a very aggressive, hot, concentrated NaCl solution gives rise to the evolution of chlorine gas as well as corrosion. Potentiostatic polarization at 1.5 V(SCE) exhibited initially a high anodic current density without chlorine gas evolution and turned the solution in the vicinity of the alloy specimen to brown color, indicating active dissolution of the alloy. Subsequently, an abrupt decrease in current density was observed with simultaneous cessation of brown coloration of the solution, suggesting passivation. At the same time, vigorous evolution of chlorine gas began to occur and lasted for a few hundred seconds. However, chlorine gas evolution gave rise to the degradation of the passive film by nascent chlorine and to resumption of active dissolution which resulted in a sharp increase in current density. This behavior was repeatedly observed. It can, therefore, be said that rapid dissolution leads to passivation even in a very aggressive environment where breakdown of passivity takes place by nascent chlorine.

Since the beneficial species for passive film formation also dissolve during active dissolution of alloys, the slow active dissolution of alloys does not result in a high concentration of the beneficial species at the alloy-solution interface, and hence such alloys hardly passivate. Consequently, when alloys which are able to passivate are exposed to an aggressive solution, the faster the active dissolution, the faster is the passivation and the higher the concentration of beneficial species in the passive film.

Figure 9 shows the comparison of active dissolution and subsequent passivation between amorphous Fe–10Cr–13P–7C alloy and type 304 stainless steel[20]; immediately after mechanical abrasion of the specimen surfaces during polarization in 1 *N* H_2SO_4 at 300 mV(SCE) where passive films can be formed on both the alloys, current decay curves were measured. The current density at zero second corresponds to the dissolution rate or activity of bare surfaces where the surface films were removed by mechanical abrasion. The current decay with time means a decrease in dissolution rate due to the formation of a passive film. Accordingly, the faster decay corresponds to the faster formation of the passive film, and the lower current density in the steady state indicates the higher protective quality of the passive film. The current density of the amorphous alloy immediately after mechanical

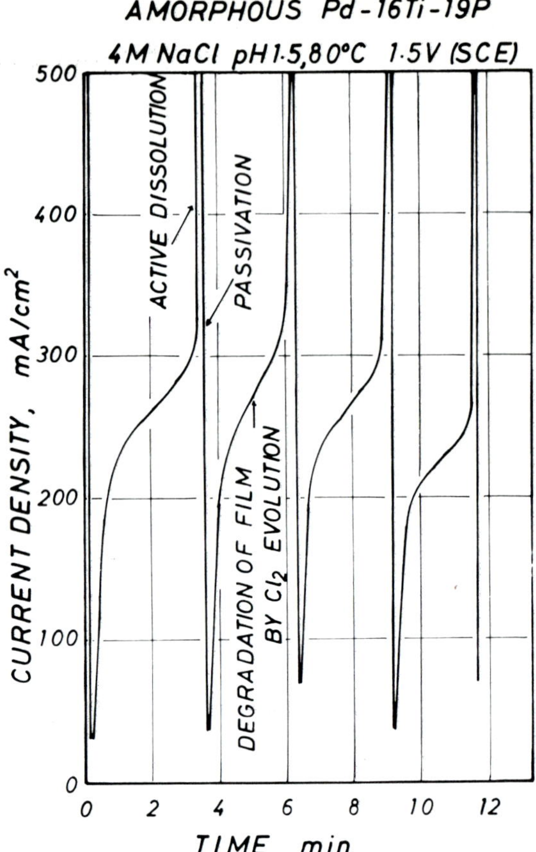

FIGURE 8. Current density-time curve measured for amorphous Pd–16Ti–19P alloy during potentiostatic polarization at 1.5 V(SCE) in 4 *M* NaCl at pH 1.5 and 80°C. (From Hara, M., Hashimoto, K., and Masumoto, T., *Electrochim. Acta*, 25, 1215, 1980. With permission.)

abrasion is higher than that of the stainless steel, indicating a higher activity of the amorphous alloy. This is in good agreement with the fact that amorphous iron-metalloid alloys without adding the second metallic elements showed a higher dissolution rate than crystalline iron, as shown in Figure 3. These results imply that, if a stable surface film is not formed, amorphous alloys generally dissolve rapidly. On the other hand, the current density of the amorphous alloy sharply decreases with time, suggesting a rapid formation of a passive film. When the alloys contain beneficial elements for the formation of a protective passive film such as chromium, the high activity of the amorphous alloys leads to rapid enrichment of the beneficial species at the alloy-solution interface and to rapid formation of the passive film which is highly protective, owing to a significant enrichment of beneficial species such as chromic ions. The high protective quality of the passive film formed on the amorphous alloys is responsible for a low current density in the steady passive state. As can be seen from these results, the high corrosion resistance of amorphous alloys in various environments is partly attributable to the high reactivity of the alloys which provides the rapid formation of a highly protective passive film.

In connection with correlation between the activity of an alloy and its passivation behavior, interesting results are obtained from a comparison of amorphous iron-, cobalt- and nickel-base alloys. Figure 5 shows changes in average corrosion rates of amorphous iron-,

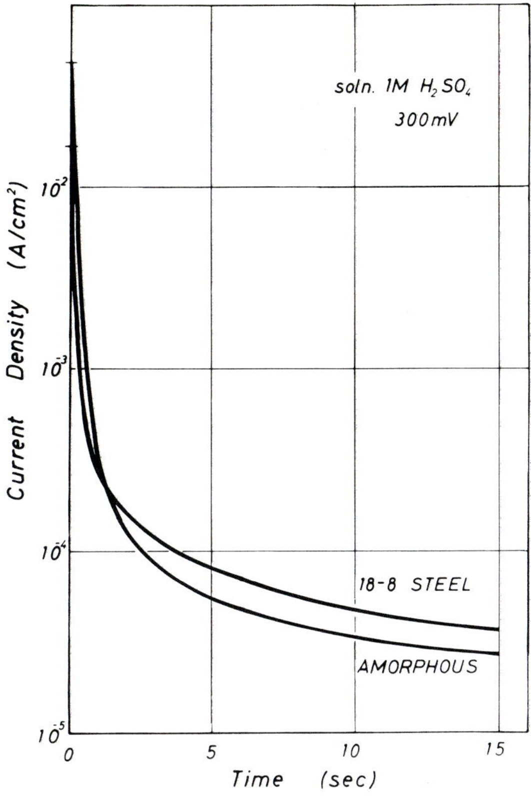

FIGURE 9. Current decay curves of amorphous Fe–10Cr–13P–7C alloy and Type 304 stainless steel after the cessation of abrasion during polarization at 300 mV(SCE) in 1 *M* H_2SO_4. (From Hashimoto, K., Osada, K., Masumoto, T., and Shimodaira, S., *Corros. Sci.*, 16, 71, 1976. With permission.)

cobalt-,and nickel-base alloys measured in 1 *N* HCl as a function of chromium content.[8] The comparison of three series of alloys clearly reveals that, if a sufficient amount of chromium is not added, the corrosion rate is mainly determined by the reactivity of alloys; when the chromium content is less than 30 at.% for boron-bearing alloys and less than 3 at.% for phosphorus-bearing alloys, that is, when a protective surface film is not formed, the most active iron-base alloys show the highest corrosion rate while the most noble nickel-base alloys possess the lowest corrosion rate. However, when a sufficient amount of chromium is added, the most active iron-base alloys passivate spontaneously by the addition of the lowest amount of chromium due to the most rapid accumulation of the beneficial species such as chromium in a high concentration in the surface film. The corrosion rate becomes undetectable by the addition of 8 at.% chromium to Fe–13P–7C alloy and 35 at.% chromium to Fe 20B alloy in 1 *N* HCl. Because cobalt is more noble than iron and shows a slower active dissolution rate than iron, the addition of 10 at.% chromium to Co–13P–7B alloy or 50 at.% chromium to Co–20B alloy is required for providing the high corrosion resistance when the corrosion rate becomes undetectable in 1 *N* HCl. Furthermore, it is difficult to accumulate beneficial species at the alloy solution interface for the most noble nickel-base alloys because of low reactivity of the alloys, and hence the corrosion rate in 1 *N* HCl is still detectable even if the chromium content of Ni–Cr–20P alloy is 20 at.%. Consequently, when an alloy contains an element with a high capability for passivation, the higher the reactivity of the alloy, the easier is the passivation.

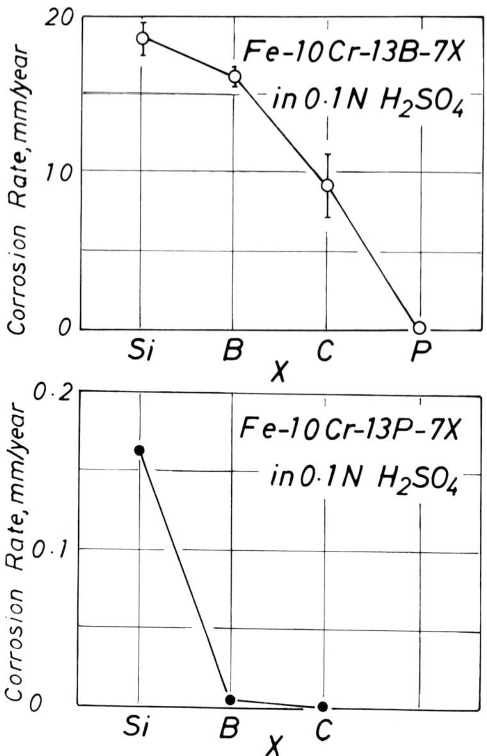

FIGURE 10. Effects of additive metalloid elements on corrosion rates of amorphous Fe–10Cr–metalloid alloys in 0.1 *N* H_2SO_4. (From Naka, M., Asami, K., Hashimoto, K., and Masumoto, T., *J. Non-Cryst. Solids*, 28, 403, 1978. With permission.)

VI. EFFECTS OF METALLOIDS

The reactivity of amorphous alloys is greatly affected by metalloid elements contained in the alloys. Figure 10 shows the change in corrosion rate of two series of amorphous Fe–10Cr–metalloid alloys as a function of minor metalloid elements.[21] A comparison of the ordinates of top and bottom figures clearly indicates that the corrosion rate of the alloys containing boron as a major metalloid element is more than two orders of magnitude as high as that of the alloys containing phosphorus as a major metalloid element. In addition, the change in minor metalloid elements in the order of silicon, boron, carbon and phosphorus decreases the corrosion rate of the alloys in that order. To clarify the role of metalloid in improving the corrosion resistance the experiment similar to Figure 9 was carried out. Figure 11 shows effects of minor metalloids, X, on the activity and passivation behavior of amorphous Fe–10Cr–13B–7X alloys after mechanical abrasion of specimen surfaces during potentiostatic polarization in the passive region.[22] The phosphorus-bearing alloy exhibits the highest current density immediately after mechanical abrasion, that is, the highest reactivity of the alloy and shows the lowest steady state current density corresponding to the highest protective quality of the film. It is evident that phosphorus is the most effective metalloid element to increase the rate of protective film formation.

In this connection, the corrosion resistance of amorphous metal-metal alloys is considerably raised by the addition of a small amount of phosphorus. For instance, the corrosion

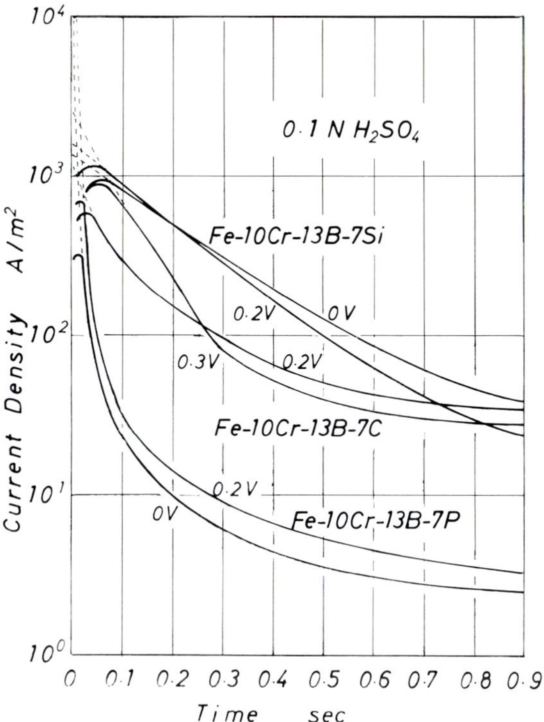

FIGURE 11. Current density-time curves of amorphous Fe–10Cr–13B–7X alloys immediately after cessation of abrading the alloy surfaces during anodic polarization at constant potentials in 0.1 *N* H_2SO_4. X and polarization potentials relative to the saturated calomel electrode are described in the figure. (From Hashimoto, K., Naka, M., Asami, K., and Masumoto, T., *Corros. Eng. (Boshoku Gijutsu)*, 27, 279, 1978. With permission.)

resistance of amorphous Ti–50Cu alloy is lower than crystalline pure titanium due to detrimental effect of copper.[9] As shown in Figure 12, the amorphous Ti–50Cu alloy shows a higher current density in both the active and passive regions in 1 *N* HCl than crystalline pure titanium.[23] Even if the titanium content is raised to 70 at.%, the corrosion potential of the Ti–30Cu alloy in 1 *N* HCl is still lower than that of crystalline titanium. However, when 5 at.% copper is substituted with phosphorus, the corrosion resistance remarkably increases, showing spontaneous passivation. In particular, the passive current density of amorphous Ti–25Cu–5P alloy measured in 1 *N* HCl is about two orders of magnitude lower than that of crystalline pure titanium. This has also been interpreted by the beneficial effect of phosphorus to form a protective passive film.

Metalloid elements contained in amorphous alloys also affect the composition of anions in the passive film formed and hence the protective quality of the film. For instance, Table 2 summarizes the molar ratios of chromic ions and polyoxyanions to the total cations in the surface films formed on amorphous cobalt-chromium-metalloid alloys by immersion in 1 *N* HCl.[15] Co–20Cr–20B and Co–25Cr–20B alloys dissolve actively in 1 *N* HCl, while Co–30Cr–20B and Co–10Cr–20P alloys passivate spontaneously in 1 *N* HCl. The cationic fraction of chromium in the surface film formed on the Co–25Cr–20B alloy is 0.81, which is significantly higher in comparison with that in the passive films formed on the ferritic stainless steels as shown in Table 1. Nevertheless, this amorphous alloy does not passivate

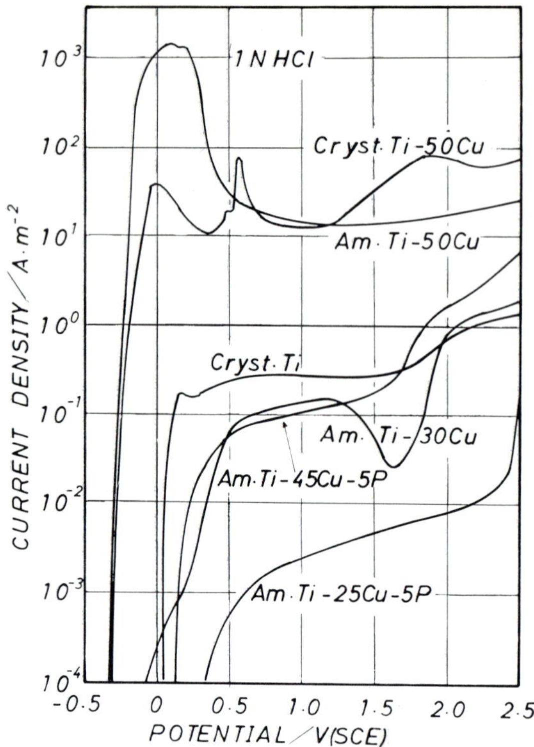

FIGURE 12. Anodic polarization curves of amorphous and crystalline Ti–Cu alloys and amorphous Ti–Cu–P alloys measured in 1 *N* HCl. (From Naka, M., Asami, K., Hashimoto, K., and Masumoto, T., in *Titanium '80, Proc. 4th Int. Conf. Titanium,* Vol. 4, Kimura, H. and Izumi, O., Eds., The Metallurgical Society of the American Institute of Mining, Metallurgical, and Petroleum Engineers, Warrendale, Pa., 1981, 2677. With permission.)

Table 2
MOLAR RATIOS OF CHROMIC ION AND POLYOXYANIONS TO TOTAL CATIONS IN SURFACE FILMS FORMED ON AMORPHOUS Co–Cr–METALLOID ALLOYS IMMERSED IN 1 *N* HCl

Amorphous alloy	State	Cr^{3+}	BO_3^{3-}	PO_3^{3-}
Co–20Cr–20B	Active	0.55	0.47	—
Co–25Cr–20B	Active	0.81	0.49	—
Co–30Cr–20B	Passive	0.99	0.52	—
Co–10Cr–20P	Passive	0.95	—	0.03

spontaneously in 1 *N* HCl. The amorphous Co–30Cr–20B alloy passivates spontaneously in 1 *N* HCl, forming the passive film in which the cationic fraction of chromium is 0.99. Accordingly, remarkable enrichment of chromic ions in the film is required for passivation of Co–Cr–20B alloys. The question has been raised why the amorphous Co–25Cr–20B alloy does not passivate spontaneously despite the fact that the chromic ions comprise 81% of cations in the surface film. The surface films formed on the amorphous Co–Cr–20B alloys are characterized by a high concentration of boron as borate; the molar ratio of boron to metallic ions is about 0.5. Similarly, if amorphous chromium-bearing alloys contain silicon as a metalloid, a large amount of silicate was found in the surface film formed.[22] On the other hand, the amorphous Co–10Cr–20P alloys passivates spontaneously in 1 *N* HCl forming the passive film whose cationic fraction of chromium is 0.95. In addition, the phosphorus content of the film is negligibly low; the molar ratio of phosphorus to metallic ions in the film is 0.03.

Even if a surface film consists exclusively of chromium as cations, the formation of chromium borate or silicate decreases the concentration of beneficial hydrated chromium oxyhydroxide in the film, and therefore, decreases the protective quality of the film. In contrast, amorphous iron-base alloys containing phosphorus and a definite amount of chromium exclusively form a passive hydrated chromium oxyhydroxide film, in which the phosphate content is negligibly low. Consequently, phosphorus contained in amorphous metal-metalloid alloys accelerates active dissolution of alloys prior to the passive film formation, and accordingly leads to the rapid enrichment of beneficial species at the alloy solution interface and to the rapid formation of passive film. In addition, phosphorus is not largely incorporated into the surface film as phosphate and hence favors the formation of hydrated oxyhydroxide films with a high protective quality.

The role of carbon in the acceleration of active dissolution is not predominant in comparison with phosphorus. However, carbon contained in the amorphous metal-metalloid alloys does not constitute the passive film as carbonate. Therefore, next to phosphorus, carbon is the most effective metalloid in improving the corrosion resistance of amorphous alloys. Silicon and boron contained in amorphous alloys do not greatly accelerate active dissolution of alloys, and hence are less effective than phosphorus or carbon in increasing the rate of passive film formation. Furthermore, they form silicate or borate films, and accordingly prevent the formation of a passive hydrated oxyhydroxide film with a high protective quality. Consequently, although the addition of boron and/or silicon to amorphous alloys is effective in improving the corrosion resistance, it is less effective in comparison with the addition of phosphorus and/or carbon.

VII. HOMOGENEITY OF ALLOY

In general, the corrosion resistance of almost all metallic materials is determined by the composition, stability, and uniformity of the surface film formed as well as the kinetics of formation. As mentioned above, amorphous metal-metalloid alloys have superior characteristics with respect to the high corrosion resistance — the rapid formation of the passive film containing a high concentration of beneficial species with a high stability. When the alloys possess such superior characteristics, the corrosion resistance will depend upon the uniformity of the passive film formed, since the passive film generally contains weak points whose activity is responsible for the passive current density in aggressive solutions. In this connection, the homogeneous nature of amorphous alloys is responsible for the high passivating ability and the high resistance against pitting. As is well known, crystalline defects are often chemically heterogeneous and tend to act as initiation sites for corrosion. Amorphous alloys are, however, free of defects associated with the crystalline state. In addition, rapid quenching from the liquid state prevents solid state diffusion during quenching and precludes segregation, precipitation, and formation of other compositional fluctuations. Accordingly, the alloys are regarded as ideal alloys with the chemically homogeneous, single phase nature, which provides the high corrosion resistance. For example, crystalline iron does not passivate in 1 *N* HCl. The addition of molybdenum to crystalline iron is not effective for passivation by anodic polarization in 1 *N* HCl and increases the passive current density in 1 *N* H_2SO_4.[6]

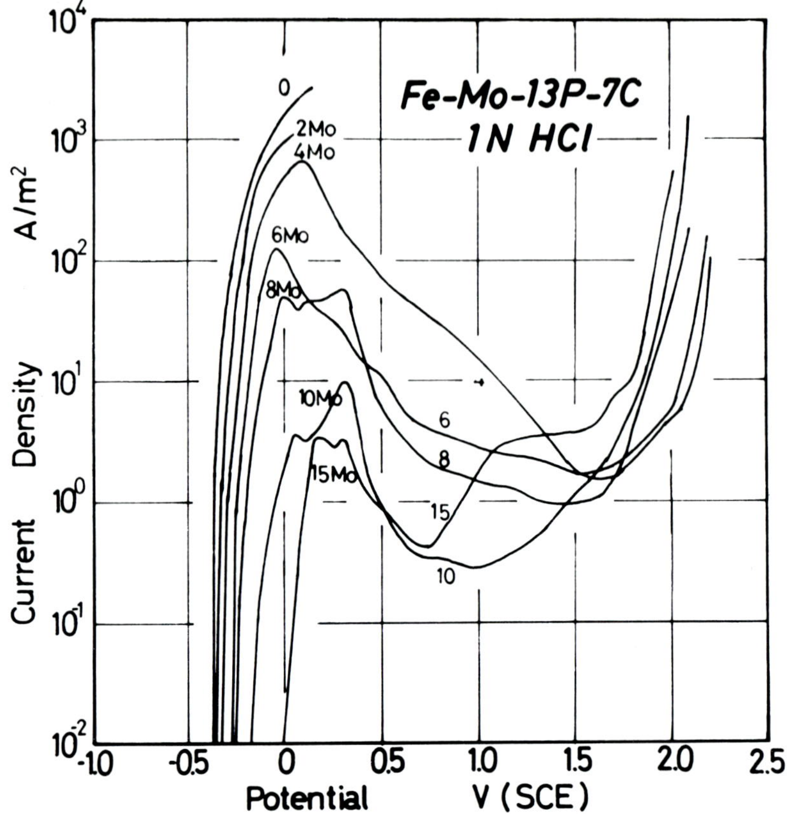

FIGURE 13. Potentiodynamic polarization curves of amorphous Fe–xMo–13P–7C alloys measured in 1 *N* HCl. x is the atomic percent content of Mo indicated by the number in the figure. (From Naka, M., Hashimoto, K., and Masumoto, T., *J. Non-Cryst. Solids,* 29, 61, 1978. With permission.)

In contrast, amorphous Fe–Mo–P–C[24] and Fe–Mo–C[2] alloys passivate by anodic polarization, even in 1 *N* HCl, showing no pitting corrosion as shown in Figure 13. The passive films were analyzed by X-ray photoelectron spectroscopy. The cationic fraction in the surface films formed on the amorphous Fe–Mo–P–C alloys is shown as a function of polarization potential in Figure 14. Molybdenum cations are significantly concentrated in the corrosion product film formed in the active region. However, the passive films formed on these alloys contain a very small amount of molybdenum cations and are exclusively composed of hydrated iron oxyhydroxide,[25,26] which is the same as the passive film formed on crystalline iron in 1 *N* H_2SO_4.[12] However, iron does not passivate in 1 *N* HCl, because the crystalline heterogeneous substrate cannot be uniformly covered by hydrated iron oxyhydroxide in the aggressive hydrochloric acid. In this manner, the composition and stability of the passive film formed on amorphous Fe–Mo–metalloid alloys are not essentially different from those of the passive film formed on iron if iron is able to passivate. Accordingly, easy passivation of the amorphous Fe–Mo–metalloid alloys is attributed to the homogeneous nature of the amorphous alloys, which provides the formation of a uniform passive film assisted by the high reactivity of amorphous alloys.

The chemical heterogeneity seems to form a high density of weak points in the passive film with respect to corrosion as well as localized corrosion attack. For instance, crystallization of amorphous alloys results in about two orders of magnitude increase in the passive current density.[20,27] The heat treatment of amorphous alloys gives rise to the formation of

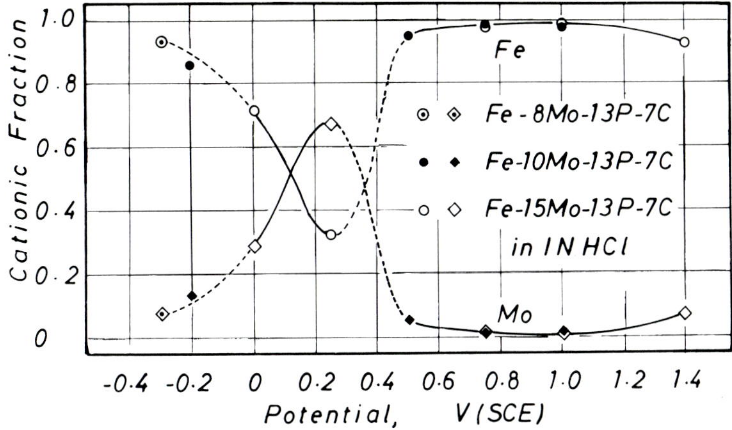

FIGURE 14. Cationic fractions in surface films formed on amorphous Fe–8–15Mo–13P–7C alloys by immersion and potentiostatic polarization for 1 hr in 1 *N* HCl. (Hashimoto, K., Naka, M., Asami, K., and Masumoto, T., *Corros. Sci.*, 19, 165, 1979. With permission.)

various crystalline metastable phases in the amorphous matrix before the formation of stable crystalline phases. Diegle and Slater[27] compared anodic polarization curves of amorphous Ni–40Fe–16P–4B and Ni–25Fe–15Cr–16P–4B alloys in 1 *N* H_2SO_4 + 0.1 *N* NaCl with those of the alloys crystallized by heat treatment at 550°C for 1 hr, and reported that the current density of the crystallized alloys in the passive region is about two orders of magnitude as high as that of the amorphous alloys. Kulik et al.[28] also studied the change in the corrosion behavior of amorphous Ni–32Fe–14Cr–12P–6B alloy by heat treatment and stated that crystallization of the alloy resulted in a decrease in the corrosion resistance to the level of ferritic 17Cr stainless steel. It is interesting to follow the change in the corrosion behavior with crystallization process of amorphous alloys and to identify active surface sites with respect to corrosion.

Figure 15 shows the temperature-time-transformation diagram of Fe–10Cr–13P–7C alloy, indicating that the crystallization of the amorphous alloy occurs through the formation of two kinds of metastable crystalline phases called MS I and MS II.[29] After the isothermal heat treatment at 430°C for different times, anodic polarization curves of the alloy were measured in 1 *N* HCl as shown in Figure 16. The amorphous alloy passivates spontaneously. However, as soon as the MS I is formed in the amorphous matrix, the anodic current density increases and the corrosion potential suddenly changes from the passive region to the active region, indicating the detrimental effect of the formation of compositional heterogeneity.

An interesting result regarding the active surface sites with respect to corrosion has been obtained from the study of rapidly quenched crystalline binary titanium-nickel alloys.[23] The ordinary crystalline Ti–40Ni and Ti–61.5Ni alloys consist of phase mixtures of Ti_2Ni and TiNi or an eutectic of TiNi and $TiNi_3$. Both rapidly quenched Ti–40Ni and Ti–61.5Ni alloys are composed of a single TiNi type phase with fine grains, indicating an expansion of the homogeneous range of the compound by rapid quenching. As shown in Figure 17, the crystalline multiphase alloys exhibit high anodic current densities in 1 *N* HCl and hardly passivate by anodic polarization. On the contrary, the rapidly quenched crystalline single phase alloys reveal significantly low anodic current densities. In particular, the rapidly quenched crystalline Ti–40Ni alloy passivates spontaneously and show a lower passive current density than that of pure titanium. Grain boundaries have been thought to act as the active surface sites with respect to corrosion. The density of grain boundaries in the rapidly quenched crystalline alloys is significantly higher than the ordinary crystalline alloys. Nevertheless, the rapidly quenched alloys possess a considerably high corrosion resistance in

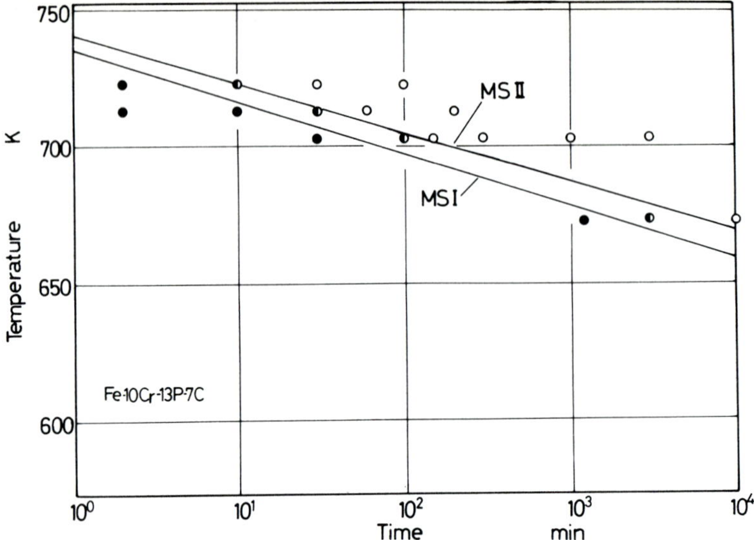

FIGURE 15. Temperature-time-transformation diagram of Fe–10Cr–13P–7C alloy. Lower and upper lines indicate borders where metastable crystalline MS I and MS II phases become detectable by X-ray diffraction. (From Naka, M., Hashimoto, K., and Masumoto, T., *Corrosion (Houston),* 36, 679, 1980. Reprinted by permission of the National Association of Corrosion Engineers, Houston, Texas.)

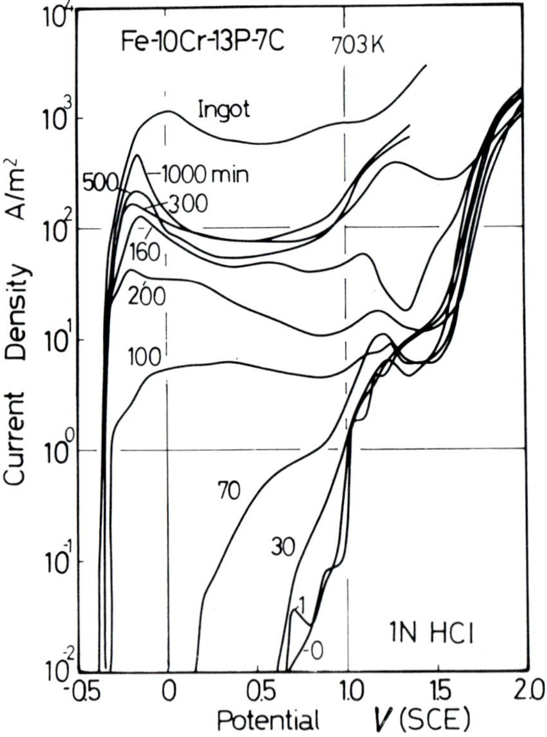

FIGURE 16. Change in potentiodynamic anodic polarization curves of Fe–10Cr–13P–7C alloy measured in 1 *N* HCl with the time of heat treatment at 703 K. The time of heat treatment is denoted in the figure in minutes. (From Naka, M., Hashimoto, K., and Masumoto, T., *Corrosion (Houston),* 36, 679, 1980. Reprinted by permission of the National Association of Corrosion Engineers, Houston, Texas.)

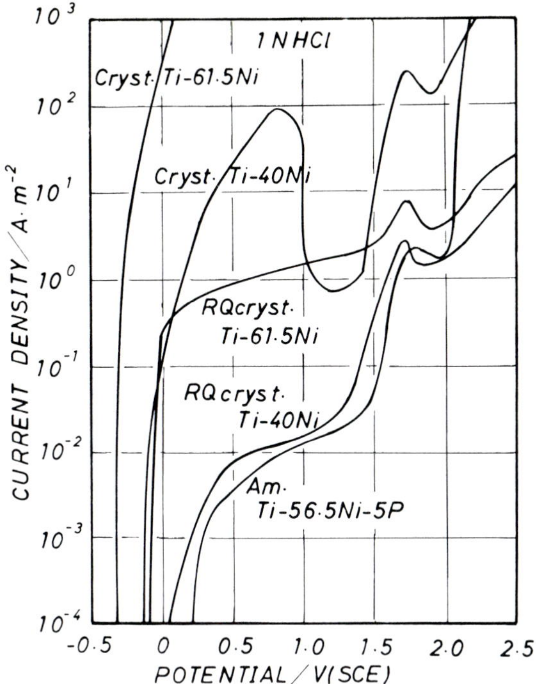

FIGURE 17. Potentiodynamic polarization curves of ordinary crystalline Ti–Ni alloys, rapidly quenched crystalline Ti–Ni alloys and amorphous Ti–Ni–P alloy measured in 1 *N* HCl. (From Naka, M., Asami, K., Hashimoto, K., and Masumoto, T., in *Titanium '80, Proc. 4th Int. Conf. Titanium,* Vol. 4, Kimura, H. and Izumi, O., Eds., The Metallurgical Society of the American Institute of Mining, Metallurgical, and Petroleum Engineers, Warrendale, Pa., 1981, 2677. With permission.)

comparison with the ordinary crystalline alloys with the same average composition, although the formation of the amorphous structure by adding a small amount of phosphorus is further effective in improving the corrosion resistance as can be seen from the comparison of polarization curves of three 38.5Ti alloys, i.e., the ordinary crystalline Ti–61.5Ni, rapidly quenched crystalline Ti–61.5Ni and amorphous Ti–56.5Ni–5P alloys.

According to Anthony and Cline,[30] rapidly quenched crystalline type 304 stainless steel after laser surface melting possesses a high corrosion resistance since various compositional fluctuations are removed by rapid quenching. On the other hand, prolonged heat treatment of amorphous alloys at temperatures lower than the critical temperature for the formation of MS I results in the formation of crystalline supersaturated solid solution phase (SS) in the amorphous matrix. The average composition of the SS phase has been thought to be almost the same as that of the amorphous matrix. Nevertheless, we found that the passive current density of the Fe–10Cr–13P–7C alloy increases two orders of magnitude due to the formation of the SS phase in the amorphous matrix as shown in Figure 18.[20]

Therefore, rapidly quenched single phase alloys show significantly high corrosion resistance, even if they are crystalline. On the other hand even when the crystalline phase formed in the amorphous matrix is not different from the amorphous matrix in the apparent average composition, heat treatment considerably decreases the corrosion resistance. As mentioned previously, rapid quenching from the liquid state prevents solid state diffusion and the formation of various compositional fluctuations. Accordingly, crystal defects, such as grain boundaries, formed during rapid quenching from the liquid state may be chemically clean

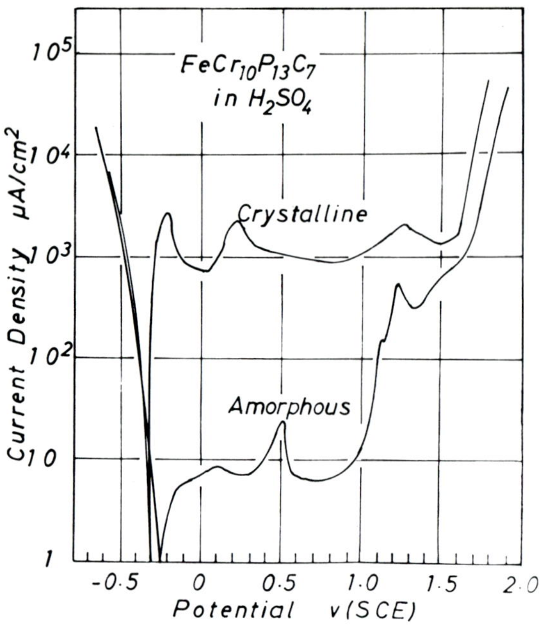

FIGURE 18. Potentiodynamic polarization curves of amorphous and crystallized Fe–10Cr–13P–7C alloys measured in 2 *N* H_2SO_4. Crystallization of amorphous alloy was carried out by heating at 350°C for 2 months. (From Hashimoto, K., Osada, K., Masumoto, T., and Shimodaira, S., *Corros. Sci.*, 16, 71, 1976. With permission.)

and may not act as preferential active surface sites with respect to corrosion. In contrast, heat treatment inevitably induces solid state diffusion and hence results in various compositional fluctuations such as precipitation, segregation, and other compositional gradients. These compositional fluctuations may act as dominant active surface sites with respect to corrosion.

VIII. HIGH RESISTANCE TO LOCALIZED CORROSION

If the high corrosion resistance of amorphous alloys is mainly attributed to the alloy homogeneity, the alloys would be quite resistant against initiation of localized corrosion attack. However, when a crevice assembly exists, they may suffer serious corrosion attack, since crevice corrosion does not require surface heterogeneities for initiation. Devine[31-33] tried to demonstrate this fact by using heavily cold rolled specimens. According to him, plastic deformation to a total strain of 3.6% had no effect on the passive corrosion rate nor on the pitting corrosion resistance of amorphous Ni–30Fe–15Cr–14P–6B alloy. However, cold-rolling the alloy to 32% reduction in thickness markedly increased the corrosion rate of the alloy as indicated in Figure 19. As shown in Figure 20, after anodic polarization in 1% NaCl solution severe pitting corrosion was found due to crevice corrosion at the intersections of surface microcracks which were formed by heavy cold-rolling.

Diegle[34] reexamined the effect of cold rolling on the corrosion behavior of amorphous alloys: cold-rolling the amorphous Fe–30Ni–16Cr–14P–6B alloy to a 38% reduction in thickness led to a significant increase in the current density in both the active and passive regions as shown in Figure 21. However, during potentiostatic polarization for 60 min at 0.6 V(SCE), which is significantly anodic to promote pitting and crevice corrosion if it would indeed occur, the current density decreased rather than increased, the latter being

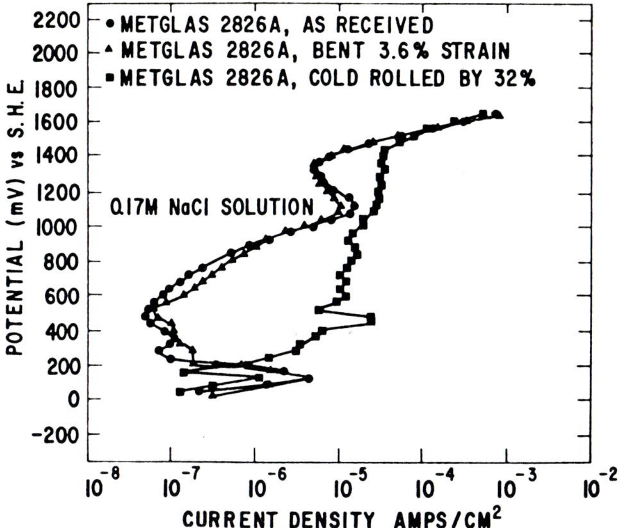

FIGURE 19. Effect of cold work on the anodic polarization curves of amorphous Ni–30Fe–15Cr–14P–6B alloy measured in 1% NaCl solution. (From Devine, T. M., *J. Electrochem. Soc.*, 124, 38, 1977. Reprinted by permission of the publisher, The Electrochemical Society, Inc.)

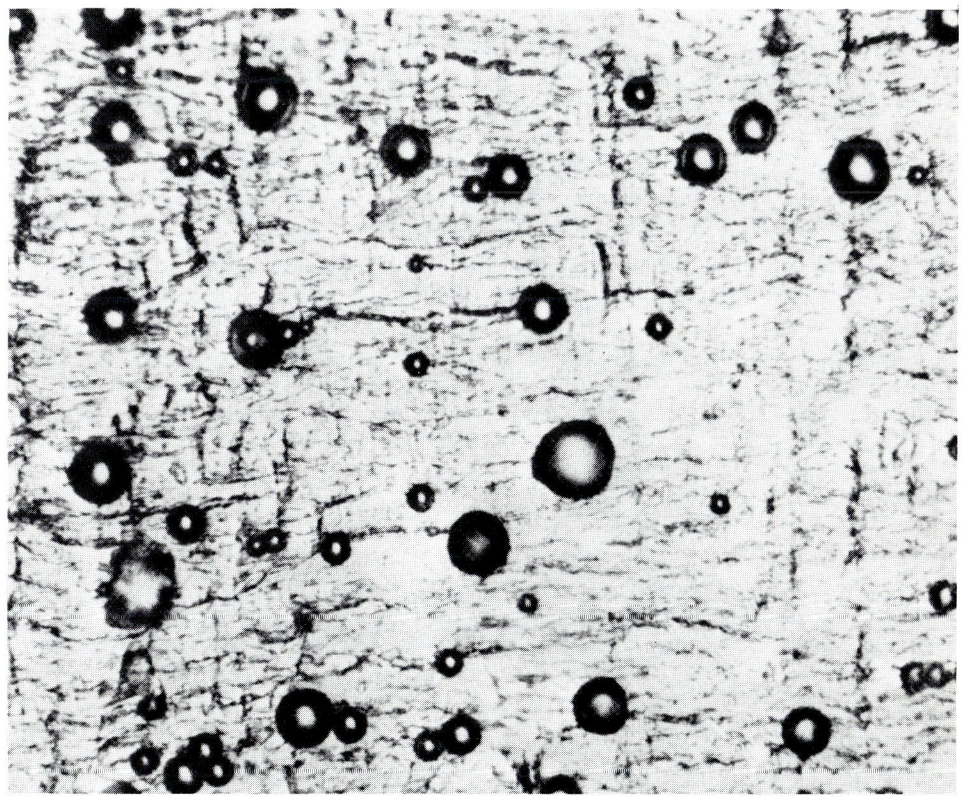

FIGURE 20. Photomicrograph of amorphous Ni–30Fe–15Cr–14P–6B alloy cold-rolled to 32% reduction in thickness and anodically polarized in 1% NaCl. (From Devine, T. M., *J. Electrochem. Soc.*, 124, 38, 1977. Reprinted by permission of the publisher, The Electrochemical Society, Inc.)

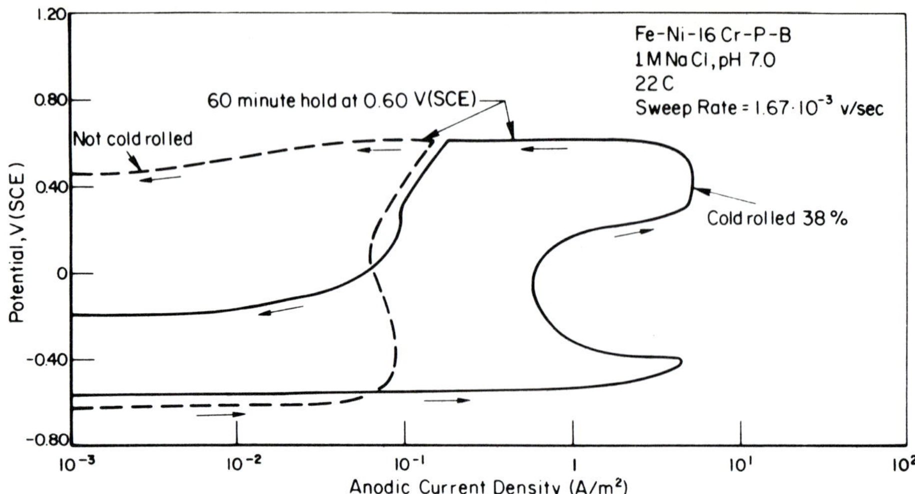

FIGURE 21. Potentiodynamic/potentiostatic polarization curves for amorphous Fe–30Ni–16Cr–14P–6B alloy in the undeformed and cold-rolled conditions measured in 1 *N* NaCl. (From Diegle, R. B., *Corrosion (Houston),* 35, 250, 1979. Reprinted by permission of the National Association of Corrosion Engineers, Houston, Texas.)

expected if pitting and crevice corrosion progressed deeply. Furthermore, after this experiment a negative hysteresis of the polarization curve was observed as seen in Figure 21. These results clearly reveal that, even if pitting and crevice corrosion initiate by anodic polarization at the intersections of surface microcracks, which act as tight crevices; potentiostatic polarization in the passive region easily repassivates such actively dissolving sites. Diegle further carried out crevice cell experiments. As shown in Figure 22, when an amorphous Fe–30Ni–5Cr–14P–6B alloy specimen placed inside a crevice was suddenly polarized at high potentials, the crevice cell current was observed but decayed instantaneously, indicating that the alloy within the crevice began to be crevice-corroded by sudden anodic polarization but was readily repassivated.

Consequently, the amorphous metal-metalloid alloys containing specific elements such as chromium possess extremely high corrosion resistance not only to initiation of pitting corrosion even by anodic polarization in strongly acidic chloride solutions due to the chemically homogeneous single phase nature of the alloys, but also to propagation of pitting and crevice corrosion owing to high passivating ability due mainly to the high reactivity of metalloid-containing amorphous alloys.

IX. STRESS CORROSION CRACKING AND HYDROGEN EMBRITTLEMENT

Stress corrosion cracking of crystalline solids is associated with grain boundaries or some crystallographic planes. It is interesting to examine whether amorphous alloys, which are free of grain boundaries and a crystal lattice, are susceptible to stress corrosion cracking. Pampillo[35] reported that amorphous Ni–29Fe–14P–6B–2Al alloy failed by immersion in a 3.5% NaCl solution at 26°C for 117 hr under an applied tensile stress, 75% of the fracture stress in air. In general, if the specimens do not have uniform thickness and width the fracture stress of ribbon-shaped thin amorphous alloy specimens is not reproducible. To obtain reproducible results, both the surfaces and edges of the specimen must be carefully polished before the mechanical test. In particular, the alloy used by him did not contain chromium and should be readily corroded in the NaCl solution. Accordingly, if a corrosion notch was formed at the edge of the alloy specimen, mechanical failure would easily occur

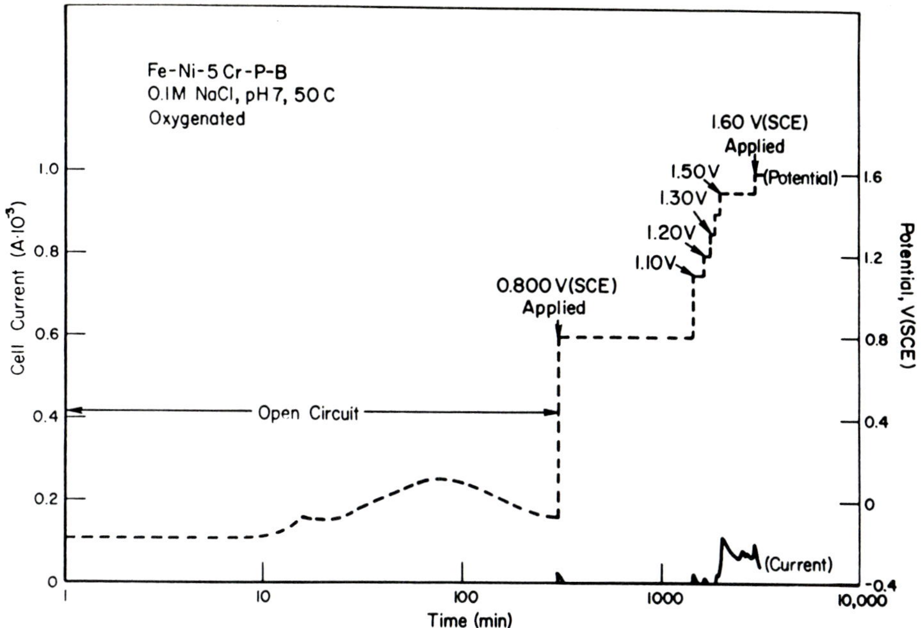

FIGURE 22. Crevice cell current and applied potential vs. time for amorphous Fe–30Ni–5Cr–14P–6B alloy in the crevice test cell. (From Diegle, R. B., *Corrosion (Houston)* 35, 250, 1979. Reprinted by permission of the National Association of Corrosion Engineers, Houston, Texas.)

under such a high applied stress. No other work reporting stress corrosion cracking for amorphous alloys has so far been published. We also have tried but failed to produce stress corrosion cracks for amorphous alloys. It can, therefore, be said that there are no reports confirming the formation of stress corrosion cracks for homogeneous amorphous alloys and that they may not suffer stress corrosion cracking because of their homogeneous nature.

On the other hand, some amorphous alloys suffer hydrogen embrittlement when they are corroded under an applied tensile stress or when hydrogen is charged to them.[36-38] As mentioned previously, when a metallic specimen is corroded by immersion in an aqueous solution, cathodic Reactions 3 and 4 take place. The sum of the cathodic reaction rates is equal to the corrosion rate of metal. Reaction 3 consists of the following elementary reactions:

$$H^+ + e \rightarrow H_{ads} \tag{7}$$

$$2H_{ads} \rightarrow H_2 \tag{8}$$

or

$$H_{ads} + H^+ + e \rightarrow H_2 \tag{9}$$

Accordingly, corrosion of metal results in the formation of hydrogen atoms adsorbed on the metal surface, H_{ads} being an intermediate species of cathodic reaction. Some adsorbed hydrogen atoms are absorbed in metals as

$$H_{ads} \rightleftharpoons H_{abs} \tag{10}$$

Consequently, aqueous corrosion leads to absorption of hydrogen in metal and often to hydrogen embrittlement.

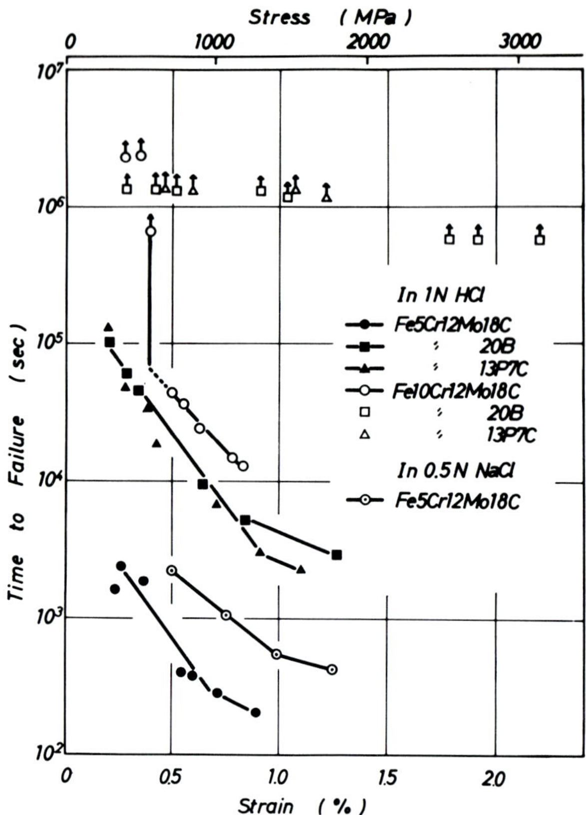

FIGURE 23. Variation of time to failure with applied strain or stress for amorphous Fe–Cr–Mo–metalloid alloys measured by bending tests in 1 *N* HCl and 0.5 *N* NaCl at 30°C. (From Kawashima, K., Hashimoto, K., and Masumoto, T., *Corrosion (Houston)*, 36, 577, 1980. Reprinted by permission of the National Association of Corrosion Engineers, Houston, Texas.)

Figure 23 shows the relation between time to failure and applied strain or stress obtained by a bending test of amorphous Fe–Cr–Mo–metalloid alloys in 1 *N* HCl and 0.5 *N* NaCl solutions at 30°C.[39] Fracture strain and fracture stress for these alloys measured in air are 2.0 to 2.4% and 2800 to 3000 MPa, respectively, and are considerably decreased by immersion in the solutions. As shown in Figure 24, the fracture stress decreases with increasing time of traction test in the aqueous solution, indicating that embrittlement results from deformation in the aqueous solution.[36] It is evident that the decrease in fracture stress and strain is due to hydrogen embrittlement, since the fracture stress of the specimen embrittled by a traction test in the solution is completely recovered by a subsequent heating in an oil. As can be seen from Figure 23, the 5Cr alloys are more susceptible to hydrogen embrittlement than the 10Cr alloys. This indicates that an improvement of the corrosion resistance greatly increases the resistance to hydrogen embrittlement. In addition, metalloids contained in the alloys significantly affect the susceptibility to hydrogen embrittlement. In particular, the alloys containing only carbon as a metalloid are most susceptible to hydrogen embrittlement. It can, therefore, be said that the susceptibility to hydrogen embrittlement of amorphous alloys can be significantly decreased when the alloy composition is modified so as to improve the corrosion resistance or to substitute carbon with other metalloid elements.

As shown in Figure 25, the most characteristic fractographic feature of embrittled amor-

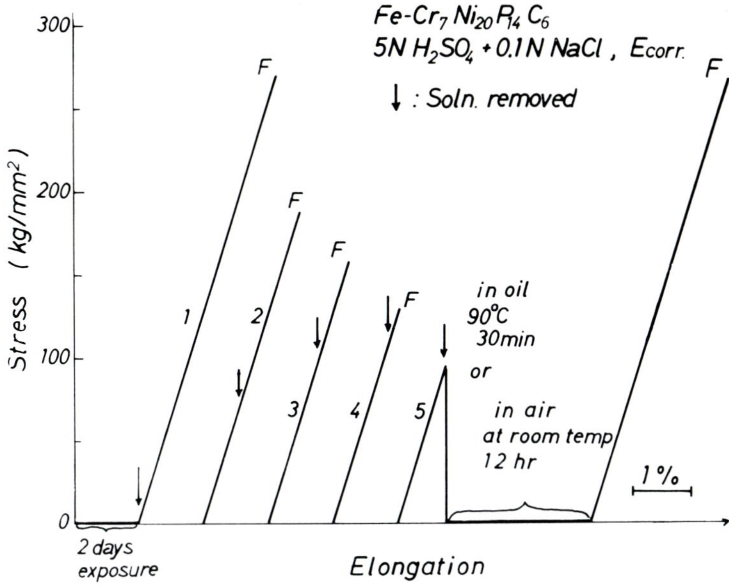

FIGURE 24. Tensile stress versus elongation for amorphous Fe–7Cr–20Ni–14P–6C alloy in 5 N H_2SO_4 + 0.1 N NaCl at the corrosion potential and a strain-rate of 5.6 $\times$ 10^{-5} sec^{-1}. Symbol F denotes fracture of specimen. Arrows indicate the time when the solution was removed from the cell and the specimen was rinsed with distilled water. Subsequently, traction test was continued in air. In Experiment No. 1 the unstressed specimen installed in a tensile test machine was exposed for 2 days in the solution. After the specimen was rinsed with distilled water the traction test was commenced in air. In Experiment No. 5, traction tests in the solution were stopped at the time indicated by an arrow and the crosshead of the tensile test machine was returned to the original position. After specimens were rinsed with distilled water and heated in silicon oil at 90°C or stored in air at room temperature for 12 hr, the traction test was then resumed in air. (From Kawashima, A., Hashimoto, K., and Masumoto, T., *Corros. Sci.*, 16, 935, 1976. With permission.)

phous alloys is a crack propagation trace including Walner lines, which are characteristic of fracture surfaces of completely brittle solids such as inorganic glasses, ceramics and intermetallic compounds.[40] The velocity of crack propagation derived from Walner lines is about 1200 m/sec, indicating a catastrophic failure. Accordingly, the time of failure is mainly governed by the time of local enrichment of hydrogen in the alloys as a result of corrosion.

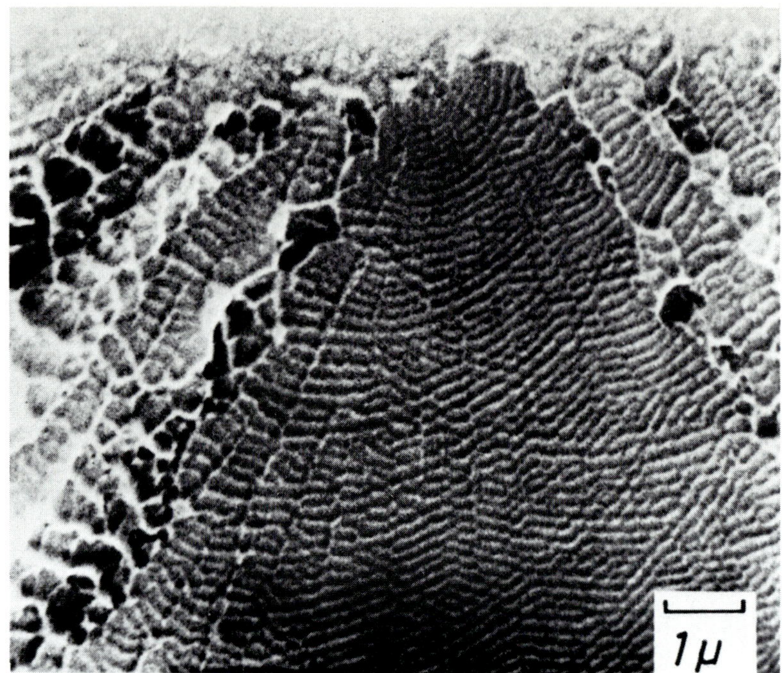

FIGURE 25. Fracture surface of amorphous Fe–10Cr–12Mo–18C alloy failed during the bending test in 1 *N* HCl solution. The upper part was tensile-stressed. The applied tensile strain at the top of the specimen was 0.52%. (From Kawashima, A., Hashimoto, K., and Masumoto, T., *Scr. Metall.*, 14, 41, 1980. With permission.)

REFERENCES

1. **Masumoto, T., Hashimoto, K., and Naka, M.,** Corrosion of amorphous metals, in *Proc. 3rd Int. Conf. Rapidly Quenched Metals,* Vol. 2, Cantor, B., Ed., The Metals Society, London, 1978, 435.
2. **Naka, M., Hashimoto, K., Inoue, A., and Masumoto, T.,** Corrosion-resistant Fe-C alloys containing chromium and/or molybdenum, *J. Non-Cryst. Solids,* 31, 347, 1979.
3. **Naka, M., Hashimoto, K., and Masumoto, T.,** Change in corrosion behavior of amorphous Fe-P-C alloys by alloying with various metallic elements, *J. Non-Cryst. Solids,* 31, 355, 1979.
4. **Naka, M., Hashimoto, K., and Masumoto, T.,** High corrosion resistance of chromium-bearing amorphous iron alloys in neutral and acidic solutions containing chloride, *Corrosion (Houston),* 32, 146, 1976.
5. **Hashimoto, K., Kobayashi, K., Asami, K., and Masumoto, T.,** Corrosion-resistant amorphous alloys in hot concentrated hydrochloric acids, in *Proc. 8th Int. Cong. Metallic Corrosion,* Vol. 1, Deutsche Gesellschaft für chemisches Apparatewesen, Mainz, 1981, 70.
6. **Sugimoto, K. and Sawada, Y.,** The role of molybdenum additions to austenitic stainless steels in the inhibition of pitting in acid chloride solutions, *Corros. Sci.,* 17, 425, 1977.
7. **Asami, K., Naka, M., Hashimoto, K., and Masumoto, T.,** Effect of molybdenum on the anodic behavior of amorphous Fe-Cr-Mo-B alloys in hydrochloric acid, *J. Electrochem. Soc.,* 127, 2130, 1980.
8. **Naka, M., Hashimoto, K., and Masumoto, T.,** Effect of addition of chromium and molybdenum on the corrosion behavior of amorphous Fe-20B, Co-20B and Ni-20B alloys, *J. Non-Cryst. Solids,* 34, 257, 1979.
9. **Naka, M., Hashimoto, K., and Masumoto, T.,** Corrosion behavior of amorphous and crystalline $Cu_{50}Ti_{50}$ and $Cu_{50}Zr_{50}$ alloys, *J. Non-Cryst. Solids,* 30, 29, 1978.
10. **Brusic, V., MacInnes, R. D., and Aboaf, J.,** Passivation properties of amorphous Fe-Si thin films, in *Passivity of Metals,* The Corrosion Monograph Series, Frankenthal, R. P. and Kruger, J., Eds., The Electrochemical Society, Princeton, N.J., 1978, 170.

11. **Hashimoto, K., Asami, K., Naka, M., and Masumoto, T.,** The role of alloying elements in improving the corrosion resistance of amorphous iron-base alloys, *Corros. Sci.*, 19, 857, 1979.
12. **Asami, K., Hashimoto, K., and Shimodaira, S.,** ESCA study on passivity of a series of iron-chromium alloys in a sulfuric acid, *Corros. Sci.*, 18, 151, 1978.
13. **Asami, K., Hashimoto, K., Masumoto, T., and Shimodaira, S.,** ESCA study on the passivity of an extremely corrosion resistant amorphous iron alloy, *Corros. Sci.*, 16, 909, 1976.
14. **Hashimoto, K., Naka, M., Noguchi, J., Asami, K., and Masumoto, T.,** Effect of alloy elements on corrosion resistance of amorphous iron-base alloys, in *Passivity of Metals*, The Corrosion Monograph Series, Frankenthal, R. P. and Kruger, J., Eds., The Electrochemical Society, Princeton, N.J., 1978, 156.
15. **Hashimoto, K., Asami, K., Naka, M., and Masumoto, T.,** Surface films formed on amorphous Co-Cr alloys in 1 *N* HCl, *Corros. Eng. (Boshoku Gijutsu)*, 28, 271, 1979.
16. **Hashimoto, K., Kasaya, M., Asami, K., and Masumoto, T.,** Electrochemical and XPS studies on corrosion behavior of amorphous Ni-Cr-P-B alloys, *Corros. Eng. (Boshoku Gijutsu)*, 26, 445, 1977.
17. **Hashimoto, K. and Asami, K.,** An X-ray photoelectron spectroscopic study of the passivity of ferritic 19Cr stainless steels in 1 *N* HCl, *Corros. Sci.*, 19, 251, 1979.
18. **Hashimoto, K., Asami, K., and Teramoto, K.,** An X-ray photoelectron spectroscopic study on the role of molybdenum in increasing the corrosion resistance of ferritic stainless steels in 1 *N* HCl, *Corros. Sci.*, 19, 3, 1979.
19. **Hara, M., Hashimoto, K., and Masumoto, T.,** The anodic polarization behavior of amorphous Pd-Ti-P alloys in NaCl solutions, *Electrochim. Acta*, 25, 1215, 1980.
20. **Hashimoto, K., Osada, K., Masumoto, T., and Shimodaira, S.,** Characteristics of passivity of extremely corrosion resistant amorphous iron alloys, *Corros. Sci.*, 16, 71, 1976.
21. **Naka, M., Asami, K., Hashimoto, K., and Masumoto, T.,** Effects of metalloidal elements on corrosion resistance of amorphous iron-chromium alloys, *J. Non-Cryst. Solids*, 28, 403, 1978.
22. **Hashimoto, K., Naka, M., Asami, K., and Masumoto, T.,** XPS and electrochemical studies of effects of metalloid additives on corrosion behavior of amorphous iron-chromium alloys, *Corros. Eng. (Boshoku Gijutsu)*, 27, 279, 1978.
23. **Naka, M., Asami, K., Hashimoto, K., and Masumoto, T.,** Corrosion behavior of amorphous titanium alloys, in *Titanium'80, Proc. 4th Int. Conf. Titanium*, Vol. 4, Kimura, H. and Izumi, O., Eds., The Metallurgical Society of the American Institute of Mining, Metallurgical, and Petroleum Engineers, Warrendale, Pa., 1981, 2677.
24. **Naka, M., Hashimoto, K., and Masumoto, T.,** High corrosion resistance of amorphous Fe-Mo and Fe-W alloys in HCl, *J. Non-Cryst. Solids*, 29, 61, 1978.
25. **Hashimoto, K., Naka, M., Asami, K., and Masumoto, T.,** An X-ray photoelectron spectroscopic study of the passivity of amorphous Fe-Mo alloys, *Corros. Sci.*, 19, 165, 1979.
26. **Hashimoto, K., Asami, K., Naka, M., and Masumoto, T.,** Effect of molybdenum on the corrosion behavior of amorphous Fe-Mo-C alloys in 1 *N* HCl, *Sci. Rep. Res. Inst. Tohoku Univ. Ser. A*, 27, 237, 1979.
27. **Diegle, R. B. and Slater, J. E.,** Influence of crystallinity on corrosion behavior of ferrous alloys, *Corrosion (Houston)*, 32, 155, 1976.
28. **Kulik, T., Baszkiewicz, J., Kaminski, M., Latuszkiewicz, J., and Matyja, H.,** The electrochemical corrosion of amorphous $Ni_{36}Fe_{32}$-$Cr_{14}P_{12}B_6$ alloy (Metglas 2826A), *Corros. Sci.*, 19, 1001, 1979.
29. **Naka, M., Hashimoto, K., and Masumoto, T.,** Effect of heat treatment on corrosion behavior of amorphous Fe-Ni-Cr-P-B and Fe-Cr-P-C alloys in 1 *N* HCl, *Corrosion (Houston)*, 36, 679, 1980.
30. **Anthony, T. R. and Cline, H. E.,** Surface normalization of sensitized stainless steel by laser surface melting, *J. Appl. Phys.*, 49, 1248, 1978.
31. **Devine, T. M. and Wells, L.,** Anodic polarization behavior of an amorphous metal, *Scr. Metall.*, 10, 309, 1976.
32. **Lee, D. and Devine, T. M.,** Deformation, fracture and corrosion behavior of an amorphous alloy, in *Proc. 4th Int. Conf. Rapidly Quenched Metals*, Section I, Grant, N. J. and Giessen, B. C., Eds., The Massachusetts Institute of Technology, Cambridge, Mass., 1976, 393.
33. **Devine, T. M.,** Anodic polarization and localized corrosion behavior of amorphous $Ni_{33}Fe_{30}Cr_{13}P_{14}B_6$ in near-neutral and acidic chloride solution, *J. Electrochem. Soc.*, 124, 38, 1977.
34. **Diegle, R. B.,** Localized corrosion of amorphous Fe-Ni-Cr-P-B alloys, *Corrosion (Houston)*, 35, 250, 1979.
35. **Pampillo, C. A.,** Flow and fracture in amorphous alloys, *J. Mater. Sci.*, 10, 1194, 1975.
36. **Kawashima, A., Hashimoto, K., and Masumoto, T.,** Stress corrosion cracking of amorphous iron base alloys, *Corros. Sci.*, 16, 935, 1976.
37. **Nagumo, M. and Takahashi, T.,** Hydrogen embrittlement of some Fe-base amorphous alloys, *Mater. Sci. Eng.*, 23, 257, 1976.

38. **Viswanadham, R. K., Green, J. A. S., and Montague, W. G.,** Hydrogen embrittlement of an Fe-based amorphous metal, *Scr. Metall.*, 10, 229, 1976.
39. **Kawashima, K., Hashimoto, K., and Masumoto, T.,** Hydrogen embrittlement of amorphous Fe-Cr-Mo alloys, *Corrosion (Houston)*, 36, 577, 1980.
40. **Kawashima, A., Hashimoto, K., and Masumoto, T.,** Fractographic study of amorphous iron base alloys embrittled by hydrogen and heat-treatment, *Scr. Metall.*, 14, 41, 1980.

INDEX

A

B

C

D

E

H

I

K

N

O

P

Q

R

S

T

U

V

W

X

Y

Z